Slide Rules

Their History, Models, and Makers

Peter M. Hopp, C.Eng. MBCS.

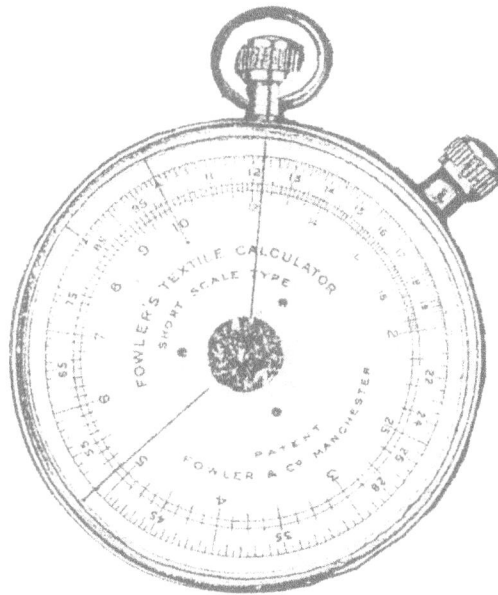

Astragal Press
Mendham, New Jersey
United States

ISBN 978-1-879335-86-8

Library of Congress Catalog Card Number 98-074740

Published by
THE ASTRAGAL PRESS
5 Cold Hill Road, Suite 12
P.O. Box 239
Mendham, NJ 07945-0239

Preface.

Any book that attempts to provide comprehensive information on the slide rule with exhaustive listings of makers, manufacturers, and types is inevitably going to be an iterative process, and never complete! Each new source provides additional information and the temptation is to continue adding information and never establishing a baseline. I thus begin by making apologies and stating categorically that errors and omissions are mine and mine alone, there is no reason other than my ignorance for these.

It would be nice for this to be the definitive work on slide rules; however with my present knowledge of slide rules and associated information, it is more appropriate to consider this a foundation which can be built on as additional information becomes available. To keep this as a living document requires input from the readers. I therefore welcome comment and criticism as well as any information that adds to the data presented. I strongly believe that it is vital to publish with the data that is available rather than waiting for more information as it is only by presenting what is known that one becomes aware of what is missing. For readers who wish to become more involved with collecting slide rules I strongly recommend that they join the Oughtred Society, a group dedicated to the advancement of information on slide rules which can be contacted via *The Oughtred Society, P.O. Box 99077, Emeryville, California 94662, USA.* It is very soothing to know that there are like- minded individuals world-wide with an interest in slide rules.

My family considers me gently lunatic but have been extremely supportive during the years that I have been compiling this study. The book is dedicated to them, and especially my wife Carol, who has been especially understanding and encouraging. It is also dedicated to all fellow collectors and those interested in slide rule history: they are all performing a vital function in preserving these delightful devices for future generations.

P.M. Hopp
1 Dorewards Ave.
Bocking, Braintree
Essex, CM7 5LT
England
April 1998

Acknowledgements.

Many people have been extremely kind with the willing and free provision of information on the makers, manufacturers and slide rule types in their collections, and without their help the book would be very much poorer. The slide rule collecting fraternity are extremely generous with their time and information, and those who have been kind enough to comment and add to the information are too numerous to mention individually.

However I must specifically mention the following:

Fellow collectors Colin Barnes and John Knott, whose generosity with information from their sources has been a great help over a number of years.

Members of the Dutch Circle of Slide Rule collectors, especially Herman an Herweijnen, Otto von Poelje, and Ysebrand Schuitema, who have been particularly helpful with providing information about their collections.

Dieter von Jezierski and Gunther Kugel in Germany, who have provided information on German rules and manufacturers.

David Rance, who was very generous with positive comments and a wealth of detail on a range of manufacturers.

Heinz Joss, who let me have the names of numerous Swiss manufacturers.

Pierre vander Meulen in Belgium, who has provided an outstanding amount of detail on the slide rules in his collection.

Cyril Catt in Australia, who has provided the bulk of the information on Australian rules and manufacturers.

Peter Kenward, who has added considerably to my collection over the years.

The production of the manuscript was in danger of foundering when Roy Williams very kindly read the first draft of the manuscript. He deserves a special mention and thanks for his positive and helpful comments, which encouraged me to continue. Mrs Julanne Arnold, my very helpful and enthusiastic editor who bashed the whole into its present shape, deserves a special thank-you.

Table of Contents.

Figures and Tables

Figures:

Tables:

Introduction.

Few, if any, technological developments in history mirror the ascendancy and demise of the slide rule. It is probably unique as an item which was developed and enhanced over a long period, was in everyday use by a large and ever increasing segment of the professional population for over 300 years, and then disappeared virtually overnight following the production of the first cheap electronic pocket calculators in the early part of the 1970's. By 1975 very few manufacturers still advertised slide rules, and at the end of the decade it was virtually impossible to buy such a device except second hand.

By the end of its development, the slide rule had become an exceedingly sophisticated calculating machine, available in many forms to cater to general as well as many specialised applications. There were a profusion of manufacturers world-wide who at their peak were producing tens of thousands of slide rules per year, with a bewildering range of capabilities in numerous styles. This variety is one of the many aspects which can both delight and frustrate collectors. The relatively recent disappearance of slide rules from common usage means that good condition items up to 50 years old are readily available at very reasonable prices at, among other places, small local sales. For the serious collector, auctions of antique mathematical and other instruments will sometimes include ivory, brass, or wooden instruments from previous centuries at an appropriately greater price.

What is a slide rule?

The slide rule is a very crude form of analogue computer which relies on the manual movement of the various computing elements (the scales) to perform the calculations desired. These scales are an interpretation of logarithms into a linear, circular, spiral or helical form, depending on the style of slide rule. Generally two identical scales are moved relative to each other to perform the calculation. A cursor enables the numbers on other parallel scales to be aligned against each other.

How does a slide rule work?

A slide rule is provided with logarithmic scales of various forms and in various combinations. The most common is a "single cycle" logarithmic scale, where the markings are from 1 to 10. The distance of each mark from the index (at the far left of the scale) is proportional to its logarithm; e.g., for a ten inch rule, the mark for 2 is placed 3.010 inches from the index, because the logarithm of 2 is 0.3010, and this becomes the proportion of the scale representing 2.

Similarly 3 is placed 4.307 inches from the index, and so on. These single cycle scales are normally placed on the bottom of the slide and the adjacent edge of the stock, and are labelled "C" and "D". Two cycle scales each occupy half the distance (5 inches on a 10 inch rule). Two of these are put together to cover numbers from 1 to 10, then 10 to 100, and are labelled "A" and "B". In some cases three cycle scales are also supplied. The logarithmic relationship between these means that if the indices of the various scales are coincident, then numbers on a two cycle scale are equivalent to the square of the numbers on the single cycle scale, and numbers on the three cycle scale are equivalent to the cube of the single cycle scale. The final scale to be considered is a reversed scale, i.e. a scale where the datum is taken from the right-hand end of the rule. This allows the inverse of numbers (reciprocals) to be calculated by reference to its neighbouring non-inverted scale.

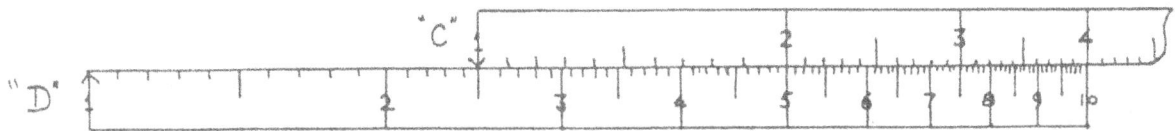

Figure 1: Logarithmic Slide Rule Scales

Logarithmic Scales.

Figure 1 shows a section of two linear logarithmic scales, labelled "C" and "D". The sections are shown in the performance of the calculation of the simple sum 2.5 x 2.0. The index (1) of the "C" scale is moved to coincide with the 2.5 on the "D" scale, whereupon the answer can be read on the "D" scale opposite the figure 2.0 on the "C" scale. This simple multiplication sum demonstrates that by adding the mantissa of the logarithm of the numbers we are able to perform the multiplication of the numbers whose logarithms they are. Division is the converse of multiplication. The set-up in **Figure 1** can be used to demonstrate 5.0 divided by 2.0. To do this, 2.0 on the "C" scale is brought into coincidence with 5.0 on the "D" scale, whereupon the answer may be read off the index (1) on the "C" scale. Thus it follows that a complex series of multiplication and division sums can be performed by the systematic movement of the elements containing these scales.

These simple examples must suffice as an introduction to the use of a slide rule. The instruction leaflets supplied with slide rules, and numerous books as well, described in varying detail and sophistication, methods of using a slide rule to perform general and specialist calculations. A number of the books listed in **Appendix 3** contain information primarily aimed at teaching the reader the many techniques for performing complex calculations with a slide rule, and it would serve little purpose to repeat this information here.

Who used a slide rule?

Two quotations from the 18th century illustrate the feelings of contemporary mathematical practitioners regarding the slide rule. The first is taken from Edmund Stone's 1723 translation of *The Construction and Principal Uses of Mathematical Instruments* from the French of Michael Bion:

> *Mathematics are now become a popular study, and make a part of the Education of almost every Gentleman...Mathematical Instruments are the means by which those Sciences are rendered useful in the Affairs of Life.*

This is mirrored approximately twenty years later, with some plagiarism, by George Adams when he published his *Micrographia Illustrata* in 1746:

Mathematical Instruments are the means by which those noble Sciences, Geometrie and Philosophy, are rendered useful in the Affairs of Life. By their assistance an abstracted and unprofitable Speculation, is made beneficial in a thousand instances: In a word they enable us to connect Theory with Practice, and so turn what was only bare Contemplation, into the most Substantial Uses.

Generations of technologists in many disciplines, as well as the general public, needed to have some knowledge of slide rule usage to enable them to do simple multiplication and division without the aid of pencil and paper or, for more complex calculation, the use of logarithmic tables. Slide rule usage was taught in most schools for a number of years, and it is probable that all but the youngest of today's population will have some memory of learning how this instrument worked, and was used. The majority of the population over the age of 40 may even have owned a slide rule at some time.

Why use a slide rule?

Gottfried Wilhelm Leibnitz, the famous scientist, wrote in 1685 :

It is unworthy for excellent men to lose hours like slaves in the labour of calculation which could be safely relegated to anyone else if machines were used.

In any profession or job which required large numbers of individual calculations or repetitive calculation to be performed, some form of automation or mechanical assistance removed the drudgery from the task. The slide rule was the source of such assistance for some hundreds of years, and was, in all probability, used in the design of all "high technology" products that we are familiar with today. It was the forerunner of today's ubiquitous electronic calculator, but whereas it is probable that most households have at least one calculator, the slide rule never achieved such popularity.

It has been argued that the advent of the calculator was the start of the widespread lack of numeracy that seems prevalent today. It is certainly true that using a slide rule required a better understanding of the meaning of numbers, and a general facility in mental arithmetic that enabled a user to check the validity of results obtained. Perhaps the slide rule will have to return to popularity if we wish to improve the general level of numeracy.

Why study slide rules?

A recent book prefaces its text on collecting with the statement:

We are destroying too much, too fast, without realising how much we are losing ...
If not now, then later, we shall all have cause to regret this loss.

This is a fair reflection on the state of information relating to slide rules and their manufacturers. If these devices are not studied and recorded now, we shall lose the opportunity to preserve an accurate history of an important part of mathematical application.

As the slide rule has only recently passed from general usage, numerous examples are available at very reasonable prices. A comprehensive collection can thus be put together for a reasonable outlay, and by specialising, a unique collection is achievable. It is doubtful whether any of the large and specialist rules still exist on the open market. It is to be hoped that large organisations that still have some of the more spectacular devices tucked away in a forgotten corner, will discover what they are and, rather than scrapping them, will consider donating them to a collector, or to a museum.

This book is intended to show the perhaps surprising range and variation of these delightful instruments. This may whet the appetite of a few more people to start looking at these now almost forgotten instruments, and to take pleasure in the ingenuity of our forefathers who regularly enjoyed using them. To quote from the 1940 instruction book for PIC slide rules:

> *If there were no slide rules I would give up Engineering. The arithmetic would be so tedious; with a slide rule the arithmetic is rather fun.*

Organization of the book.

This book has been divided into two periods: first from the invention of the slide rule in the early 17th century to 1850; then from 1850 to the beginning of the end of significant slide rule manufacture in 1975, when the cheap electronic calculator made its first impact, and continuing with limited slide rule manufacture thereafter. This separation is not arbitrary. as will be explained later.

The first 80 years of development in the 17th century were effectively the formative years when the "shape" of the slide rule was set. In the United Kingdom the first standards were set primarily by Coggeshall and Everard, as there were no obvious makers or manufacturers elsewhere. This period also showed increasingly refined design and fabrication techniques, as the mathematical practitioners were more specialised: *"...astronomers, surveyors, military engineers, navigators, schoolmasters, instructors..., nevertheless one and all remained in close touch with the instrument-makers: there was an exchange of ideas about new designs and new methods on equal terms"*. Taylor, 1966.

In the 18th century *"...the great upsurge of scientific achievement in the 18th century, particularly in regard to navigation, was the result of a truly remarkable collaboration and understanding between some of the world's finest scientists, some superb craftsmen, and some remarkable naval officers, together with a host of teachers, writers, practical navigators and the 'lesser' men..."* (D.H. Sadler, Astronomer Royal, in the forward to E.G.R. Taylor's *Mathematical Practitioners of Hanoverian England, 1714 - 1840*). Professor Taylor identified this period as one where all worked together on practical solutions for the outstanding mathematical problems of the day, and the instrument making craftsmen (there were a very small number of women) were a vital, essential and equal part of the scientific community.

By the middle of the 19th century, technology and communications had moved on to the stage where it is possible to look at world-wide movements and inventions. Choosing 1850 as a breaking point has significance in the world of slide rules; soon after, the invention of the Mannheim rule was, I believe, the start of true world-wide standardisation. In Professor Taylor's view this was also the point where *"... the scientist had now disappeared into the laboratory, the craftsman into the factory, the private teacher had become a schoolmaster or a professor, the best surveyors were in the army or navy."* By 1850 it was virtually impossible to differentiate between the instrument maker and the retailer; and the manufacturer was just beginning to appear. The tables in Chapters 4 and 6 attempt to show that separation. Before 1850 the listings cover instrument makers, and after this point they deal with instrument manufacturing companies. This separation point emphasises the change between the craft of instrument making and the industry of slide rule manufacture. There is some inevitable repetition and overlap between the instrument makers and the manufacturers in the two periods, and this allows the transition points to be demonstrated and identified.

The 20th century is the period when the greatest numbers of slide rules were manufactured world-wide, and also the era that has seen the slide rule replaced virtually overnight by the electronic calculator, which rapidly became cheaper and easier to manufacture. It was also much easier to use.

The book considers slide rule types within these two periods, initially as specialist designs made by instrument makers in the earlier period, and then slide rule manufacturers producing a range of world-wide standard designs in the latter period. The appendices give supporting information and additional detail.

Chapter 1.

The History of the Slide Rule

The slide rule's early history and development are not comprehensively documented. It is now generally accepted that William Oughtred invented the slide rule between 1620 and 1625, but even this was not established until after a particularly bitter argument between Oughtred and a pupil of his, Richard Delamain. There were many claims and counterclaims as to the invention of many of the early devices, the form they took and so on. There were also a number of parallel developments within England and in other countries. This is not surprising considering how relatively slow and inefficient communications were in the 17th century.

Richard Delamain's Mathematical Ring of 1730 was apparently granted Royal Letters Patent (an early form of protection), but no details of this are known. The first English patent on a slide rule was only issued in 1753, to Suxpeach, some 130 years after the slide rule's invention and about 140 years after the modern patent system was adopted. There is thus over a century of development during which no patents were registered. Why this should be is not known.

The early development of the slide rule was mirrored by parallel development streams for other calculating devices, such as the sector, Gunter's line and Napier's bones. At about the same time came the early mechanical adding machines, the first being Pascal's Arithmetical Machine of 1642. These will not be considered in any great detail here; they are mentioned in order to put the slide rule into a comparative historical context. **Table 1** (p.17) compares the evolution of the various calculating devices and the milestones in their development from the beginning of the 17th century. **Appendix 8,** at the end of the book, looks at some major historical milestones, starting around 1000 A.D., in order to put the development of the slide rule into a wider historical context. To begin, we need to look at pre-slide rule calculation.

Pre-slide rule calculation.

The numerical notations of the ancient peoples, and the mathematical outlook of our ancestors, were not suitable for the mechanisation of calculations. There was little or no commonality of number bases used in most parts of the world, and even as these gradually standardised it took some time for a common numbering system to evolve. Most day to day calculation was performed using counting tables, pebbles, sticks and other variants of the abacus, which is one of the few "instruments" capable of coping with different numbering schemes. Indeed our word "calculate" stems from the Latin *ponere calculos*, to lay the stones, from the counting boards and stones used by Roman merchants for calculating. These counters, in rows, on strings and in frames, can be seen as the embryo of a number of mechanical calculator types.

The need for some form of mechanisation in calculation stemmed primarily from an increase in the calculations carried out in various fields of science, particularly in astronomy. The demand for more accurate navigation, the need for better and more accurate use of artillery, the requirement to survey lands, and the increased requirement for sophisticated calculation as research into the natural sciences began to escalate, all added to the need for more efficient methods of performing calculations.

A mathematician, or natural philosopher, was able to select from a considerable array of mathematical tools and calculating instruments. Immediately before the invention of logarithms, the abacus and the sector were used for most calculation; they both continued in use for some time and were further developed. After the invention of logarithms there was Gunter's line; then the slide rule and also Napier's bones, though these were unrelated to the invention and development of logarithms. The relative dates and simple chronology of these various instruments are shown in **Table 1**.

Individual preference and bias must inevitably have played a considerable role in the selection of a particular instrument by a mathematician. Each of the instruments evolved to some extent; however the development of the slide rule was inexorable. As improvements were made to the slide rule, it gained in popularity and gradually spread, first throughout Europe and then to America and Japan, until by the beginning of the 20th century, there were few parts of the "civilised" world that were not using slide rules to some extent.

The Sector.

The forerunner of the slide rule recognised as the earliest and most commonly used form of mechanical calculator was the sector (see **Figure 2**). It was used and developed for many years, specialist versions being produced for a number of applications, particularly for architects and gunners.

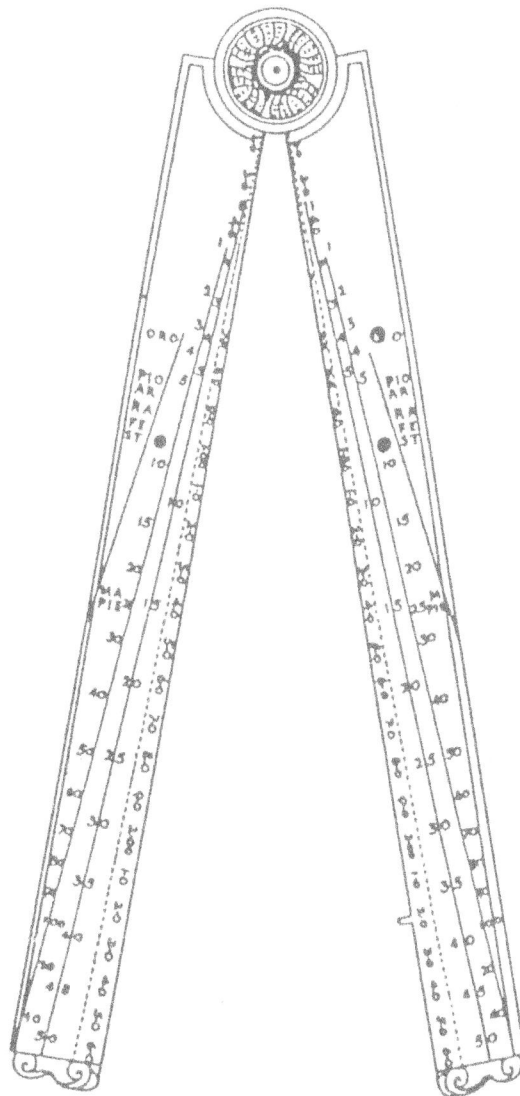

Figure 2 : The Sector

Sectors were still supplied in the late 19th century, and were in common use by mathematicians, scientists, natural philosophers and other professionals.

Invention of the sector is generally attributed to Galileo Galilei in approximately 1597; however this attribution was violently disputed by another Italian, Baldessare Capra, who claimed that he had invented the sector at about the same time. Later it was found that the sector was actually invented in 1568 by a friend of Galileo's who also made instruments for him, one Guidobaldo del Monte, and that Galileo only enhanced and popularised the device. It is known that in England both Hood and Gunter had each invented a sector in the late 16th century. It was also claimed that a version of a sector was produced in Belgium at about the same time. However, this latter invention was found to be a set of tables inscribed on a metal plate, and not a sector. The inventor had been aware of Galileo's work during his life, but had never claimed to have preceded his work.

The sector is a hinged and graduated rule which uses the theory of similar triangles, with the aid of a pair of dividers, for the calculation of natural numbers, squares, cubes, reciprocals, chords, tangents etc. It was popular until the late 18th century, when it began to be displaced by the slide rule. It is claimed that only the regular adoption of the cursor (runner or indicator) in the mid 19th century finally gave the slide rule true pre-eminence. A practised user of the sector with dividers was able to work faster and more accurately with his instrument than a slide rule user without a cursor. In 1866 W.F. Stanley wrote:

> *"The sector is a kind of twofold rule, commonly supplied with a case of mathematical instruments, as a kind of established ornament."*

Even so, a sector was supplied as part of a drawing set and in most boxes of navigational equipment for another 40 or 50 years, well into the 20th century.

Napier's Bones.

Another method of multiplying, dividing and extracting squares and roots was Napier's bones (see **Figure 3,** next page), invented slightly earlier than the slide rule and produced in many different forms. Because of its intrinsic design, the bones could not be developed in any comprehensive way, but were incorporated into some mechanical calculators.

Napier's bones, invented in 1617 by John Napier, who was also responsible for natural logarithms, were a pocket set of multiplication tables which were able to perform quite complex multiplication. They were most usually supplied as a number of rods in a case of some sort. The rods were marked with numbers in the form of an ancient Arabic series, which allowed multiplication to be performed quite simply by the addition of numbers. Variants of the simple rods, including rotating columns in a box, very fancy gear driven posts, and a number of other very ornate versions, were also available. Napier's bones remained in use for approximately 150 years; however the slide rule replaced them as the prime method of performing multiplication much earlier, probably as a simple result of the slide rule's convenience and ability to perform strings of multiplication and division much more easily.

Gunter's Line.

Gunter's Line of Proportion was the first interpretation of logarithms into a mechanical form. Edmund Gunter, who was professor of astronomy at Gresham College in London, was a friend of Napier. He produced a two foot rule on which the distance along the rule was marked with the logarithm of that distance, that is, as the "B" and "C" scales of a slide rule as we now know it. This scale could be used to perform multiplication and division by using a pair of dividers or calipers to step distances along, and adding or subtracting the distances (adding or subtracting the logarithms) as necessary.

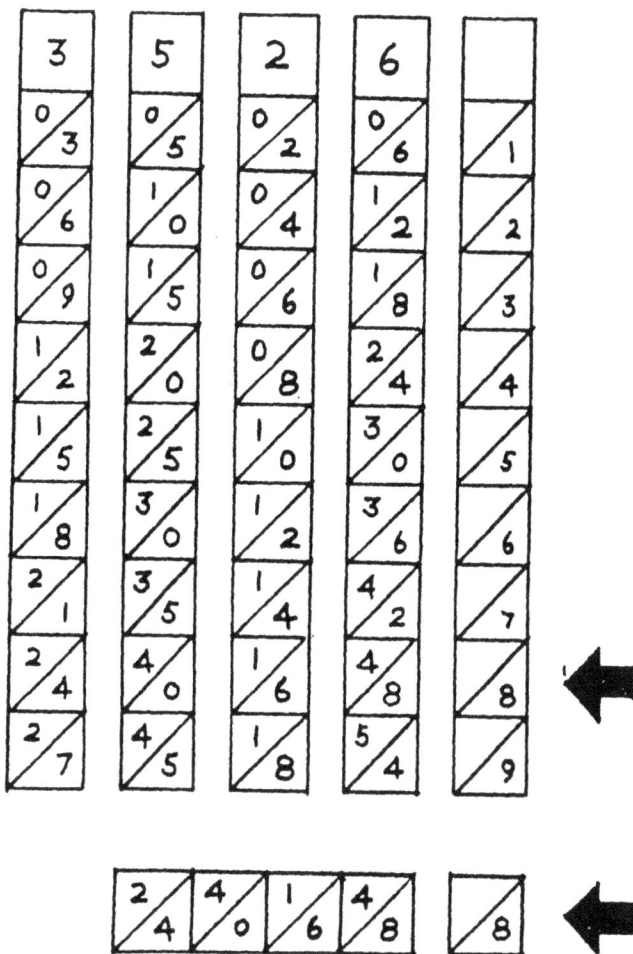

3526 x 8

Working from the right, add diagonally across the lines:
The last digit is 8, the next is 6 + 4 = 0, carry 1
then 0 + 1 + 1 (carried) = 2
then 4 + 4 = 8
first digit = 2, ANSWER is 28208.

Figure 3 : Calculating with Napier's Bones

Gunter was a prolific inventor and writer, with inventions of astrolabes, the surveyor's chain, and a sector to his name, in addition to the original logarithmic scale. Early Gunter's rules had few scales on the rule, generally a "line of numbers" equivalent to the "C" or "D" scales on a slide rule, a "line of proportion" equivalent to the squares or "A" scale, and a "line of cubes" equivalent to the "K" scale. There were also scales for gunnery calculations.

This form of calculator was capable of great accuracy and the Royal Navy continued to use Gunter's line as their standard form of navigational calculator until well into the 19th century. It was a relatively cheap, extremely robust and simple device, which was probably not as convenient, or as capable of increased functionality, as the slide rule which followed it. Later "Gunter's" (as they came to be called) had additional "lines" or scales added for some increased functionality. These included sines, tangents, equal parts, sines and tangents of rhumbs and other lines, depending on the device's intended use.

Gunter's line can be seen as an obvious first step in the direction of the slide rule, two Gunter's lines in immediate juxtaposition being effectively the "C" and "D" scales of a slide rule. A number of contemporary descriptions speak of such a pair of rules as the first slide rule.

Figure 4 : John Napier, Baron of Murchiston

The Slide Rule.

Napier invented natural logarithms in 1614. This must be taken as the seminal invention which would ultimately lead to the slide rule. It is interesting to note that Napier produced his bones after his invention of the logarithm. At about the same time, Jost Burghi in Switzerland also invented a form of logarithm, though this was never developed and exploited.

It took a number of other people to take the steps from logarithm to slide rule. The first step was the translation of Napier's "natural" logarithms into the more commonly used and convenient logarithms to the "base 10", also called Briggsian logarithms. This step was made by Henry Briggs, Professor of Mathematics at Oxford, in 1617 - the same year as the bones were invented. Briggs was also a friend of Napier; contemporary accounts describe Briggs' visiting Napier in Scotland and being full of praise and admiration over the invention of logarithms, as well as Napier's involvement with the translation to base 10.

The use of logarithms in a mechanical form for calculation was the vital next step, and is attributed to Gunter as described above. His vital contribution was the interpretation of the Naperian/Briggsian ratios into a two foot straight line, the earliest Line of Numbers.

It is now generally accepted that the inventor of the slide rule was The Reverend William Oughtred (see **Figure 5**), though there has been considerable discussion as to who had the basic idea originally. The accepted version is that Oughtred, a mathematics teacher, believed that his pupils should be able to work with numbers without the aid of any "new fangled" mechanical aids, and he put two of Gunter's lines together (side by side) for his own use, and later produced circular logarithmic scales.

Oughtred never bothered to publish or publicise his design in any way immediately after he invented it. It is believed that one of his pupils, Richard Delamain, saw the circular logarithmic scales turned into a circular slide rule, either at Oughtred's school or at his instrument makers. Delamain went on to publish his *Grammelogia or The Mathematical Ring* in about 1630 in which he described a circular slide rule that he claimed to have invented. This was actually the first published description of a slide rule, in both flat circular and spiral forms, with either a flat compass or an index/cursor.

Following this in 1631, Oughtred published, in Latin, his *The Circles of Proportion and the Horizontal Instrument &c.,* which described his circular slide rule, the Circles of Proportion. Another student of Oughtred's, William Foster, begged Oughtred for permission to translate his Latin script into English, and then went on to publish it with an Appendix. The Appendix, *To the English Gentrie ... the just Apologie of W. Oughtred against the ... insinuations of R. Delamain in a pamphlet called Grammelogia or the Mathematical Ring,* was published with Foster's translation of Oughtred's Latin manuscript in 1632 and attacked Delamain's claim to have produced the first slide rule. A later book, *An Addition unto the Use of the Circles of Proportion for the working of Nautical Questions,* was published in 1633. Professor David Bryden's article on *Grammelogia,* entitled *"A patchery and confusion of disjointed stuffe ... ",* gives a fascinating insight into what appears to have been a strong case of plagiarism, not only of Oughtred's design, but also of Thomas Browne's design of a spiral scale circular slide rule.

From the above, and from a statement Oughtred made in 1633 that as early as 1621 he had put two Napier-Gunter lines together to do away with the need for dividers on Gunter's line, we can conclude that Oughtred invented the slide rule in London sometime between 1620 and 1625. In his book he described a slide rule in its earliest form as including concentric logarithmic scales on a circular plate with a rotatable index. (Hutton in 1815 wrote that Oughtred had proposed a circular rule in 1627; we are not sure of his sources.)

Oughtred is also credited with producing a logarithmic rule for the Vintners Company in 1633. This is described in *The New Artificial Gauging Line or Rod: Together with Rules concerning the use thereof : Invented and written by William Oughtred,* published in London.

Oughtred's design of the original slide rule was fabricated by Elias Allen, a well-known instrument- maker whose shop was near St. Clement's Church in the Strand, and with whom Oughtred had a long and excellent relationship. Allen heavily influenced the design, which we can see in various museums. Oughtred's original design had two discs which rotated against each other. Each disc had concentric logarithmic scales, which would have been extremely difficult to make accurately. It was Allen who advised that the opening indexes on a single scale would be easier to make, fabricating what we now know as the Circles of Proportion (see **Figure 6**).

GULIELMUS OVGHTRED ANGLVS
ex Academia Cantabrigiensi, Ætat. 73. 1646

Figure 5 : The Reverend William Oughtred

In common with a number of other instrument makers, Allen had a workshop that was a meeting place for mathematical practitioners, natural philosophers and other scientists. Here information was exchanged and designs reviewed and admired. Allen was a mathematical instrument-maker in the Grocers Company, with a number of young men apprenticed to him. He founded a dynasty whose work would become famous world-wide.

Oughtred's high regard for Elias Allen and Foster was such that he bequeathed his designs for his Circles of Proportion to them. Oughtred *wished to reward a favourite pupil and assistant William*

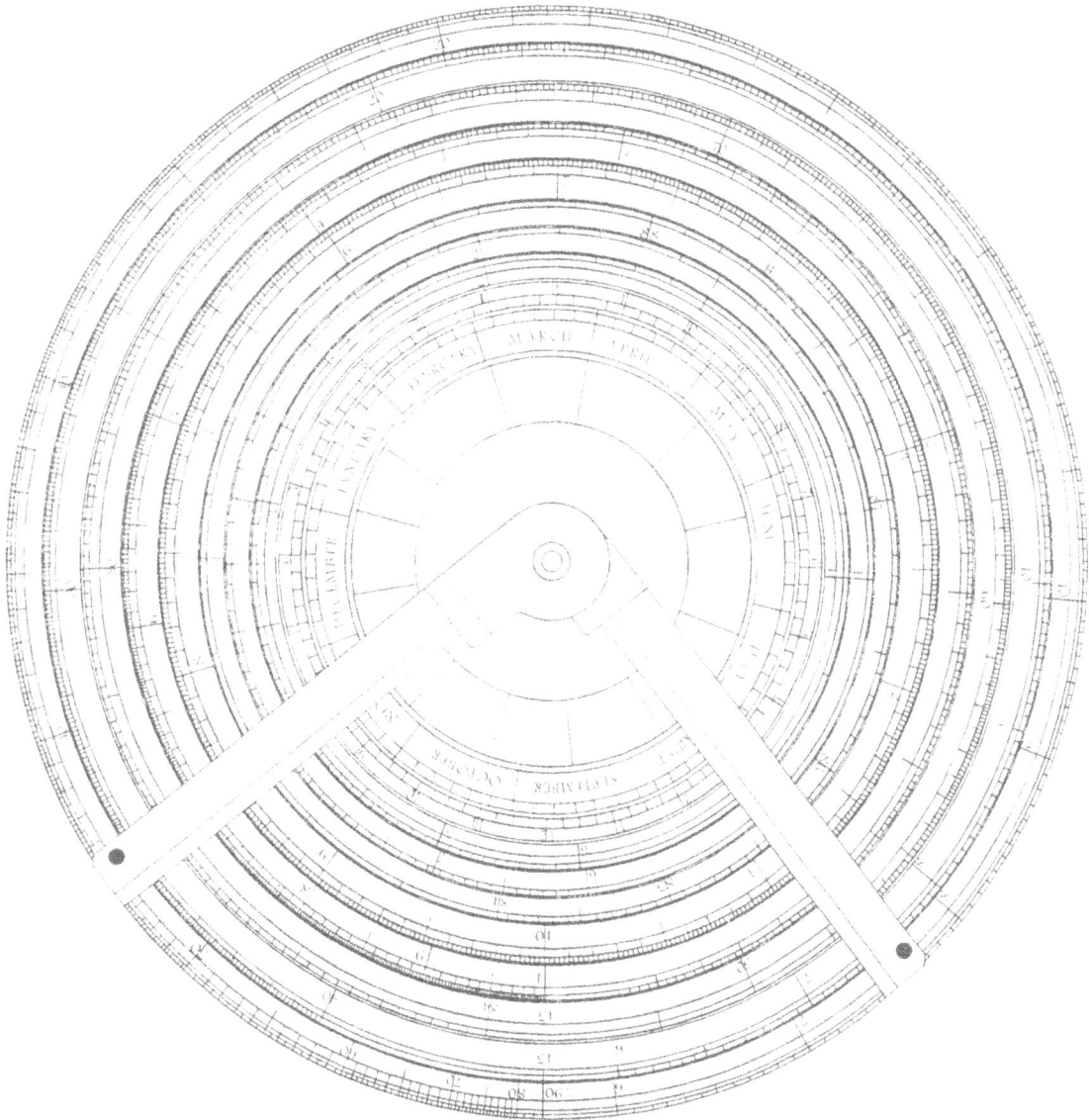

Figure 6 : William Oughtred's "Circles of Proportion"

Forster, who, with Allen who made the instruments, were to have the 'advantage', that is, any profit on the sales of the instruments.

Allen is known to have made a number of Circles of Proportion. The only known examples still in existence are one in the Whipple Museum, Cambridge; one in the Museum of the History of Science in Oxford; one in the Royal Museum of Scotland (made by Robert Davenport); and a recent model in the Science Museum, London.

The acrimony between Oughtred and Delamain could have stemmed from their common use of Allen as their instrument maker. The voluntary interchange of information would not have resulted in any acrimony. It was the plagiarism and priority claims by Delamain that resulted in the considerable debate and ill-feeling that caused Delamain to change his instrument maker to John Allen, whose relationship to Elias is not clear: he may have been son or nephew, or perhaps not related at all.

There were further priority disputes between Oughtred and Delamain on Oughtred's Horizontal Instrument, a plane projection of the sphere. This dispute also continued without resolution for some considerable time. Delamain went on to be Instrument Maker to the King under Royal Warrant.

The dispute as to who invented the slide rule continued to this century. Florian Cajori, in 1909, tried to show that Edmund Wingate was the inventor of the slide rule. It was only in the final appendix of the book that he acknowledged Oughtred. Other authors credited Bissaker, Everard and various other inventors. In some texts, Wingate was credited with the introduction of a slide to a fixed body in 1626. He was supposed to have produced his Lines of Proportion, or Rule of Proportion, by drawing log scales on two rules which were capable of being placed in juxtaposition. It is now recognised that Wingate, who was a friend of Gunter, actually modified the Gunter's rule and called it the Line of Proportion. This device was taken to France by Wingate when he went to be tutor to a French princess. He wrote prolifically on the Line of Proportion, or Echélle Anglaise as he called it, and its use, as well as on other mathematical instruments, but not the slide rule.

The next major development in the slide rule evolutionary saga is attributed to Robert Bissaker, who was credited with the introduction of a slide between fixed stocks in 1654. In it we can identify an instrument that bears more than a passing resemblance to the one which continued in use for the next 330 years. The Science Museum in London has a Bissaker rule dated 1654; it is the earliest known rectilinear slide rule in existence.

It should be noted that some attribute the introduction of a slide between fixed stocks to Seth Partridge. In 1658 he wrote an account of constructing a slide rule, the "Double Scale of Proportion", when he put together two Rules of Proportion (Gunter's). This was the final development of the basic rectilinear slide rule; it was to continue in this form to the end of its life.

Early instruments bore very grand and different names that attempted to describe in some way the mathematical function of the calculating device. Although Thomas Brown produced the first spiral scale slide rule in 1632, it was not called a slide, or sliding rule, until 1662 when John Brown, Thomas Brown's son, used the term in a book. It is difficult to understand why this particular name should have found favour as it is probably the least descriptively accurate. However, the name was gradually adopted and by the end of the 17th century was in general usage.

From this we can see that "Gunter's lines", that is, a mechanical interpretation of a logarithmic scale, were used in both straight and circular forms, on straight rules or discs of various material. Initially the straight forms were held against each other by hand. On the circular rules, dividers were replaced by an opening index, effectively dividers pivoted at the centre of the discs. We can identify the idea of a fixed stock with a moveable slide, and we can certainly see Oughtred's two lines as the forerunner of the "C" and "D" scales that are the basis of all slide rules to this day. It is justified thus that Oughtred should be credited as the inventor of the slide rule.

The importance of the slide rule continued to grow after the middle of the 17th century. Many inventors produced slide rules for general and specialist applications, and a number of designs are covered in detail later. Two of the most popular and effective designs were produced by Coggeshall and Everard. These became standards and were sold in large quantities for many years.

Coggeshall and Everard rules, in common with all rules at this time, were extremely crude and rather inaccurate. Even though Hooke invented his first gear-cutting engine in about 1670, it was to take another 100 years before dividing engines by Ramsden and others were regularly used, and the accuracy of rules radically increased.

The 150 years between 1700 and 1850 saw further gradual improvement in the accuracy of the slide rule, particularly following work by Boulton and Watt on the accuracy of the gradations. They produced what became the standard Soho engineer's slide rule, which they used for their work on steam engine designs. The increase in accuracy was mirrored by an increase in the functionality of slide rules. An example of this was the introduction of the log-log scale by Roget.

This was probably the single most important invention of the 19th century, even though it did not find favour initially, probably because there was no requirement for the type of calculation that the log-log scale enabled. Coulson, Hoare, Routledge and others also produced engineer's slide rules with additional facilities and in different styles. The Coulson slide rule had two slides, separated, within a dual-sided, open frame construction stock. Hoare's slide rule on the other hand had two slides side by side within a closed frame stock. The Routledge engineer's slide rule is a form of the Coggeshall carpenter rule.

Amédée Mannheim, a French artillery officer, is acknowledged as the inventor of the modern slide rule as we know it today. Mannheim's original design of about 1850 included the rearrangement of the scales into the present day grouping, with the general adoption of a cursor or indicator coming slightly later. The cursor is said to have been the final improvement, ultimately ending the popularity of the sector by producing a slide rule that was as convenient to use as the sector, with added functionality.

Mannheim rules were the earliest true standard produced from 1859. Initially very popular in France, they were produced in large quantities, and were imported to Great Britain in about 1880, where they were made by a number of manufacturers. They were imported into the United States in about 1890, where they also became a standard design manufactured by a number of companies. The final major milestones in the design during the 19th century came with Dennert and Pape's 1886 invention, in Germany, of celluloid scales. This was followed in 1890 by William Cox's invention of the dual-faced slide rule, with scales accurately correlated between the two sides, and a glass double cursor to enable transfer of figures between the two sides.

The years after 1850 saw great increases in accuracy and sophistication. Other inventions at this time primarily produced specialist long scale slide rule designs to increase accuracy, and other improvements such as the Rietz and Darmstadt scale layouts. It is interesting to note that prior to 1890, slide rules were manufactured in quantity only in England, France and Germany. In the early 20th century came the invention of the final major standard, the Rietz disposition of scales. This evolved from the Mannheim layout; seen by many as an improvement, it was probably the standard for the largest number of slide rules produced world-wide. Countless thousands of slide rules were manufactured in the 20th century; it is often said that the mark of a technologist or engineer during this period was ..." *the slide rule at the belt!*"

Both the mechanical calculator, available for many years, and the electronic calculator, available for a shorter time, were used in parallel with the slide rule. It was only the availability of cheap and reliable electronic pocket calculators from Sinclair in England in the early 1970's, and then from other manufacturers in different countries soon after, that effectively stopped the development and use of the slide rule virtually overnight.

As with the slide rule, the calculator was not immediately recognised as the boon it turned out to be. There are many tales of older professional people insisting that all calculations performed on an electronic calculator be checked either manually or with a slide rule!

Re-inventions.

As with the slide rule itself, several people are credited with the invention of its various elements. The cursor is a particular case in point. In about 1677, Newton was using log scales for calculating cubic and higher order equations and in so doing had to lay a straight line - a hair, hence the term hairline - across several scales. This was probably the first use of a runner or indicator. Some 100 years later the cursor was again claimed to have been invented, this time by John Robertson when he modified a rule for navigation and mentioned a runner or index. And yet again, in 1850 Amédée Mannheim claimed to have produced the first cursor.

There were various similar re-invented developments in the area of indicators; whether these "reinventions" were simply rediscoveries of existing devices is not known. It is probable, as with many inventions, that an idea or piece of equipment may have lapsed in usage and then been re-introduced at a later date.

The log-log scale provides a similar case. Roget first postulated the idea at the Royal Society in London in 1815. It was re-invented by Thomson in 1881. Yet other sources credit Professor Perry with the same invention in about 1900, confirmed by his patent of 1901. The probable scenario is that first Thomson and then Perry re-introduced the log-log scale with improvements, and Boardman made further improvements in 1933. These ideas came as a result of the increasing need to be able to perform calculations in thermodynamics, physics and electrical engineering, which were much simplified with a log-log scale. All of these important extensions were undoubtedly based on the original concept as proposed by Roget.

A summary of some of the key dates in the development of the slide rule is given in **Appendix 2**. These dates are put into a historical context in **Appendix 8**. In this, it can be seen that America had only been discovered for just over 130 years when Oughtred and Delamain were engaged in the verbal recriminations as to who had first introduced the slide rule to the world. It is likely that for many years few people actually cared, since so few used the slide rule until the early 1700's, almost 100 years later.

TABLE 1 : COMPARISON OF THE USAGE OF MATHEMATICAL CALCULATING INSTRUMENTS

	SECTOR	LOGARITHMS	NAPIER'S BONES	GUNTER'S RULE	SLIDE RULE	CALCULATOR	ABACUS
1575	1568 Guidobaldo I del Monte						
1600	1597 Galileo	1614 Napier					
1625		1617 Briggs / 1620 Log tables	1617 Napier	1620 Gunter	1620/5 Oughtred	1623 Schickard's Calc Clock	
1650						1642 Pascals Pascaline	
1675			1685 Coyle		1677 Coggeshall	1660 Morland	1665 English Exchequer
1700					1683 Everard	1670 Leibnitz	
1725							????
1750							
1775					1775 Robertson		
1800							
1825				1842 Bateman		1822 Babbage	
1850	1866 Stanley				1851 Mannheim		
1875			1885 Genaille / 1888 Blater				
1900					1902 Reitz		
1925					1935 Darmstadt		
1950							
1975						1972 Sinclair	

Chapter 2.

Descriptions and Definitions.

From the invention of the slide rule until production ceased, slide rule manufacturers were constantly searching for greater accuracy, robustness, simplicity and convenience of use, and inevitably a gimmick which would enable their particular brand to sell in larger numbers than those of their competitors. A large number of developments came during the Victorian era, a period that produced so many clever, but often less than practical, ideas. Many of these ideas, and the slide rules they engendered, are now known only by descriptions in the literature of the day, rather than by actual examples.

Most slide rules were manufactured in the rectilinear or rectangular format. The next most popular type was circular. This is somewhat surprising as Oughtred's earliest design was circular, and it can be postulated that this should have developed as the most popular construction. However, the circular slide rule was difficult to make, and Oughtred's earliest design with two discs was changed to the single disc with two cursors for ease of construction. There were also many designs of the tubular type and other splendid examples of the long scale rule. These were not made in such large numbers as the two main types.

It is possible to classify the various types of rules into the three basic forms; all others are derivatives of, or variations on, these.

Linear Slide Rule.

This shape is ideal for the essential features of a slide rule which are shown in **Figure 7** (next page). All rules have a stock in which a slide is able to move smoothly. The scales are engraved or otherwise placed on the stock and slide so that, in the simplest rules, identical "A" and "B" scales will be on the upper part of the stock and slide respectively, and identical "C" and "D" scales will be on the lower part. Early rules (prior to 1860) were not supplied with a cursor or other form of indicator which ran on the stock (in grooves or other tracks) to enable a better translation of figures between the various scales. With the development of the duplex (double-sided) rule, accurate cursors capable of translating between the two sides of the rule became essential.

Greater accuracy was possible by increasing the length of the rule. Linear slide rules of various lengths between 3 inches and 40 inches can be found. There are also longer (up to 7 foot) teaching slide rules. The sheer impracticality of a 40 inch rectilinear rule is apparent. During a calculation with the slide nearly fully extended at either end, approximately 8 feet of space is required. The additional accuracy provided by such rules would have been the major benefit when used in an office; the inconvenience of using them would have made them unlikely contenders for use in the home. Some special rules were made which were even larger; these will be dealt with later.

The shape of the linear rule also enabled greater numbers of special scales to be added, and the duplex rule enabled scales on the front and back faces of the rule. In some cases over 30 scales were available on one rule.

Prior to the availability of duplex rules, a number of manufacturers supplied multiple slides for use in the one stock, some of them with ingenious clips and mountings enabling them to be used simultaneously.

SLIDE

CURSOR

a) SOLID FRAME

STOCK

SLIDE

STRAP

CURSOR

b) OPEN FRAME

STRAP

STOCK

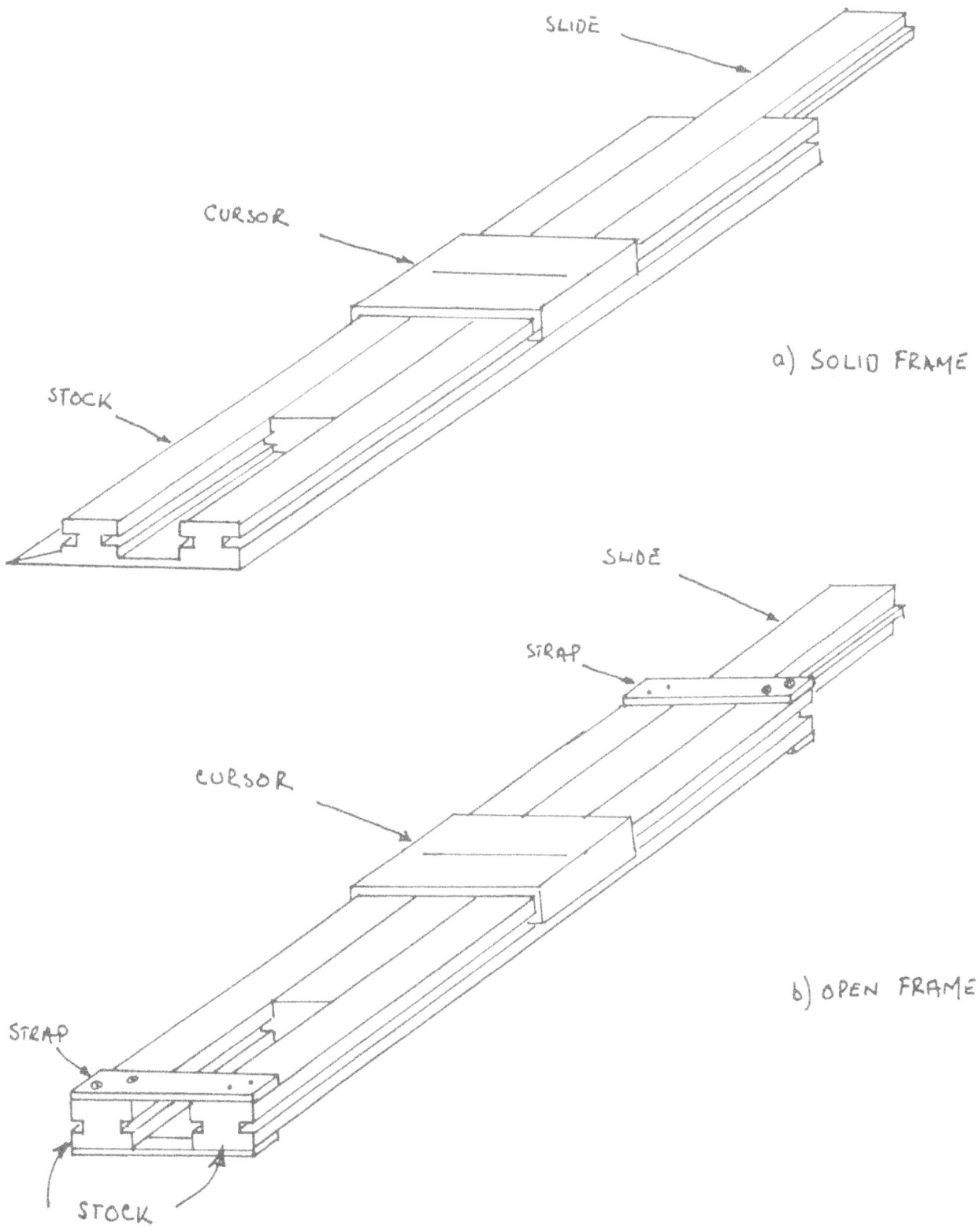

Figure 7 : Linear Slide Rule

The robustness of these rules is obvious. The materials used in the early ones included brass, silver and ivory as well as the more common hardwoods. Excellent examples of brass-bound wooden rules can still be found. The advent of plastic enabled cheap, robust rules to be produced in huge quantities. The late 19th century also saw the advent of celluloid-faced wooden rules, which had clearer scales on a very strong rule, a type which continued in production into the 1970's. Various plastics replaced the glass cursor, and in some makes it is possible to trace the development of the cursor from glass in metal, plastic in metal, and finally to all-plastic.

Circular Slide Rule.

This was the earliest form of slide rule, but probably due to the difficulties of marking or dividing circular scales, it never gained in popularity until the availability of plastic. Plastic was cheap and easier to mark, allowing economical circular rules to be made in quantity. Small circular slide rules were able to provide accuracy similar to that in a much longer rectilinear slide rule. A 3 inch diameter circular rule has a similar length scale on its outer edge to a 10 inch linear rule, and would certainly be much more "pocketable". Circular rules have an advantage over all other forms because the scales are continuous, enabling multiplication with fewer movements of the scales and cursor. The proponents of circular rules made much of this supposed advantage over the rectilinear rule. There is a modern equivalent to the discussion; we have the debate between the traditionalists who prefer the normal algebraic notation used in most electronic calculators, and the advocates of reverse Polish notation used in other types of calculators.

Figure 8 (next page) shows the component parts of the main types of circular slide rules. The most common is the multi-disc variant, which has an inner and outer disc and a cursor or cursors. The other type is the multi-cursor rule, which requires some method of enabling the cursors to be locked together and then moved relative to the discs.

Excellent examples of circular "pocket-watch" calculators from a number of manufacturers can be found. These are undoubtedly the top-of-the-range devices, with clever and complex gears and winders used to move the scales and cursors. Many of these are double-sided devices with some of the elements locked together.

Circular plastic slide rules were popular advertising gifts and examples are available in all countries. Examples of circular rules in all materials are available. Unless the scales are etched or engraved, painted or photo-anodised scales can lose detail as a result of careless handling.

Long Scale Slide Rule.

These designs, the result of the search for increased accuracy, extend scale length without the attendant increase in length which would be necessary with a linear rule. Long scale rules can be further classified into two sub-categories, the helical or spiral scale slide rule, and the multiple scale slide rule.

Helical scale rules are the more common form of long scale rule, with examples still found on sale at the end of the slide rule era. This type of rule has the scale produced as a helix on a tubular stock, with the cursor incorporated on a separate tube running on the stock. The helix format allows a scale many feet long to be put on a rule that is only a number of inches long. **Figure 9** (p.22) shows the component parts of such a rule.

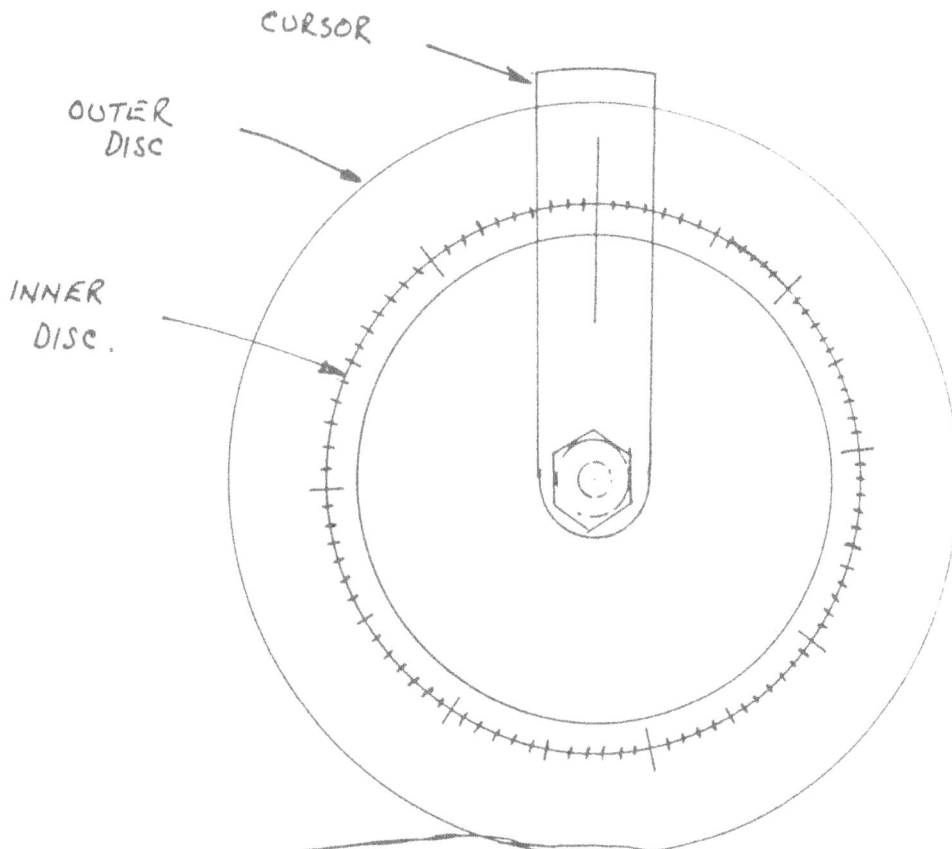

CURSOR

OUTER DISC

INNER DISC.

a) TWO DISC

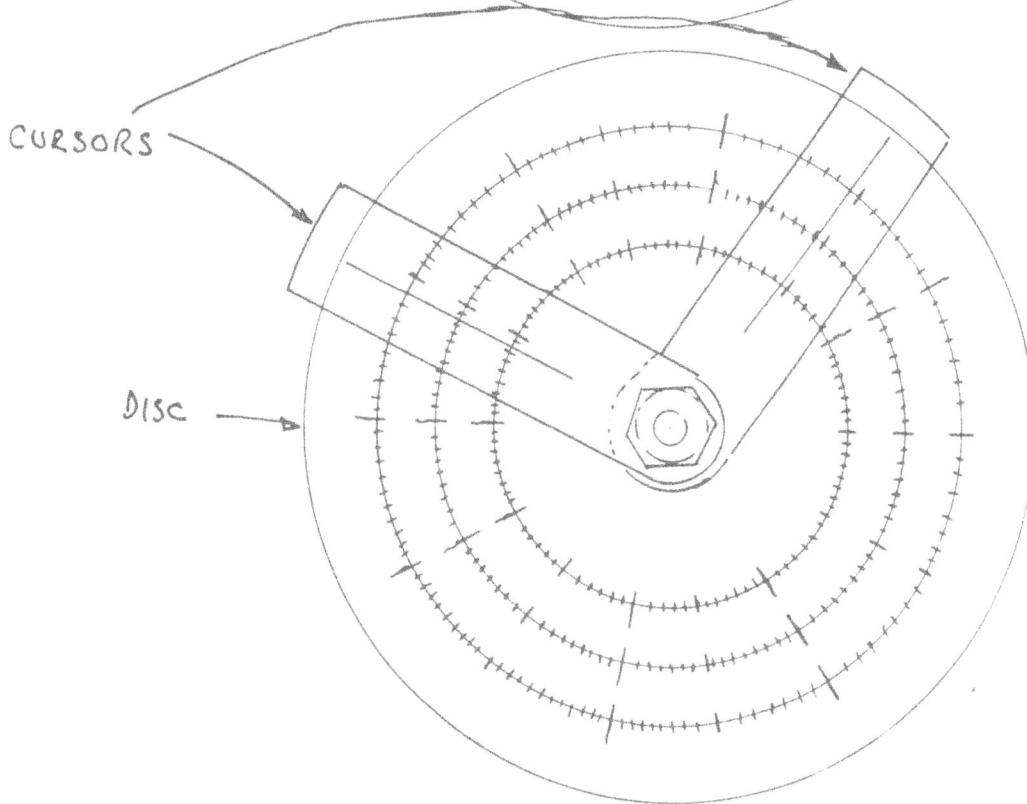

CURSORS

DISC

b) TWO CURSOR

Figure 8 : Circular Slide Rule

THE CYLINDER

CURSOR

THE HOLDER

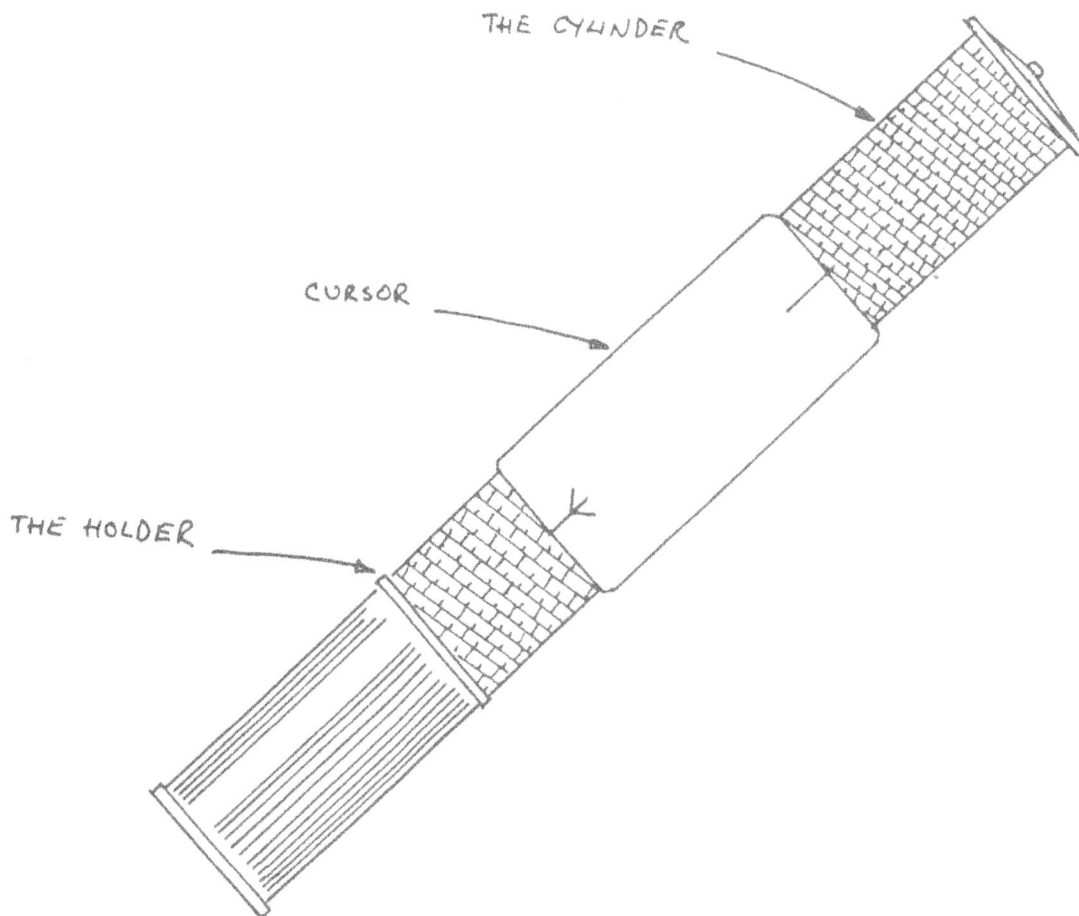

Figure 9 : Tubular/Helical Scale Slide Rule

Spiral scales were used to extend the scale length of circular slide rules, and examples are known with various numbers of turns in the spiral. Five turn spiral scales were relatively common.

Multiple scale or grid-iron rules came in many forms. All broke the total scale length into a number of shorter lengths with more or less overlap. These shorter lengths were then mounted in various ways so as to allow them to be turned to, or used with, some form of cursor or indicator during the performance of calculations. Equivalent scale lengths of many feet were possible in rules which were between 12 inches and 20 inches long, and a few inches in diameter. These were fitted with various devices to enable the scales to be locked in place or to enable the whole device to be moved linearly or rotated.

Special Features.

Some slide rules are provided with unique features and do not easily fit into any of the previous categories, either as a result of a constructional feature, a novel way of enabling a particular set of functions to be used, or a particularly clever piece of design which carries out all of the above. In no particular order we find the following:

Braille slide rule.

Keuffel & Esser produced a Braille slide rule, the model 4081-3 Log Log Duplex. This was available to special order, and was provided with a number of features for people with impaired vision. These included Braille marks on all scales as well as a large plastic holder with a wire cross hair which was mounted to the standard cursor frame. The holder could pass over the end plates of the rule to allow the wire cross hair to be aligned at both ends of the slide rule.

Scales in the well of the stock.

The well of the stock was an obvious place to add additional scales, and using these scales required some form of additional cursor on one end of the slide.

Faber-Castell Electro slide rules.

The slide on this series of rules was fitted with a metal, chisel-edged fork at the left-hand end. This enabled the electrical scales placed in the groove of the rule to be read and used with the scales on the front of the rule.

A.G. Thornton Electrical.

There was a similar feature on the Thornton Electrical rules. These, however, had a rounded end and a hairline marked on the stem of the extension to the slide which enabled the scales to be read as with a cursor, rather than the chisel edge of the Faber rules. The rounded end was fitted on slide rules just before World War II. The literature implied that the design of the rounded end resulted from user complaints relating to the sharpness of the chisel-edged fork.

Movement adjusters.

No matter how carefully the slide rule was made and the materials selected, because of wear or an inherent problem with the materials, movement of the slide within the stock may be either too loose or too tight. A movement adjuster provided a method of altering the stiction between slide and stock.

Note that the traditional method of regulating the movement of the slide within the stock was by fitting a curved cross-section piece of spring steel along the back of the closed frame. This applied pressure on the edges of the slide as a result of a slot cut in the back of the stock. In some cases this slot was cut through the complete thickness of the back of the rule, the stock being held together by the spring steel; in others the cut was only partial. Other methods included the insertion of a wavy shaped piece of spring steel into the slot to provide the pressure on the slide.

John Davis.

Some versions of these closed-frame slide rules were fitted with a sophisticated screw arrangement, within the width of the stock, to enable movement of the slide to be regulated.

Keuffel and Esser.

These were similarly fitted with a screw adjustment. However they moved only one side of the stock on their closed-frame design. The movement was up and down relative to the thickness of the rule.

Blundell, Blundell Harling.

These were fitted with screw adjusters within the width of the stock, and a small screwdriver was supplied for adjustment. Later versions had no adjuster other than the usual spring steel insert between the two halves of the slide.

Stanley.

Adjustment was provided within the thickness of the stock, one half of the stock being moved relative to the other by adjusting screws set through the base of the stock.

Stands.

Slide rule stands were available for a number of makes of slide rules, one of the most effective coming from Aristo (Dennert & Pape). This design was rather clever in allowing a duplex slide rule to be easily used on either side while it was lying on a desk.Other manufacturers had some form of stand built into the rule, either as a set of rubber feet, or as a feature of the plastic bridges joining the two halves of the stock

Cursors.

The cursor, also known as a runner or indicator, is an essential part of any rectilinear slide rule. Indeed it is what transformed the slide rule into a tool more popular than the sector and Gunter's rule. It is also a very distinctive part of all slide rules. To those people who have used a slide rule it is perhaps surprising that early users were able to perform calculations for so long without the benefit of a cursor. However, it was not until the early 1900's - some 50 years after Mannheim - that the large majority of slide rules, particularly of traditional design and for normal mathematical calculations, were fitted with a cursor as standard. Some special slide rules never had a cursor, these being deemed capable of sufficient accuracy in their reading without the benefit of this device.

The cursor in its simplest form is a piece of glass, mica, or plastic with a single hairline engraved on it, and held in some form of frame so that the hairline is at a right angle to the scales on the slide rule. The complete cursor can be moved along the body of the slide rule in grooves or slots. It is held at a right angle to the scales by a simple spring which also provides the necessary friction to stop the cursor from moving other than when so required by the user.

The construction and design of the cursor can provide an indication of the date of production when compared with pictures in dated catalogues. Some dates are as follows:

Basic metal with knife edges:	pre-1920
Wooden frame:	pre-1930
Metal and glass:	pre-1940
Multi-line:	post-1940
All plastic:	post-1960

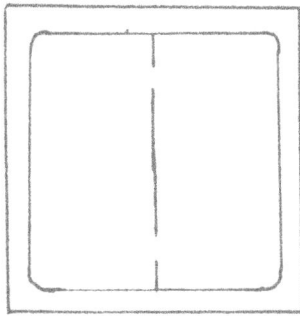

BROKEN-LINE CURSOR.

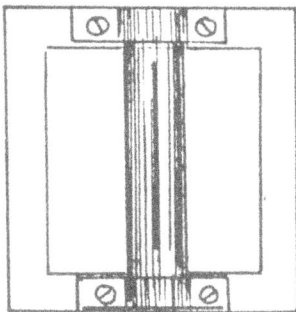

MAGNIFYING CURSOR.

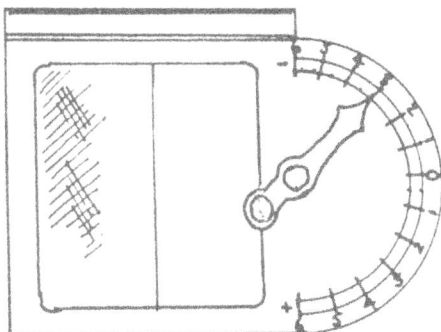

DIGIT REGISTERING CURSOR.

Figure 10 : Cursors

It must be stressed that these dates are very approximate, and because the cursor is transferable between rules as well as easily replaced after loss or breakage, it should not be used for definitive dating. Metal and glass cursors are seen on post-1960 slide rules; however the quality of workmanship of these is better than for the earlier designs, and the finish (chroming etc.) may give a better indication of the age of the cursor. Examples of various cursors are shown in **Figure 10**.

The cursor is essential on a duplex slide rule, as it is the best and simplest way of enabling figures to be transferred from scales on one side of the rule to the other. There are examples of some Aristo duplex rules where the cursor has to be taken off and placed on the reverse of the rule, a piece of penny-pinching which cannot have been popular with rule users.

The cursor was an area where considerable ingenuity was expended to increase functionality as well as accuracy. The multi-line cursor was produced with a second line (post-1930) and numerous other lines were scribed on them. These were a specific and constant distance from the basic centre hairline. These distances often varied according to manufacturer and function. The best-known examples are probably the kilowatt and horsepower lines on a number of electrical engineer's rules. Other examples relate to the diameter of a circle, and others have the main gradations from 95 to 105 on the respective scales repeated on the cursor. This is very useful for adding and subtracting small percentages.

The broken line cursor (see **Figure 10**) was seen on some early slide rules. This was an attempt to ease setting of the cursor, but it did not prove to be particularly popular.

The pointed cursor, which made its appearance in the 1930's, was a method whereby additional scales off the main faces of the rule could be brought into use. Some early examples extended the pointer over an inches scale on the rule to enable plotting lengths representing the logarithms of numbers for use in graphic calculations and for indicator diagrams.

The Goulding cursor (see **Figure 11**) was an ingenious device intended to provide a third or fourth significant figure to a reading from the scales. A metal frame carrying a plate with a triangular scale fitted into the normal cursor grooves. Chisel edges were used to read figures from the scales. These chisel edges were free to move independently from the frame, and were connected to a pointer pivoted on the frame so as to provide magnification of the movement of the chisel edges. The pointer could then be used to read off additional figures from the triangular scale. Whereas the Goulding cursor was a sophisticated form of magnifying cursor, more traditional magnifying cursors included a strip of plano convex glass along the hairline, the shaping of the complete cursor in a magnifying glass, or indeed a supplementary lens carried by arms from either the top or bottom of the cursor so that it could be used to read the figures with greater accuracy. A magnifying cursor with a magnifying power of about 2 will enable a 5 inch slide rule to be read with the accuracy of a 10 inch rule, and a 10 inch rule with the same accuracy as a 20 inch rule. One is shown in **Figure 10.**

The radial cursor (see **Figure 12**) was produced to enable complex powers and roots to be found using a slide rule. It utilised the theory of similar triangles to enable the complex power or root to be found from the normal and square scales.

Other special cursors included the A.W. Faber digit registering cursor (see **Figure 10**) which enabled the operator to register the number of digits to be added or subtracted as part of the calculation.

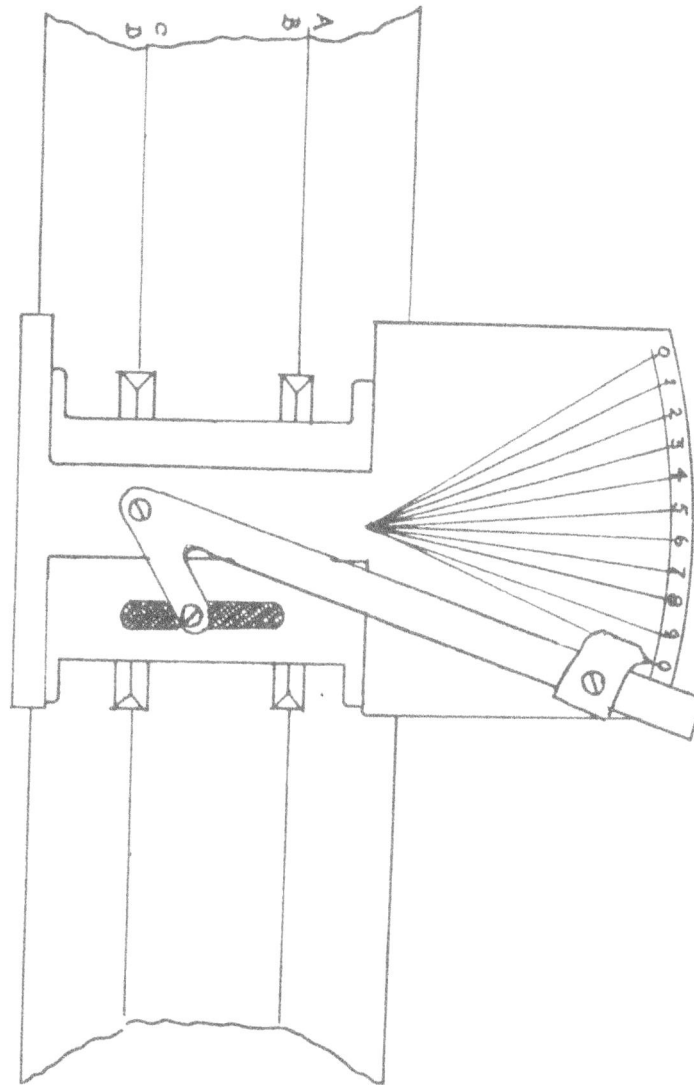

Figure 11 : Goulding Cursor

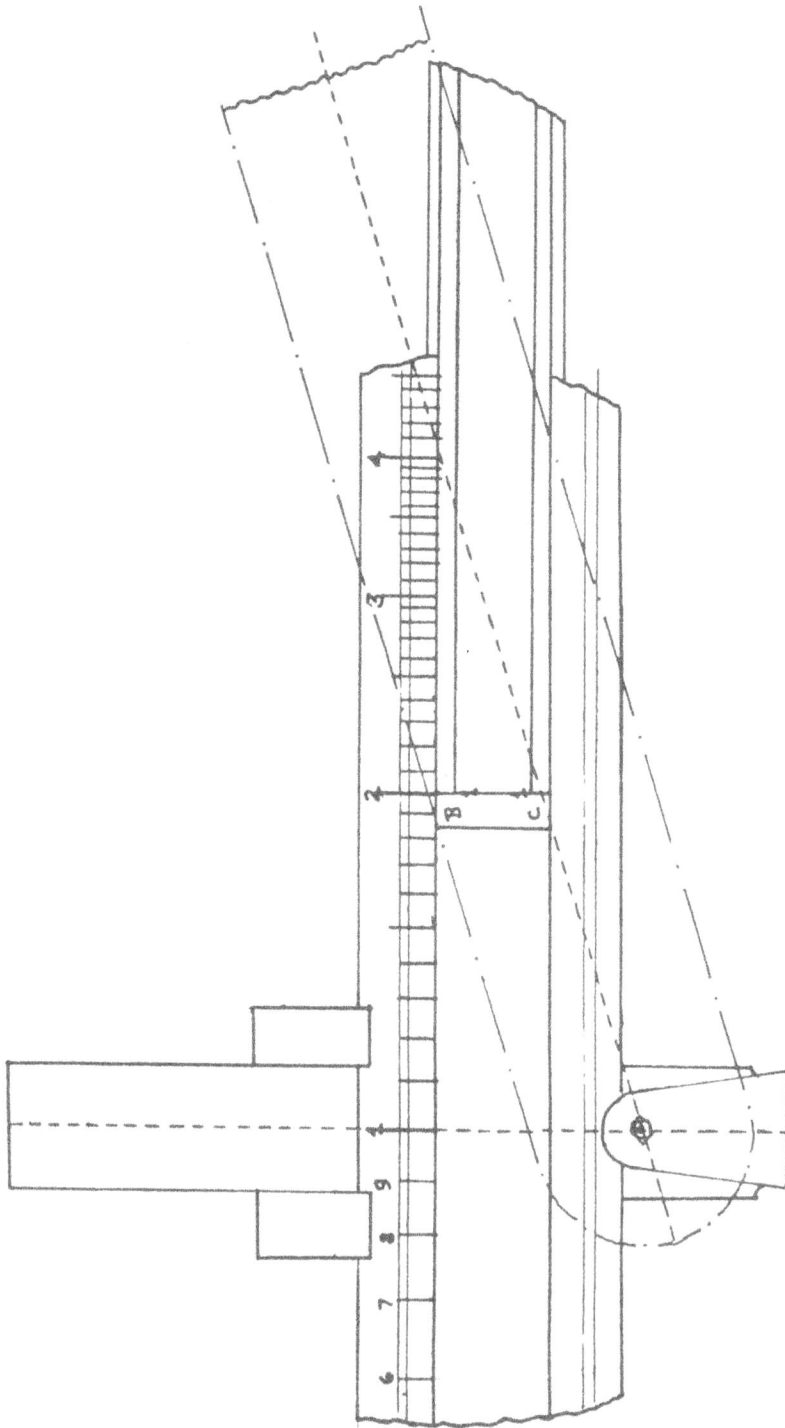

Figure 12 : Radial Cursor

Packaging and presentation.

An interesting sidelight in any study of slide rules is the packaging (the cardboard box that the slide rule comes in); the case, wallet or other tube (plastic, cardboard etc.) provided to protect the slide rule; and the presentation in the instruction leaflet or book. These provide an insight into the perceived quality of the product.

Packaging and cases.

The cardboard box originally provided with the slide rule was primarily there as a marketing tool that provided some minor advertising space. Very few items of the original packaging remain, as most users would have thrown them away and relied on the case or wallet provided to protect the device while in a briefcase or on a desk. These ranged from rather flimsy plastic wallets to more durable and high-quality leather wallets, from fragile cardboard boxes to more substantial high-density cardboard tubes, and numerous types of plastic containers. Most of these provided a high degree of protection. The ultimate containers were the leather and lined mahogany cases for tubular and other long scale devices at the top of the range. They must have been a considerable proportion of the total manufacturing cost if provided as part of the product, or an expensive special purchase for a proud owner.

The circular slide rule by its very shape is harder to package and the majority were provided with some form of wallet.

Of particular note:

– Unique slide rules were generally available in extremely flimsy cardboard containers, the ends of which were prone to collapse. Some versions were supplied in a hinged cardboard box, which was fragile and required a rubber band, or similar, to keep it closed in transit.

– Faber and Faber-Castell provided a variety of cases, from heavy duty cardboard tubes (many of which are still providing protection after 50 years, albeit with some deterioration of the ends), to plastic pull-apart tubes and a clever plastic rectangular tube with a hinged plastic end. Although the hinges are often beginning to deteriorate, these slide rules are still well protected. Some of the smaller models were supplied in plastic wallets which had a high survival rate, the Addiator models being in a zipped plastic wallet which included a stylus for its operation.

– Earlier Thornton rules were supplied in a heavy duty cardboard tube and later models in either a plastic wallet with cardboard reinforcing (which are generally in fairly poor condition by now), or in a very distinctive and durable plastic tube.

– Early Aristo models came in heavy duty cardboard tubes; later models were packaged in extremely distinctive red and white plastic tubes that were very durable.

– Fowlers Watch calculators were sometimes supplied in a small baize-lined hinged tin box, also in a variety of very sturdy leather wallets.

– Haldens Calculex watch type calculators were supplied in leather wallets, a lined tin case, and a rather splendid lined wooden case.

Instructions.

The format of the instruction leaflet for most makes of slide rule was such that by folding it lengthways in half or thirds, it would fit in the case or box of the larger (> 10 inch) rules and thus be readily available for the user. This also gave it a degree of protection. While this may have been a laudable aim, most users would seem to have found it an inconvenience and removed the instructions to a safe place, never to be seen again. Later models of some slide rules carried instruction leaflets which were long and thin to mirror the shape of the rule.

The majority of slide rules available now are thus without instruction leaflets, and any instructions still in existence are often in poor condition as a result of having had a slide rule pushed past in returning it to its case.

The instruction leaflet is a good source of information as to the original parentage of a slide rule that was bought in from a large manufacturer. While it may have been worthwhile reprinting the information leaflet with amendments to include the new supplier's name and model number, it would hardly have been worth changing or improving the text content. There are a number of examples where a comparison of the content of the instruction manuals has revealed such similarities that it has been possible to ascertain that the slide rules supplied under the Boots label were very probably Faber-Castell, and those supplied under the W.H. Smith label were more than likely from the Blundell Harling factory(subsequently confirmed by Blundell Harling). This is not always the case, however. Hemmi and a number of other Japanese suppliers often were not obvious as the suppliers of a number of other manufacturers' rules in the United States, and Dennert & Pape was supplier to many manufacturers in both the United Kingdom and the United States of America. Note that, in some cases, a look at the rule itself will reveal its parentage. In the case of D&P, the presence of either the initials or a German patent is a good indicator.

The major manufacturers also separately supplied more comprehensive books relating to their slide rules, the various editions of these booklets provide another source of dating information.

Descriptions of slide rules.

A format has been developed for describing slide rules in this book. Each slide rule is covered by a one- or two-line description. The one-line description covers "Simplex" slide rules where there are scales either on one side only, or where the slide may also have scales on the reverse. "Duplex" slide rules have a two-line description, the reverse scales being on the second line. Details of the format are given in Chapter 6. An example from the Aristo range is as follows:

1	2	3	4	5	6	7
Duplex Mod 968 Studio	25L	Dup	Pl	pl	LL01,LL02,LL03,A/B,L,C/LL3,LL2,LL1 A/B,C/D	

This shows that the Aristo Duplex Model 968 Duplex slide rule is a 25cm Linear slide rule in Plastic with a plastic cursor, with LL01,LL02,LL03,A/B,L,C/LL3,LL2,LL1 scales on the front and A/B,C/D scales on the reverse.

Chapter 3.
Slide Rule Types 1620 - 1850

The period from 1620 to 1850 encompasses the basic development of slide rules, starting with Oughtred's slide rule about 1620 through to the introduction of the Mannheim rule in about 1850. During this time the large majority of slide rules were made by individual specialist mathematical instrument makers who were recognised craftsmen plying their trade. These instrument makers advertised their capability and made instruments to order rather than speculatively. A mathematician or engineer who had designed a slide rule, or who had produced an improvement to an existing standard, would either commission his favourite instrument maker to produce the rule to his design, or would find someone with the relevant skills to do the appropriate work. In some cases the resulting rule would have come from a number of craftsmen, each having contributed his particular skills.

For the first 50 years of this period, the mathematical instrument makers jealously guarded the secrets of their trade, and each had some specialisation, either as the maker of a particular range of instruments or by having such specific expertise as engraving, woodwork, metalwork, etc. In 1677 Moxon published his *Mechanick Exercises,* which described the secrets of the instrument makers trade in some detail. This allowed the expansion of the trade and accelerated improvements in the technology by cross-fertilisation.

Oughtred's slide rule was hand-made and hand-engraved by Elias Allen from verbal information or from simple drawings. The mathematical instrument maker of this period was as important a man as the mathematical practitioner, and the relationship between them was synergistic. Mathematical instruments continued to be hand-made until well into the beginning of the 19th century. A statement from 1738 says that Maupertais "*...set out scales with his own hand.*" The resulting instruments are thus unique, and often will also have engraved on them the name of the customer. For example Bissaker's rule in the London Science Museum was made for T.W., whoever he may have been.

The major users of calculating instruments reflected the "industry" of the period, namely navigators, astronomers, gunners, people involved in the timber trade, and of course the excise man who was responsible for the taxation of alcohol. These last two trades, in particular, favoured the slide rule and were responsible for a large number of improvements in rule design, as well as a number of designs that became early standards in production. The timber and carpenter's rules, as well as gauging rules, were the earliest standards available from a number of instrument makers.

The earliest standard slide rule design came from Henry Coggeshall, who lived in Suffolk and was an expert in mensuration and timber gauging. The Coggeshall slide rule for the timber trade was invented in 1677. His first rule was a relatively standard linear design; a later design in 1682 used as its basis the established carpenter's two-fold, two-foot rule. This latter rule demonstrates how existing designs were used to good effect, in this case by incorporating a "sliding Gunter" in one leg of the rule for simple calculations relating to timber volume. One face of the sliding Gunter had the normal slide rule type scales with a number of gauge marks intended for calculating volumes of timber. The reverse of the slide had an inch scale so that when the rule was opened out and the slide extended, a yard could be measured. See **Figure 13** for an illustration of a later Coggeshall rule.

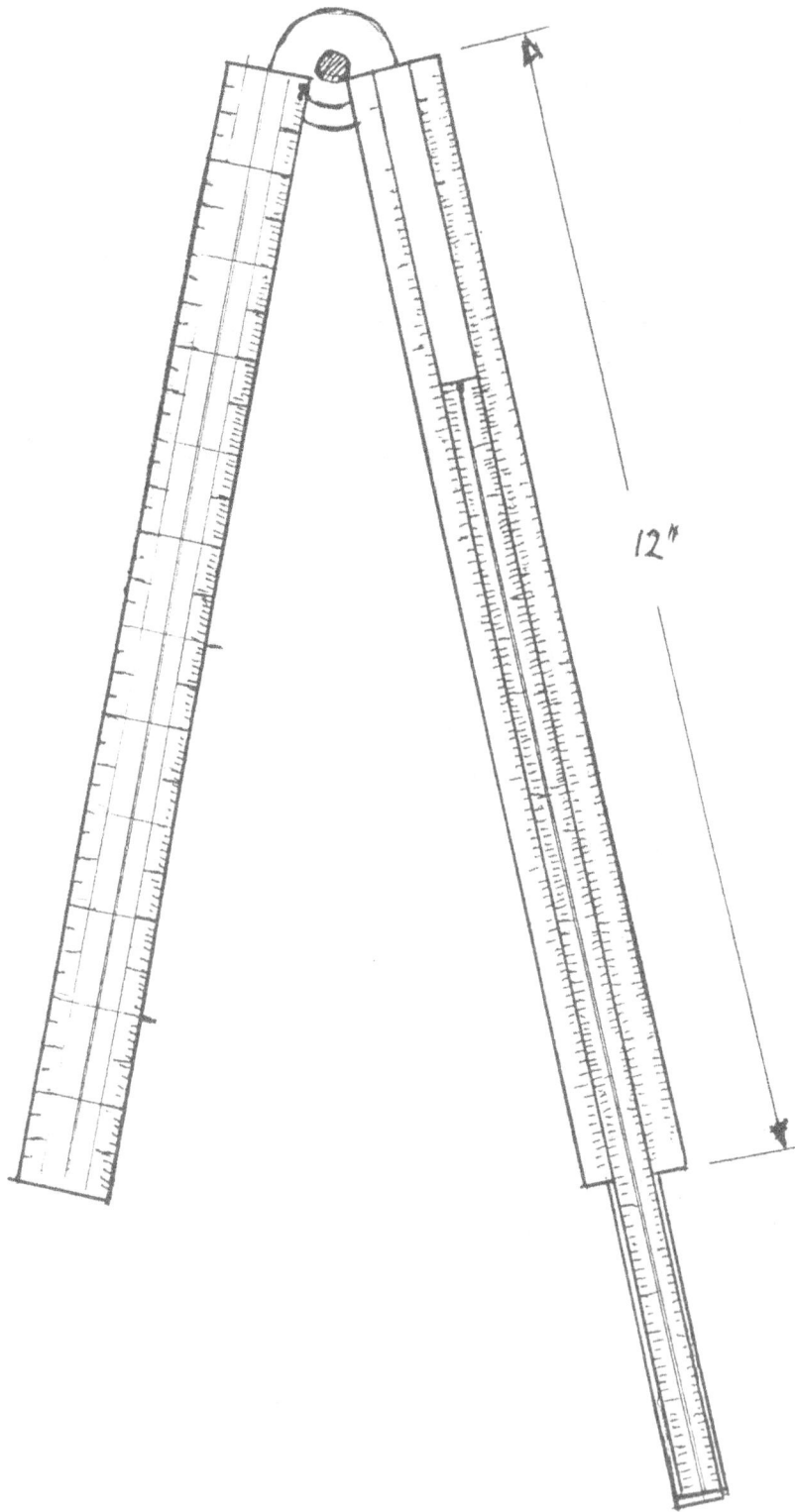

Figure 13 : Coggeshall Slide Rule

Several people produced improvements and variants to the Coggeshall rule, and it remained in use for over 200 years. Stone showed both versions of Coggeshall's rule. The Coggeshall rule was to become the earliest standard slide rule, and it continued to be made and used until 1874, even after the invention of the Mannheim rule.

Little is known of Thomas Everard other than that he was a gauger (customs officer) working in the excise department in Southampton. In 1683 he developed a slide rule, which also became a standard, with particular application to the gauging trade. This was a more typical rectilinear rule in the style of Bissaker, some two feet long, which had slides on the opposite faces of a square section stock. This rule continued to be developed over a number of years; Leadbetter added a third slide, and Shirtcliffe added the "SR" scale for ullaging. Vero, and a number of others, added more scales and gauge points as well as additional slides, until the original two-sided (non-duplex) slide rule finished up with slides on all four sides of the square section rule. **Figure 14** shows an Everard slide rule. Everard wrote in 1705 that many thousands of his rule had been sold and were in use.

In parallel, a number of different designs of slide rule were produced, some of which became minor standards in their own right. These included spiral scale circular rules and tubular rules with non-helical scales. Indeed by 1850 most designs that were to become common in the next 150 years had been seen in some early implementation. The majority of these designs were aimed at improving the accuracy of the calculations which could be achieved from a device that was normally less than 2 feet long at its most useable and convenient size.

The earliest spiral slide rule is credited to Brown in 1631, and another design to Milburne in 1650. Stone (1723) mentions Brown's design but makes no mention of Milburne. It may be that Milburne's design was made by Brown, though why there was the difference in dates is not known. Little is known of either design and they do not appear to have been popular.

William Nicholson, a prolific inventor, was responsible for a large number of slide rule improvements and ingenious designs in the latter half of the 19th century. Not many of the designs were actually produced in any large quantity (if at all), though many of his designs and improvements were re-invented several times in Great Britain and elsewhere during the following decades.

The use of the slide rule for engineering only achieved any sort of prominence after Boulton and Watt modified a number of rules from John Rowley in London and ultimately produced their Soho rule (named after their factory in Birmingham) for engineers in 1779. The Soho rule was a traditional closed-frame design with four scales, a standard square scale (two radius) occupying the "A", "B" and "C" positions, and a standard single radius scale occupying the "D" position. There were also inch scales graduated in various fractions. (See **Figure 15** p.36.)

There followed a number of other engineer's rules in various formats. Routledge produced a design utilising the two-foot two-fold rule; a similar design from Carrett and Coulson produced a two-slide rectilinear design with separate slides. There were also designs from Bevan, Hoare and others, but Boulton and Watt were the main protagonists who improved the graduation of rules rather than looking for more esoteric designs as the way forward.

By 1850 dividing engines had been in existence for some 100 years but were still virtually unknown in the manufacture of slide rules. The majority of rules were still hand scribed from a master pattern, and were still basically a craft produced instrument. It was another 50 years before slide rules were commonly produced with the assistance of automatic machinery.

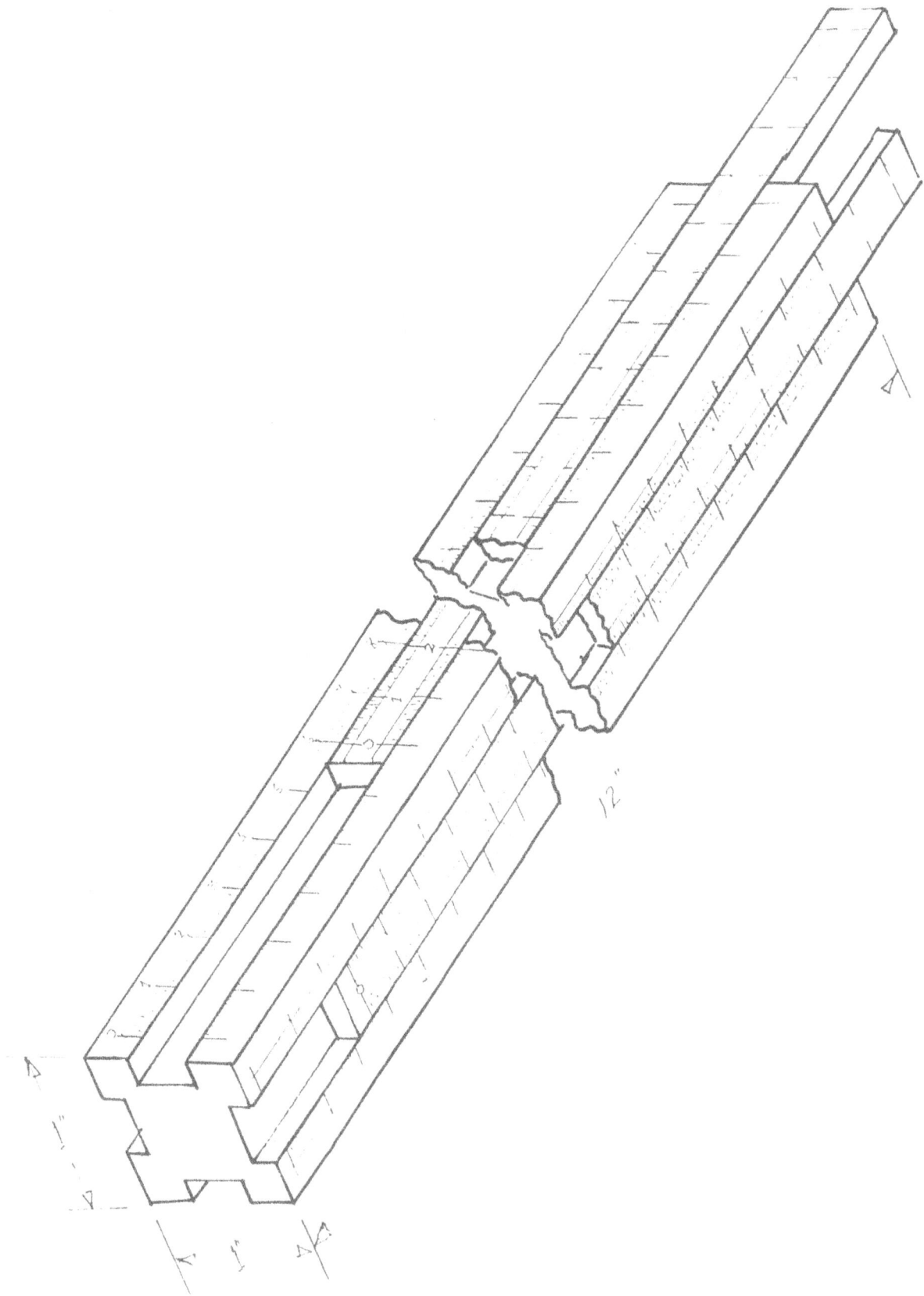

Figure 14 : Everard's Slide Rule

Before 1770 slide rules were used in large numbers only in Great Britain. This fact has been attributed to the English education system, which taught decimal fractions at school, this being particularly apposite to the use of not only the slide rule but also other forms of mechanical calculator.

France, as England's nearest neighbour, benefited from the transfer of slide rule technology. Wingate moved to France soon after the publication of the earliest books on slide rules, and he wrote prolifically on the Gunter's scale. Gunter's rule, and then the slide rule, began to find common usage there, particularly after the French Revolution when, in 1795, there was a declaration that all public servants would have to be familiar with the use of slide rules as part of their entrance exams. There was also, around this time, a change of measuring units in France; the adoption of the metric system permitted many advances.

Great Britain's continuing use of £.s.d and lbs/ozs was a burden for many years to come and resulted in a range of very special slide rules with unusual scales dealing in UK quantities. The year 1824 was an important date for slide rules in England, as this was when the imperial system of units replaced the earlier Winchester system. The 20% larger imperial gallon was introduced to replace all liquid gallons. In gauging, this replaced the Wine Gallon, and led to different gauge points being scribed on rules. A gauge point scribed at 2150 for MB (Malt Bushel) denotes a pre-1824 rule, and one at 2219 for MB is post-1824, though there are examples of pre-1824 rules with the scales planed off and re-scribed, illustrating that slide rules were expensive tools not readily replaced. Due to difficulties in implementation, the changes in gauge points took place mainly in 1825 and 1826. Another date to note is 1818 when duty on verjuice was removed; any rule with verjuice gauge points predates that year. The arrival of imperial measures rendered the W, AG, MB and WG points redundant, to be replaced by IM G and IM B on the appropriate scales.

The use of Gunter's rule by the Royal Navy also continued well into the 19th century. It was only when the cursor came into general use with the Mannheim slide rule, thus making the slide rule much easier to use, that it gained popularity with navigators.

Inevitably, the need to be able to lob cannon balls more accurately also added to the requirement for quicker and more accurate calculation. The earliest mechanical aid for the artilleryman was the sector; it is thus wholly appropriate that the first standardised slide rule should have been invented by an artilleryman, Mannheim. Following Mannheim's rule we see a plethora of artillery and other gunnery slide rules, particularly during World War I with its huge reliance on artillery.

Probably the single most important invention after that of the slide rule itself was the invention of the log-log scale by Roget in 1815. Initially this invention did not find any tangible use, as there was no real need for it. However, with the mathematical requirements generated by electrical engineering and thermodynamics some 80 years later, these scales were re-invented several times, notably by Thomson in 1881 and Perry in 1901.

The period 1620-1850 saw the advent of many designs, few of which became true standards or were made in any large quantities. **Table 2** (p.37) lists a number of well-known designs and their inventors, as well as the makers where known. The list is comprehensive to 1850; after this date further research may well be required. Cajori lists some 250 types of slide rules invented in the 19th century, with an additional 90 designs in the first decade of the twentieth century. Designs after 1900 are covered in **Chapter 5.**

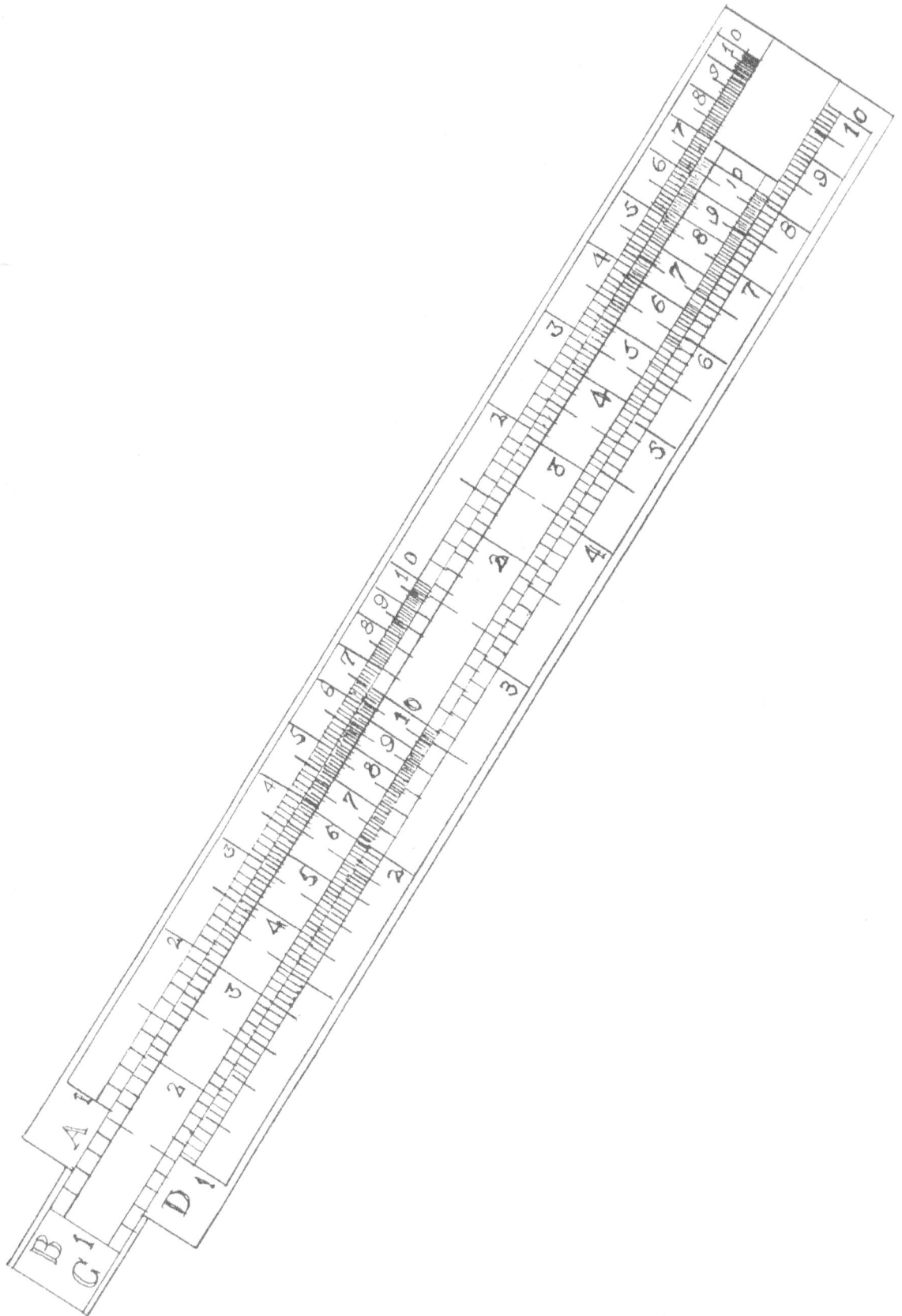

Figure 15 : Soho Slide Rule

Table 2 : Slide Rule Designs and Makers 1620 - 1900.

DATE:	SLIDE RULE:	MAKER:
1624	William Oughtred's rule	Elias Allen, London.
		Robert Davenport, Edinburgh.
1630	Richard Delamain's Mathematical Ring	Elias Allen, London.
		John Allen, London
1631	Spiral slide rule	Thos. Brown, London.
		John Bleyghton, London.
1650	Millburne's spiral slide rule	
c1650	Horner's folded shorter rules	
1654	Robert Bissaker's slide rule	Bissaker, London.
1657	Seth Partridge rule.	Walter Hayes.
c1660	Mr Brown's spiral 5, 10 or 20 turn rule	Brown, London.
1664	Brown's slide rule for Samuel Pepys,	Brown, London.
	the scales engraved by Cocker	
1677	Henry Coggeshall's first slide rule	
1670	Boxwood spiral slide rule	John Brown, London.
1682	Coggeshall's second slide rule,	
	a hinged two-foot, two-fold rule	Wood & Withnoll, Birmingham.
1683	Thomas Everard's slide rule	Isaac Carver, Horseydown.
1685	Encyclogium for Mathew Norwood	Walter Hayes, London.
		Anselm Jenner, Bristol.
1697	W. Hunt's slide rule	
1699	W. Hunt's slide rule	Philip Lea, London.
1700	Saveur slide rule, similar to Robertson's design,	Gevin and Le Bas.
	the earliest French slide rule	
1707	John Ward's rule	John Rowley, London.
1717?	Vero's modification to Everard's slide rule	
1722	John Warner's modification to the Coggeshall rule	Thomas Wright, London.
1727	Clairault slide rule, 21" dia. cardboard.	France.
1733	Benjamin Scott's circular slide rule	Benjamin Scott, London.
1734	Thomas Hill's new sliding rule	John Coggs Jnr., London.
1739	Brannan's 4' gauging rule (joint rule)	
1739	Charles Leadbetter adds a third slide to Everard's rule	
1740	Shirtcliffe adds the SR line to Everard's rule	Coggs & Wyeth, London.
1741	Charles Camus' slide rule	France
1748	George Adams' spiral slide rule	George Adams, London.
1750	Leadbetter's improved (Everard) sliding rule	Heath and Wing, London.
1750	Segner slide rule	Gottingen, Germany.
1753	Suxpeach's patent slide rule (Catholic Organon)	Benjamin Parker, London.
1761	Lambert's slide rule, wood & metal	G.F. Brander, Augsburg.
1774	John Robertson's improved slide rule with runner	Nairne & Blunt, London.
1777	Segner's improved slide rule	Gottingen, Germany.

SLIDE RULES

DATE:	SLIDE RULE:	MAKER:
1779	Watt & Boulton Soho rules for engineers	William Jones, Birmingham.
1780	Gauging rule with 4 slides	Joseph Stutchberry, London.
1795	A.S. Leblond circular slide rule	France.
1797	Horton's new improved rule	
1797	William Nicholson's various slide rules	
1798	Ewart's cattle gauge	
1799	Francois Gattey circular slide rule	France.
1805	J. Routledge engineer's slide rule. D scale 1-10.	Wellington, London.
1810	Gattey's Arithmographe	France.
1814	Woolaston's slide rule for use in chemistry	W. Cary, London.
1815	Roget's slide rule for involution and evolution,	Rooker, London.
	Roget's logometric circular rule, and	
	Roget's logometric chart	
1815	Silvanus Bevan's engineers rule	W. Carey, London.
c1815	Jomard règle à calcul	M. Collardeau, France.
1816	Hoyau, bôites à calculeur (cylindrical)	Hoyau, France.
1817	Silvanus Bevan's improved slide rule, 6"	Bate, London.
1818	The Bate Rule (imported into France by Benoit)	Bate, London.
1818	Benoit graduates English rules	W & S Jones, London.
1818	Henderson's double slide rule	W & S Jones, London.
1818	Thomas Donn's circular interest tables	
1821	Clouet's règle à calcul	Clouet, France.
1821	Lenoir's règle à calcul, copper 35cm.	Lenoir, France.
1822	Benjamin Bevan's circular accountant's rule	W. Carey, London.
		Stanley, London.
1823	Thomas Young's sliding rule for gauging casks	
1825	Schneider's rule	F. Dubler, Berlin
1827	Farey's Improved 'Soho' slide rule, 25".	Rooker, London.
827	Woolgar's calculator	Rooker, London.
1829	Lamb's circular slide rule	Lamb, London.
1829	Downing's slide rule	
1830	Saddington's slide rule	Thomas Saddington, London.
1832	Robert Hawthorne's railway slide rule	G.T. Bramwell.
1839	Thomas Kentish's compound slide rule	Dring & Fage, London.
1840	Coulson's slide rule (boxwood)	Dring & Fage, London.
1840	Arithmometre	Mme Dondey Dupre, Paris
1840	Lalanne, Arithmoplanimetre	
1842	Coulson's slide rule, 2 slides.	
1842	Henderson's double slide rule	Jones, London.
1842	James Chesterman circular cattle gauge	Chesterman, Sheffield.
1842	Woolgar's pocket-book rule	Rooker, London.

DATE:	SLIDE RULE:	MAKER:
1842	MacFarlane's circular interest rule and MacFarlane's cylindrical slide rule	
1843	L.C. Schulz von Strassnitzki, first Austrian rules	Vienna.
1844	Ewart's improved cattle gauge	J. Cail, Newcastle on Tyne. James Tree, London.
1844	Palmer's endless self-computing scale and pocket scale	
1845	Bradford's sliding rule	
1845	John E. Fuller's time telegraph	
1846	Lalanne, tableau graphique	
1846	J. Fuller's computing telegraph	J. Fuller
1847	Eschmann-Wild's Tachymeterschieber	J. Stambach, Aarau.
1847	C. Hoffmann slide rules	T. Baumann, C. Dorffel, C. Grunow, Berlin.
c1850	Sir William Armstrong, carpenter's slide rule, D from 4-40.	J. Rabone, Birmingham.
1850	Porro's scale logaritmiche centesimali	J. Porro, Italy.
1850	Altmutter's Rechenschieber (cardboard)	G. Altmutter, Vienna.
1850	Werner's Rechenschieber (boxwood)	W. Werner, Vienna.
1850	Werner's Tachymeterschieber	W. Werner, Vienna.
1851	Lalanne, règle à calcul à enveloppe verre	
1851	Regolo calcolatorio di oesterle	E. Sedlaczek, Vienna.
1851	Règle Mannheim (Tavernier-Gravet slide rule)	Tavernier-Gravet, Paris.
1851	Regolo calcolatorio di Schwind	E. Sedlaczek, Vienna.
1851	Regolo di Sedlaczek per i calcoli d'interpolazione	E. Sedlaczek, Vienna.
1851	E. Sedlaczek	Vienna.
1851	Nystrom's calculator	Wm. Young, Philadelphia. George Thorsted, New York.
1853	Sargent's règle à calcul	
1854	Mannheim règle à calcul cylindrique	
1854	Prestel's Arithmetisches Schiebe	Hanover.
1856	Regolo di Higgison	E. Sedlaczek, Vienna.
1856	Regolo per iscopi costruttivi e geodetici del Prof. L.C. Schulz von Starssnicki	E. Sedlaczek, Vienna.
1857	Bouche, Helice à Calcul	
1859	Quintino Sella	Italy.
1860	Dubois, Arithmograph	
1860	J.Fuller's telegraph computer	J. Fuller.
1861	W.E. Carrett's slide rule	E. Preston, Birmingham.
1861	W.H. Bayley's "One-Slide" rule and W.H. Bayley's "Two-Slide" rule	Elliot, London. Elliot, London.
1861	Routledge's slide rule (Coggeshall type)	W.F. Stanley, London.
1863	Delaveleye's règle à calcul. 2.3 m rule	

DATE:	SLIDE RULE:	MAKER:
1863	Delamoriniere règle à calcul. 1.2 m rule	
1864	Burdon règle à calcul. A log-log rule	
1864	Sonne's Rechenschieber	Landsberg & Parisius, Hannover
1866	Everett's universal proportion table, or Grid-iron slide rule	
1866	Everett's cylindrical slide rule	
1867	Hoare's slide rule, 2 slides.	
1868	Moinot, règle logarithmique pour la tacheometrie	Tavernier-Gravet, Paris.
1871	Mannheim règle à calcul cylindrique (made in wood)	
1871	Heather's cubing rule, 3' rule	
1871	Kentish slide rule	Dring & Fage, London.
1871	The Timber-contenting rule, 2 sided, 2'	
1871	Ipsologista di Paola de Saint-Robert	Italy.
1871	Regolo Soldati per i calcoli di celerimensura	Italy.
1872	Weber's Rechenkreisen	R. Webber, Aschaffenburg.
1872	Clouth Rechenschieber	Leipzig.
1873	Mannheim règle à calcul cylindrique (made in metal)	
1873	Goulier, règle pour les levers tacheometriques	Tavernier-Gravet, Paris.
1874	Rechenstab von Dennert und Pape	Dennert & Pape, Germany.
1874	Coggeshall's slide rule (still being made at this time)	
1875	Darwin's slide rules. 2 small slide rule designs.	
1875	Culmann's Rechenschieber fur Distanzmessern	C. Culmann, Zurich.
1875	Pestalozzi's Rechenschieber	C. Culmann, Zurich.
1876	De Montrichard's, règle pour le cubage des bois	Tavernier-Gravet, Paris.
1876	Lala Ganga Ram's slide rule	W.F. Stanley, London.
1877	Puscariu's Stereometer	Johann Ritter von Puscariu, Budapest
1877	Hawthorn's slide rule	W.F. Stanley, London.
1877	Hermann's Rechenknecht	Wiesenthal u. Cie, Aachen
1878	G. Fuller's spiral slide rule	W.F. Stanley, London.
1878	Boucher calculator	W.F. Stanley, London, et al.
1880	Cherry's calculator	Henry Cherry.
1880	A. Wust, Taschenrechenschieber (printed on card)	Halle.
1881	Thacher's calculating instrument	Keuffel & Esser, New York.
1881	Ruth's Rechenschieber	Augsburg.
1881	Scheve's Artillerist Rechenschieber	Berlin.
1881	Thomson's log-log rule	Dring & Fage, London.
1882	Charpentier calculator	UK/Morin, France.
1883	Cherepashinskii's slide rule (one example)	Tavernier-Gravet, Paris.
1883	M. Kloth's Apparat	
1884	Hannyngton's extended slide rule	Aston & Mander, London.
1885	Lala Ganga Ram's second slide rules	Holtzappfel & Co, London.
1885	Tower's slide instrument	Beauchamp Tower.
1885	Dixon's rules - triple radius double-slide rule	

DATE:	SLIDE RULE:	MAKER:
1886	Rechenschieber von Zellhorn (celluloid faced)	Dennert & Pape, Germany.
1886	Lebrun, règle pour les calculs de terrassements	Tavernier-Gravet, Paris.
1886	Regolo psicrometrico di A. Prazmowski	Warsaw, Poland.
1886	Beyerlen's Rechenrad	Tesdorpf, Stuttgart.
1886	Regolodi F. Stapff	Italy?
1887	Burt's patent measuring and cubing slide rule	Dring & Fage, London.
1888	Sanguet, règle pour les levers tacheometriques.	Tavernier-Gravet, Paris.
1889	Paulin, règle pour les calculs terrassements	
1889	Hart's Proportior	Leipzig.
1890	Pouech, Echelles Enroulees en Spirales pour les Racines Carrees et Cubiques	Italy?
1890	Hasselblatt's Rechenschieber	St Petersburg, Russia,
1890	Kern règle à calcul pour la stadia topographique	Kern & Cie, Switzerland.
1890	F. Blanc's Rechenschieber	
1891	Lala Ganga Ram's third slide rule	Stanley, London.
1891	K&E. patented adjustable slide rule (Mannheim)	Keuffel & Esser, New York.
1891	Crane's sewer slide rule	Keuffel & Esser, New York.
1891	J. Billeter, Rechenwalze	
1892	Pollit's hydraulic slide rule	Elliot Bros, London.
1892	Bosramier, règle pour les levers tacheometriques	Tavernier-Gravet, Paris.
1892	Wingham's slide rule for calculating blast-furnace charges	A. Wingham, UK Mint.
1892	Scherer's Rechentafel	Germany.
1893	A. Steinhauser's Rechenschiebe	Munich, Germany
1893	J. Crevat, Ruban Logarithmique	
1893	L. Torres, machine for solution of equations	Spain.
1893	Renaud-Tachet, règle circulaire	
1894	W.H. Breithaupt's reaction scale and general slide rule	E.G. Soltmann, New York.
1894	Faber's Rechenstab	A.W. Faber. Germany.
1894	Faber's improved calculating rule for electrical and mechanical engineers.	A.W. Faber, Germany.
1894	Johnson's Rule for unit strains in columns	United States
1895	Regle pour les vitesses, poids et calibres des projectiles	
1895	Wichmannn's Rechenschieber (paper scales on wood)	Gbr. Wichmann, Berlin.
1895	Colby's slide rule for stadia reductions	Keuffel & Esser, New York.
1896	E.A.P. Burt's slide rule	W. Rider, London.
1896	F.W. Lanchester's rule for solving equations	
1896	Scheuermann's computing instrument	United States
1897	Omnimeter	T. Altender, Pa.
1897	Cox's duplex slide rules	Keuffel & Esser, New York.
1897	Gallice. Règle pour les calculs nautiques.	Tavernier-Gravet. Paris
1897	Jordan Rechenschieber	Dennert & Pape, Germany.

DATE:	SLIDE RULE:	MAKER:
1897	Landsberg's Rechenschiebe	Lansberg, Hannover.
1897	F.A. Meyer's Taschenschnellrechner	
1897	Colby's sewer computer	Keuffel & Esser, New York.
1897	Gallice. Règle pour les calculs nautiques.	Tavernier-Gravet. Paris
1897	Jordan Rechenschieber	Dennert & Pape, Germany.
1897	Landsberg's Rechenschiebe	Lansberg, Hannover.
1897	F.A. Meyer's Taschenschnellrechner	
1897	Colby's sewer computer	Keuffel & Esser, New York.
1898	Naish's Logarithmicon	Dublin.
1898	Sickman scale	United States
1898	Sexton's Omnimetre (circular)	
1898	Piper's Logarithmische Skale	Lemgo, Germany.
1898	Mehmke Rechenschieber fur Komplexe Grossen	R. Mehmke, Stuttgart.
1898	Beghin's règle à Calcul	Tavernier-Gravet. Paris.
1898	Fowler's circular calculator	Scientific Publishing, UK.
1899	F.J. Vaes, règle pour la traction des locomotives.	Holland.
1899	Hamann's Proportional-Rechenschieber	
1899	Furle's Rechenschieber	Berlin.
1899	Cox's strength of gear computer	United States
1899	G. Charpentier's Calculimetre	
1899	Universal slide rule	A. Nestler and K&E.
1900	Lallemand, règle à calcul	Leipzig
1900	Puller's Rechenschieber	E. Puller, Saarbrucken.
1900	Rother's Rechenschieber	Rother, Bayern.
1900	Cox's stadia computer (circular rule)	W & L.E. Gurley, New York.
1900	Favourite slide rule (Mannheim Type)	Keuffel & Esser, New York.
1900	Herrgott, règle à deux reglettes	

Chapter 4.
Instrument Makers 1620 - 1900

In 1851 Charles Babbage was quoted as saying, "*It is not a bad definition of a man to describe him as a tool-making animal.*"

This is high praise from a man who struggled for many years to produce the ultimate calculating machine. The statement sets in context the role of the instrument maker at that date and the perception of the technological community at that time.

The mathematical instrument makers played a key role in the development of the slide rule in the period to about 1800. Many of the improvements in design were directly attributable to an individual maker. Other improvements and new designs came from mathematical teachers, mathematicians and the natural philosophers of the time. Over this period we can trace the move from individual makers working alone, to collectives and small companies or partnerships, a number of which ultimately became the first of the great instrument companies. The move from fabrication to manufacture mirrored advances in technology. It started with hand crafting and finished with fully automated processes, including the use of dividing engines. The growing requirement for more effective and accurate instruments, including slide rules, was matched by the move to more automated manufacture with the resultant increased availability.

Instrument making is an old and valued craft which was originally established in Vienna and Nuremberg in the 14th century. The migration of workers from the Low Countries to England in the middle of the 16th century started the craft in England. Instrument making grew in importance and expertise until English work was world renowned. By 1760 to 1780 it was held in especially high regard on the continent, but it was probably at its ultimate peak at the Great Exhibition of 1851 in London - the end of the period covered by this chapter. Generations of craftsmen slowly pushed forward the boundaries of their technological capability, mostly as individual experts with a particular specialisation.

Instrument making was also an art. Many very ornate instruments were produced at this time. Some of the most beautiful and sophisticated astrolabes and similar instruments associated with the study of heavenly bodies dating from the early centuries A.D. are not only highly functional, but are very beautiful in design. Sadly the slide rule is more appropriate to functional design than the artistic.

In England, and particularly in London, pursuing a trade and training apprentices for the furtherance of that trade were governed by the guilds, later called Livery Companies. All master craftsmen had to complete an apprenticeship within a guild before being allowed to practise on their own, employ journeymen or take on youngsters to train as apprentices. By the early 17th century instrument makers were to be found in a number of different guilds, many not readily associated with this particular trade. These included the Blacksmiths (formed before 1494 and chartered in 1591), the Spectacle Makers (formed in 1629 and granted Livery in 1809) and the Clockmakers (formed in 1631). One of the least immediately obvious was the Grocers Company (formed in 1373 and chartered in 1428), where large numbers of instrument makers were to be found. Whereas we can understand why there should be instrument makers in the guild of Spectacle Makers, say, it is not so obvious why the Grocers Company should have gathered this know-how. The argument is made that grocers required accurate weighing machines in the pursuance of their trade, and thus a number of instrument makers congregated in this company. **Table 4** (see page 49) shows the Master Instrument Makers within the Grocers Company from 1600 to 1800. The lines within the family tree show who worked for whom, either as apprentice or journeyman, prior to setting up on their own.

ELIAS ALLEN.
Apud Anglos Cantianus iuxta **Tunnbridge** natus, Mathematicis
Instrumentis ære incidendis sui temporis Artifex ingeniosissimus.
Obijt Londini prope finem Mensis Martij anno a Christo nato 1653 suæque ætatis

Figure 16 : Elias Allen, Master Instrument Maker

Instrument makers had to wait until well into this century before forming their own Guild, the Company of Scientific Instrument Makers, formed in 1956 and granted Livery in 1964. By the end of the 18th century the influence of the instrument makers had begun to spread outside London and into the provinces. The major cities in England and Scotland had their own instrument makers, many of whom were at least as capable as their counterparts in the capital. A number of regional instrument makers became well known; with some producing Oughtred's designs. It is not surprising that Birmingham, as a centre of engineering excellence, had makers who produced the Soho, Routledge, Fowler, Fuller and other engineers' slide rule designs later.

A number of other instrument makers within the Grocers Company, the Spectacle Makers, the Clockworkers, the Drapers and other guilds also made slide rules in addition to other instruments for mathematicians to their specific designs. The growth of instrument making from 1551 to 1851 is illustrated in the following table which compares the total number of instrument makers in England and the number of those who practised in London. Of special interest is the fact that, by the beginning of the 18th century, at least 50 of the 151 instrument makers in the British Isles were producing slide rules to a variety of designs.

As discussed previously, Elias Allen of the Grocers Company (**Figure 16**) produced the first slide rule for Oughtred sometime between 1620 and 1625. From this we can trace the growth of a new product line and of specialist instrument makers, first in London, then nationally and ultimately world-wide. The particular expertise of rule making moved abroad via Edmund Wingate and a number of other instrument makers to France, where there was already an embryonic instrument making capability. However, it is said that French craftsmen did not make instruments of any great merit until some time between 1750 and 1770. The French industry grew, and over a period of time became capable of making accurate, robust instruments, including slide rules, which were highly regarded. An example was the instrument maker Lenoir, who with a number of different partners was the founder of the French firm Tavernier-Gravet that made the world's first true standard design, Mannheim's original slide rule.

Year	British Isles Inc. London	London
1551	3	1
1601	14	8
1651	43	30
1701	151	123
1751	232	161
1801	584	297
1851	837	498

Table 3 : Instrument Makers in London and England, 1551 to 1851

Following a visit to England by the czar of Russia in 1772, a number of instrument makers moved to St. Petersburg. Several well known English makers were involved with this activity, the most famous being Benjamin Scott.

At this stage of the technology it is not right to call these instrument makers "manufacturers", as this implies large scale production, with the inherent standards of quality, quantity and reproducibility; it is more accurate to call them "makers". Even so, many of these makers produced large quantities of some types of rule, particularly as the "standard designs" of the day became popular, i.e. Coggeshall, Everard, Soho, Mannheim, etc. However neither the repeatability of the designs nor the accuracy of these early devices could in any way show all the hallmarks of a manufactured item.

A major change in emphasis and the expansion of technology interchange came with the publication of Moxon's *Mechanick Exercises* in 1677, which effectively opened up the secrets of the instrument makers. Prior to this date each instrument maker jealously guarded his instrument making skills from all but his apprentices. The publication catalysed the sharing of knowledge with many more craftsmen, and the previously slow spread of technology quickened. Later in 1723, Nicholas Bion published *The Construction and Principal uses of Mathematical Instruments* (translated in England by Edmund Stone, instrument maker to George I). This showed that French and English instrument makers were able to produce highly sophisticated and accurate instruments, and described in some detail all the processes involved. Bion provided a fascinating insight into the instrument making business at the beginning of the 18th century, and this was obviously a vital reference book in its day with reprints in 1726 and 1758, each with interesting updates.

However, the relative rarity of books and the slowness of communication generally meant that the pace of improvement was still slow. Each instrument maker concentrated on his specialisation and the range of instruments he was making for his own clientele. There was some cross-fertilisation, and a number of instrument makers worked together.

Samuel Pepys in his diary for late August 1677, mentioned his efforts to obtain a "sliding rule." He noted that he went out to find someone ...

> *... to engrave my tables upon my sliding rule with silver plates, it being so small that Brown, who made it, cannot get one to do it. So I got Cocker, the famous writing master, to do it, and I set an hour beside him to see him design it all, and strange it is to see him, with his natural eyes, to cut so small at his first designing it, and read it all over, without any missing,when, for my life, I could not, with my best skill, read one word or letter of it.*

We know that the next day Cocker brought the completed rule, as Pepys writes:

> *Comes Cocker with my rule, which he hath engraved to admiration for goodness and smallness of work. It cost me 14/- the doing.*

The 14 shillings it cost to engrave the scales is equivalent to about £150 at today's rates. The resulting slide rule *"was very pretty for all questions of arithmetik."* We also know that in 1673 Flamsteed, later the Astronomer Royal, purchased a Double Rulers of Proportion for 10 shillings, roughly equivalent to £100 today. Sadly we do not know who provided the rule at this price. These figures illustrate that slide rules were an expensive item only affordable by the rich. A further comparison of the relative value of a slide rule comes from the writing of Nicholas Hilliard, the goldsmith and miniaturist in 1600: "... *best Ultramarine of Venice ... he has paid £11.10.- the ounce."* These sums are put into perspective when we realise that the annual salary of a farm bailiff at the time was £2.

The start of the Industrial Revolution in 1759/60 heralded the increase in mechanisation and automation that would require many new scientific instruments to take full advantage of the modern techniques that grew from this date. The increase in the use of steam power generated the first designs of engineer's slide rule, as well as a number of general purpose slide rule designs. This period also saw the increase in scholarly study which required accurate measurements and calculations by scientists, and also the beginning of the need for artisans to start calculating. Slide rules could simplify as well as enable practical calculation. The sheer effort required to perform any sort of calculation is not to be underestimated. Some 60 years later in 1821 Charles Babbage, recalling the beginning of his work on calculating machines, wrote of an occasion when he was working with Sir John Herschel:

> *My friend Herschel, calling upon me, brought with him the calculations of the computers, and we commenced the tedious process of verification. After a time many discrepancies occurred, and at one point these discordances were so numerous that I exclaimed 'I wish to God these calculations had been executed by steam'.*

Note the use of the word computer here to describe the people who performed the calculations for inventors and engineers such as Babbage. In 1842 Herschel commented on some of the problems with the logarithmic tables available at that time. Logarithms were required for particularly complex problems, and the slide rule was still a relatively inaccurate instrument.

> *An undetected error in a logarithmic table is like a sunken rock at sea yet undiscovered, upon which it is impossible to say what wrecks may have taken place.*

One of the real needs in the 18th century was to be able to navigate more reliably and simply. This effectively translated into a requirement to accurately work out the longitude at any time anywhere in the world. The announcement in 1714 of an award for a practical method of finding longitude catalysed a flurry of ideas and new instruments to support these. The competition resulted in a number of less than practical suggestions, and a number of concepts that involved the accurate measure of time. This gave a boost to the instrument makers who concentrated on the ability to tell time accurately and reliably on the high seas and produced the first chronometers. Chronometers required accurate gear making to be repeatable. This led to the production of some of the earliest dividing machines. As a by-product, the first elements of automation for the production of slide rules and other measuring instruments were achieved.

The specialists, also called mechanics, who made mathematical instruments gradually disappeared into factories and other cooperative ventures until it was virtually impossible to differentiate between the maker and the retailers. By 1800 the trade directory for London showed only nine people or firms who could be called 'Instrument Maker'.

The considerable range of calculating devices available to anyone wishing to perform mathematical operations during the period to 1850 did not help sales of slide rules. There had always been some misgivings about slide rule usage, the general public perceiving it to be a complicated instrument. There probably was not much to choose between sector, Gunter's rule and the slide rule for simplicity of operation until the cursor came into common usage in the middle of the 19th century. A number of mathematicians and authors of books on slide rules from the earliest days tried to show that it was not a complicated or difficult instrument to learn or use as shown in the following quotations:

In 1768 the Reverend William Flower wrote *A Key to the Sliding Rule* ... where he remarked on

> *...the hopes I had of...removing that **indifference** or rather **dislike**, which I have observed to have appeared in many **artizans** to the use of this...instrument, and which must arise from their **apprehension** of some great **difficulty** in the right management thereof, or some **uncertainty** or **inaccuracy** in its performance...*

The emphasis is Flower's and is interesting.

This vein of thought continues through the life of the slide rule, as Burns Snodgrass wrote as late as 1955 in Snodgrass's *Teach Yourself the Slide Rule*:

> *It is unfortunate that, for some reason not easy to see, the slide rule is sometimes regarded as a difficult instrument with which to become proficient. There is a tendency for some people to become facetious in their references to this simple instrument. Journalists and Broadcasters are great offenders in this respect and some of their references are unbelievably absurd and show a lack of elementary knowledge.*

It is interesting to note which of the instruments available to the calculating public during the first 100 years of the slide rule survived, and for how long. The Gunter's rule continued to be used in the Royal Navy until well into the 19th century, and the sector featured as a part of all cases of drawing instruments until the late 19th century.

Instrument makers were able to produce instruments in wood for the low end of the market, brass and ivory for the middle, and German silver and other inlaid metals for the high or exhibition end. Some of the most exquisite items were made for royalty here and abroad, as well as for international exhibitions in later years.

Tables 5 and **6** give a comprehensive list of makers from 1600 to 1900. **Table 5** (p.50) is arranged chronologically, whereas **Table 6** (p.83) is alphabetical, giving dates for cross-referencing. Names are taken directly from instruments in various collections, as well as from trade directories and catalogues. The works of Joyce Brown, Cajori, Baxendall & Pugh, Daumas, Eva Taylor, Anthony and Gerard Turner, and Gloria Clifton deserve special mention as source documents covering instrument makers in general. Gloria Clifton's *The Directory of Scientific Instrument Makers, 1550 - 1851* is probably the best available reference on *all* mathematical instrument makers, and in it readers may find makers that I have missed. Her book covers all types of instrument maker, but does not separately identify slide rule makers as such.

TABLE 4 : MASTER INSTRUMENT MAKERS IN THE GROCERS COMPANY, 1600 - 1800

Table 5 : Chronological List of Slide Rule Makers, 1600 - 1900

The date in the left-hand column is when it is estimated that the maker started producing slide rules. The dates to the right of the maker's name indicate the span of activity starting either when the maker reached the age of 21 or other known start date and ending with retirement or death.

Asterisked dates are taken from documentation which gives that date as when the maker claimed to have been "established".

1600   **Charles Whitwell** 1591 - 1606

Near St Clement's Church, Strand, London.
Was apprentice to Humphrey Cole, Allen (sic) was his apprentice.
made: Thomas Hood's sector.

1601   **Arthur Hopton** 1588 - 1614
Clement's Inn, London.
made: designed a sector (made by Allen) which was almost immediately superseded by Gunter's sector, also made by Allen.

1607   **Elias Allen** 1607 - 1653
London: 1606-1611: Blackhorse Alley, near Fleetbridge.
1616: Strand.
1618: At the Bull's Head over against St Clement's Church.
1623-1639: Without Temple Bar, over against St Clement's Church, Strand.
1653: Horseshoe without Temple Bar, Strand.
1645: The Blackmoor without Temple Bar.
Originally admitted to the Clockmakers Company, later the Grocers Company. He was the most famous instrument maker of his day, specialising in instruments in silver and brass, his peer John Thompson doing the similar work in Wood. He was closely in touch with Oughtred and many other mathematical practitioners who used his establishment as a meeting place. Succeeded by Ralph Greatorex.
made: 1614, Hopton's sector • 1614, Gunter's sector • 1624, William Oughtred's rules • 1630, Richard Delamain's rules.

1624   **Daniel Browne** 1614 - 1634
Christ's Hospital, London.
May have been a kinsman of Thomas Brown who made the spiral slide rule.

1624   **Edmund Wingate** 1596 - 1656
Queen's College, Oxford, England — Gray's Inn, London — Paris, France.
Wingate went to France in 1624 to tutor Princess Henrietta. There he designed a version of Gunter's rule, the Line of Proportion, which was initially made by Melchior Tavernier.

1625   **B. E.**
made: a folding rule (sector) in the British Museum.

1626   **Thomas Brown** 1626 - 1657
London: 1627: The Globe, Fenchurch Streeet; Minories
1631: Fenchurch St, near Aldgate.
Sometimes spelled Browne. He had a son, John, also mentioned by most of the mathematical practitioners of that time. He worked in wood, and worked closely with Bleyghton.
made: Napier's bones • 1631, Spiral slide rule, drawn and made by him.

1629   *John Thompson* 1610 - 1662
        1611-1645: Hosier Lane, West Smithfield, London.
        Peer of Allen's, worked in wood.
        made: Gunter's scale ● Sector.

1630   *John Bleyghton* 1626 - 1654
        Near the Bull Head Tavern, Tower Street, London.
        1630-31: Tower Street, London.
        The name is also spelled Blaton or Bladon. He worked in silver or brass, and worked closely
        with Thomas Brown.
        made: Thomas Brown's Spiral Instrument or Serpentine (logarithmic) scale.

1631   *Andrew Wake(r)ly* 1631 - 1665
        Redriff (Rotherhithe) Wall, near Cherry Garden Stairs, London.
        made: Gunter's scales ● Sectors ● Carpenter's rules.

1632   *John Allen* 1632 - 37
        Near the Savoy, Strand, London.
        Son or nephew to Elias Allen. Member of the Clockmakers Company in 1653.
        made: Delamain's Mathematical Ring (in silver or brass).

1632   *William Milburne* 1620 - 1643
        Brancepath, Co Durham, England.
        May have been the inventor of the spiral (serpentine) rule often ascribed to Thomas Brown.
        Note that this instrument is referred to variously as the Spiral Instrument, Spiral Slide Rule
        and the Serpentine Rule.

1636   *Baptist Sutton* 1636 - 1653
        Holborn end of Chancery Lane, London.
        The father or uncle of the famous Henry Sutton, specialised in scales (logarithmic).
        made: 1639, Oughtred's Circles of Proportion ● Gunter's Line of Numbers.

1638   *Anthony Thompson* 1638 - 1665
        1645-1664: Hosier Lane, Smithfield, London — 1648: Gresham College (working for
        Samuel Foster) — 1663: Smithfield, London.
        Succeeded John Thomson, his shop was also a well known meeting place of mathematicians
        etc.
        made: Foster's modification to Gunter's sector.

1642   *Robert Bissaker* 1642 - 1664
        Near the Red Lion Tavern, Ratcliff, London.
        made: The earliest slide rule in existence, dated 1654, in the Science Museum, London, is
        made by Bissaker. The T.W. for whom it is made may be Thomas White, whose *Logarithmic
        Ruler was much appreciated by the gentlemen in Essex.* It is 2' long and has 19 scales.

1644   *Henry Coggeshall* 1623 - 1690
        Suffolk, England.
        An expert in timber gauging, best known for his slide rule which became a long standing
        standard.
        designed: 1689, Coggeshall's slide rule.

1646   *Ralph Greatorex* 1625 - 1712
        At the Sign of the Adam and Eve, Strand, London.
        One of Elias Allen's apprentices, also a personal friend of William Oughtred.

1647   *William Leybourne 1626 - 1716*
Monkswell Street, Cripplegate, London; also Southall, Brentford, England.
Better known as a prolific writer, some of whose books were in print for over 100 years.

1647   *Robert Davenport.*
1635: London
1647: Edinburgh.
made: Oughtred's Double Horizontal Dial • Oughtred's Circles of Proportion.

1648   *John Brown 1648 - 1695*
At the sign of the Sun Dyal, Dukes Place, Aldgate, London.
Sphere and Dial, Minories, London.
Son of Thomas Brown, specialised in joint rules (sectors). He made a slide rule so small that he could not engrave it, and Pepys had to get Cocker to engrave the scales on silver at a cost of 14 shillings.
made: 1664, a slide rule for Pepys.
1670, boxwood spiral slide rule with astronomical quadrant on reverse.

1648   *Joseph Moxon 1627 - 1700?*
At the Sign of the Atlas, Parnassus Hill, near St Michael's Church, Strand, London.
1670: At the Sign of the Atlas, Great Russell Street, Bloomsbury, London.
1673: At the Sign of the Atlas, Ludgate Hill, near Fleet Bridge, London.
1694: Westminster Hall, near Parliament Stairs, London.
Best known for his publication *Mechanick Exercises,* published in 1677, which opened up the secrets of the mathematical instrument making trades.
made: Cheap instruments of printed paper on board.

1648   *Walter Hayes 1648 fr - 1692*
Birchin Lane, Cornhill, London.
1653-1684: Sign of the Cross Daggers, near the Pope's Head Tavern, Bethlem Gate, Moorfields, London.
Also known as Haynes, worked in silver, brass and wood.
made: 1657, Partridge's rule • 1685, Encyclogium for Mathew Norwood.

1649   *Henry Sutton 1637 - 1665*
Threadneedle St, near St Christopher's Church, London.
1649: Tower Hill near Postern Spring, London.
1658: Behind the Royal Exchange, Threadneedle Street, London.
The most famous and accurate engraver of scales. Many of his instruments survive.
made: Logarithmic scales • Oughtred's Circles of Proportion.
• 1663, Brass spiral scale slide rule.

1650   *Christopher Flower*
Near the Bulwarke, Tower of London, London.
made: Logarithmic rules of various types.

1652   *Christopher Packwood 1652 - 1664*
1663-64: By Allhallows Barking Church, London.
made: Brass slide rules.

1654   *John Seller 1654 - 1697*
1660: Mariner's Compass & Hour Glass, Hermitage Stairs, Wapping, London.
1669-96: Mariner's Compass, Hermitage Stairs, Wapping, London.
1671-75: Exchange Alley, Cornhill, London.
1678-81: Pope's Head Alley, Cornhill, London.

1682-86: West Side of the Royal Exchange, London.
made: Slide rules.

1654   *Henry Wynne* 1654 - 1709
Near the Sugar Loaf, Chancery Lane, London.
Member of the Clockmakers Company.
made: Sectors.

1656   *Hilkiah Bedford.* 1656 - 1680
London: The Globe near Holborn Conduit; — 1663-66: Hosier Lane.
— 1666: Fleet St. 1674: Fleet Street, Fetter Lane, Londini. (Note the use of the Latin name
for London).
made: Oughtred's Double Horizontal Dial.

1656??   *Leonard Digges*
made: Carpenters rule (maybe 1556)

1658   *Seth Partridge* 1603 - 1686
Hemel Hempstead, England.
Partridge joined two Lines of Proportion together to make a "true" slide rule, and published
the information in 1658. There is debate as to whether Bissaker's slide rule predates his.

1659   *Richard Assheton*
made: Carpenter's joint rule in pearwood.

1663   *Isaac Newton* 1642 - 1727
Trinity College, Cambridge, England.
The famous mathematician, who was responsible for many discoveries.
designed: 1675, an instrument for solving equations consisting of a combination of two or
more logarithmic rulers. This also used a "hair" as a runner or cursor to be able to read off
results across different scales - maybe the first such example.

1663   *Joseph Hone* 1663 - 1704
1663-64: In the Bulwark, Tower Wharf, London.
1670-77: On Tower Wharf, London.
made: Scale designed by Henry Bond for dimensions of ships cordage and masts.
1704, Sliding rules (2) for the same use.

1664   *Christopher Coles*
same as Cole?
made: Carpenters slide rule.

1664   *Robert Jole (Choule)* 1664 - 1704
London: Sign of the Cross, Durham Yard, Strand — The Crown, over against Durham Yard
in the Strand — Sign of the Globe, Fleet Street — At ye Crown near ye New Exchange —
1672: Fleet St, London — 1701: St John's near Clerkenwell, London.
Admitted to the Clockworkers Company 1667.
made: Gauging rods ● Oughtred's Lines (logarithmic rules) ● Wingate's Lines.

1666   *Philip Lea* 1666 - 1700
1683-86: Atlas & Hercules in the Poultry over against the Old Jewry, London.
1687-1700: Sign of the Atlas and Hercules, Old Jewry, Friday Street, London.
1689-1695: Shop in Westminster Halls near the Court of Pleas, London.
Made and sold over 100 mathematical instruments.

made: c1699, Hunt's slide rule ● c1700, Gunter's slide rule ● c1700, Coggeshall slide rule ● c1700, Everard slide rule.

1667   *James Atkinson (Snr.) 1667 - 1717*
Redriff Wall, near Cherry Garden Stairs, Rotherhithe, London.
1673: East side of St. Saviour's Dock, over against the Griffin, London.
1673: Cherry Garden Stairs on Redriff Wall, Rotherhithe, London.
made: 1694, Sliding Gunter.

1667   *Isaac Carver 1667 - 1713*
1697-1708: Globe Dial, Horsleydown, Southwark, near London.
succeeded by his son Jacob, approx 1715.
made: c1673, Hunt's slide rule ● 1683, Everard's rule ● 1697, Gauging slide rule.
1700, Boxwood slide rule ● 1713, Boxwood slide rule.

1667   *Edward Fage 1667 - 1673*
At the Sign of the Sugar Loaf, Hosier Lane, West Smithfield, London.
made: Logarithmic scales of all types ● Slide rule.

1667   *Joseph Wells 1667 - 1690*
London.
Admitted to the Clockworkers Company en bloc 1667.
made: Logarithmic rules.

1668   *William Walgrave 1668 - 1681*
Two White Posts, Newton Street, St Giles, London.
made: Joint rule (sector) with "sights."

1669   *John Whiblin 1669 - 1676*
made: Scale rules ● Gunter's line.

1669   *John Wingfield 1669 - 1671*
By St Olave's Church, Crutched Friars, London.
made: Same instruments as John Brown.

1673   *William Hunt 1673 - 1698*
designed: 1673, slide rule (made by Carver) for use in geography, astronomy, navigation and dialling. Also sold by Philip Lea.

1674   *Michael Butterfield 1674 - 1722*
London and Paris.
Better known for his Universal Pocket Dials known as "Butterworths"
made: Surveyors sector ● Folding brass rule.

1679   *Anselm Jenner 1679 - 1692*
Bristol, England.
made: 1685, Norwood's Encyclogium (Encyclologium) with Hayes (he engraved the plates with Hayes).

1682   *John Worgan 1686 - 1714*
London: 1685: George Alley, Fleet Ditch — 1686-91: St Dunstan's Church, Fleet Street — 1693: Fetter Lane against Clifford's Inn back gate — 1699: Under St Dunstan's Church, Fleet St — 1700: Fleet St.

made: He advertised making ... *any other mathematical instruments whatever.*

1683   **Thomas Everard**
Excise Officer, Southampton, England.
designed: 1683, Gauging slide rule made by Carver and sold by Philip Lea.

1684   **John Warner** 1684 - 1722
London: King's Arms and Globe, Little Lincoln's Inn Fields, end of Portugal Row —
Near Great Lincoln Inne Fields, end of Portugal Row next to Lincoln's Inn by the Booksellers
1722: The Kings Arms & Globe, Little Lincoln Inn Fields, end of Portugal Row.
Instruments dealer.
made: 1722: A modification to Coggeshall's slide rule. Coggeshall's 1722 book notes that
Warner had made the improvement, and that all instruments carrying this improvement were
made and sold by him.

1690   **John Coggs (Snr.)** 1690 - 1740
1729: The Globe and Sun, near St Dunstan's Church, Fleet Street, London.
Worked for John Rowley, later in partnership with William Wyeth. This address was used by
a number of well known instrument makers. Worked from 1690 - 1740.

1690   **William Deane** 1690 - 1748
London: Garden House, Crane Court, Fleet Street — 1718: Crane Court, Fleet St.
1718: Golden Sphere in Three Crane Court, Fleet St. — 1726: Garden House in New St,
near Fetter Lane — 1728: West Harding St, London.
made: A sector.

1693   **Robert Yeff** 1693 - 1720
Bristol, England.
made: Gunter's scale.

1694   **Joseph Selden**
Tunbridge Wells, England.
made: 1694, jointed slide rule which he sold wholesale.

1695   **Benjamin Cole** 1695 - 1750
London: Royal Exchange — 1726: Fleet St. — 1739: The Grand Orrery, Poppings Court,
Fleet St. — 1744-48: Ball Alley out of George Yard, Lombard St. —
1747-58: The Orrery, next the Globe Tavern, Fleet St. — 1765: (Residence?) Water Lane,
Near Westminster Bridge, Surrey.
Son of Benjamin Cole of Oxford (cartographer), apprenticed to Thomas Wright.
made: Full range of instruments.

1695   **Thomas Tuttell** 1695 - 1702.   1700, Instrument maker to the King.
1695-1702: Sign of the King's Arms and Globe, Charing Cross, London.
1695-1702: Royal Exchange, Cornhill, London.
Included Pepys as a customer.
made: Ivory Gunter's scale ● Sector ● Slide rule.

1697   **John Rowley** 1697 - 1728
1691: Behind the Exchange, Threadneedle Street, London.
1702-15: The Globe, under St Dunstan's Church, Fleet St, London.
1710-27: (Residence?) Johnson's Court, Fleet Street, London.
Probably retired from active business in 1715, but was Master of Mechanics (Instrument
Makers) to George I. He was succeeded by Thomas Wright, who had been his assistant.
made: Sector ● Scale for Robert Anderson's gunnery tables ● John Ward's slide rule.

1697   *Charles Price* 1697 - 1730
Mariner and Globe, Strand, London — 1705: Hermitage Stairs, Wapping, London.
1713: Lisbon Coffee House, behind the Royal Exchange, London.
1727-30: Hammersmith, London — 1727-30: In Westminster Hall, London.
Associated with Senex and Benjamin Scott.
made: Slide rules.

1700   *Thomas Arnold* 1700 - 1730
1730: Parish of St Bartholomew the Great, London.
made: 1700, Gunter's scale

1700   *James Atkinson* 1700 - 1728
St Saviour's Dock, 21 Cherry Garden Stairs, Rotherhithe Wall, London.
Father of the same name was also a mathematical instrument maker (1668 - 1717) and teacher.
made: 1724, Coggeshall slide rule.

1700   *William Earle*
Annadale, Edinburgh, Scotland.
made: Rotula Arithmetica.

1700   *Gevin & Le Bas*
France.
made: Saveur made Partridge type rules for G & Le B.

early 1700's *B. Donkin*
made: Donkin's slide rule, 9".

1703   *John England* 1703 - 1708
Charing Cross, London.
made: Napier's bones ● Sector.

1704   *T(I). Cooke* 1704 - 1744
London: 1704: Threadneedle St. — 1704-08: Old Jewry, near the Excise Office — 1744: Old Jewry, London.
There were too many Cookes to be able to resolve the complete family relationships.
made: Boxwood slide rule.

1706   *John Good* 1706 - 1733
Seething Lane, St Olave's, Hart Street, London.
made: Slide rules, also (1717) published a pamphlet on Coggeshall slide rules which was edited and corrected by Atkinson. Later edition of 1751 is post Good.

1707   *William Collier* 1707 - 1745
London: Ye Atlas, the end of Wood St, facing Cripplegate — 1720: Chick Lane, West Smithfield — 1731: Ye Atlas next the Fountain Tavern without Newgate.
made: Sector rule.

1712   *Benjamin Scott* 1712 - 1733
1712: Exeter Change, Strand, London.
1718-33: Mariner and Globe, Exeter Exchange, Strand, London.
To Russia 1733, St Petersburg Academy of Sciences, St Petersburg.
Died in St Petersburg in 1751.

c1733, it was advertised that he ... *makes and sells all mathematical instruments in geometry, navigation & dyalling.*
made: 1733: Scott's circular rule.

1700   ***Edmund Culpepper*** 1660 - 1738.
Old Mathematical Shop, Black and White Horse, Middle Moorfields, London.
Cross Daggers, Moorfields, London. (The Old Mathematical Shop).
Under the Piazza at the Royal Exchange, London.
1700-31: Cross Daggers, Moorfields, London.
1737: Near the Royal Exchange, Cornhill, London.
He had a son, Edmund Culpepper (1758-9), whose work is difficult to differentiate from his father's. He was apprenticed to Walter Hayes whom he succeeded. His trade card was the "Cross Daggers" with the instruments he made. He was one of the first wholesalers.
made: Slide rules, sectors and other instruments.

1702   ***John Sennex*** 1702 - 1740
London: Hemlock Court near Temple Bar — 1702: Against St Clement's Church in the Strand — 1703-07: Next the Fleece Tavern in Cornhill — 1707-10: (With Price) Whites Alley, Coleman St. — 1710-21: The Globe near Salisbury Court, Fleet St. — 1710: (Residence) Salisbury Court — 1721: The Globe & Star, Opposite St Dunstan's Church, Fleet Street — 1722: Fleet St. — 1724-40: Over against St Dunstan's Church in Fleet St. — 1727: Parish of St Dunstan's in the West — 1738-40: The Globe over against St Dunstan's Church, Fleet St.
Geographer to the Queen, mathematical instrument maker and globemaker.
designed: Engraved all the plates in Stone's translation of Bion's book.

1713   ***Thomas Sarjant (or Sariant)***
London.
made: 1713: dated slide rule.

1704   ***Edmund Blow*** 1704 - 1739
London: 1736: Golden Quadrant, Plow Alley, Union Stairs, Wapping.
1738-39: Virginia St. — 1739: Half Moon Court by ye Hermitage, Wapping.
made: 1715: Napier's bones.

1713   ***George Graham*** 1713 - 1751
London: Water Lane, Fleet Street — Dial & Crown, Fleet St.
1713-1720: Dial and Three Crowns, Water Lane, Fleet Street.
1720-1751: Next door to the Globe & Marlborough's Head Tavern.
1720-1751: Dial & One Crown, Fleet St.
Nephew and successor to Thomas Tompion, unequalled skill in scale division.
The most famous maker of the day.
made: Divided rules.

1714   ***Samuel Cunn***
Lichfield Street, Newport Market, London.
Cunn, with Heath, divided anew a Brass pattern for a Foot Sector.
designed: Sector, made by Thomas Heath.

1715   ***Charles Leadbetter***
Hand and Pen, Cock Lane, Shoreditch, London.
Leadbetter was born in Croton, Lancashire, but had moved to London by 1715. He was a mathematical teacher who produced various improvements to Sectors and slide rules. He was a gauger to the Royal Excise.
designed: 1750: improved (Everard) sliding rule made by Heath and Wing.

1715   *Gabriel Stokes* 1715 - 1768
Essex St, Dublin, Ireland.
made: Slide rule.

1717   *Robert Taylor*
High Market Place, Berwick on Tweed, England
made: Gunter scale ● Slide rule.

1718   *Thomas Wright* 1718 - 1747
1720: Orrery & Globe in Fleet St, London.
1720-31: Orrery & Globe near Salisbury Court, Fleet St, London.
1734-38: Orrery & Globe near Water Lane, Fleet St, London.
1747: Orrery & Globe next the Globe & Marlborough Head Tavern in Fleet St, London.
1767: (Residence in retirement) Hoddesdon, Herts.
Instrument maker to HRH Prince of Wales. Successor to John Rowley. His great rival was
Thomas Heath in the Strand.
made: Coggeshall rules.

1718   *John Coggs, Jnr.* 1718 - 1733
1729: Globe and Sun, between St Dunstan's Church and Chancery Lane, Fleet Street, London
— 1759: 136, Fleet Street, London. With E. Cushee.
Successor and son of John Coggs senior, apprentice to Benjamin Cole.
made: 1734, Thomas Hill's new sliding rule.

1719   *Thomas Heath* 1719 - 1753
1719-1747: Hercules & Globe, near the Fountain Tavern, Strand, London.
1734-1746: Hercules & Globe near Beaufort Buildings in the Strand, London.
1750: Hercules and Globe, near Exeter Exchange, Strand, London.
Took as a partner Tycho Wing (1758), Thomas Newman succeeded Heath & Wing and took
the Exeter Exchange address.
made: c1719, assisted Cunn in graduating a sector ● c1734, slide rules.

1719   *William Wyeth* 1715? - 1741
London: Globe and Sun, near St Dunstan's, Fleet Street —
Orrery and Globe, Fleet Street — Orrery, near Water Lane.
Worked for John Rowley at the Globe and Sun, was associated with John Coggs (1733 - 40),
and in 1740 was a partner of Thomas Wright. He died in 1741
made: Slide rules.

1721   *John Fowler* 1721 - 1750
1728-38: The Globe, Swithin (Sweetings) Alley, by the Royal Exchange, London.
1750: The Globe by the Royal Exchange, London.
made: Gaugers slide rules (he is mentioned by Leadbetter in 1739).

1722   *Rees Powell*
made: Slide rule.

1724   *Jonathan Eade* 1724 - 1771
1724: Near King Edward Stairs, Wapping, London.
made: Slide rule.

1724   *Matthew Loft* 1724 - 1747
       1730: The Golden Spectacles, the Backside of the Royal Exchange, London.
       The Golden Spectacles, the North side of the Royal Exchange, London.
       made: Slide rules

1725   *Benjamin Parker* 1725 - 1753
       Isaac Newton's Head, next to the Great Turnstile, Holborn, London.
       Fulwood's Rents, Holborn, London.
       made: 1753, Suxpeach's patent slide rule, combined with a telescope.

1733   *John Coggs & William Wyeth*
       1740: near St Dunstan's, Fleet St. London.
       made: Shirtcliffe's modification to Everard's rule.

1733   *George Adams* 1707 - 1773.
       London: 1733-38: Shoe Lane (Residence) — 1734: 4 doors east of Shoe Lane, Fleet St.
       1735: Near the Castle Tavern, Fleet St, London.
       1736-57: Tycho Brahe's Head, Corner of Racquet Court, Fleet Street, London
       1767: 61 Fleet Street, London.
       George Adams (Snr.), instrument maker to the Prince of Wales and George III. His son of
       the same name had the same appointments.
       made: 1748, Adams spiral rule.

1734   *Thomas Hill*
       London.
       designed: Thomas Hill's new portable slide rule, slid out to 6'. Made by Coggs.

1735   *John Bennet(t)* 1735 - 1770
       The Globe in Crown Court, between St Ann's, Soho, & Golden Square, London.
       Crown Court, Little Pulteney St, London.
       The Globe, Crown Court, St Ann's Soho, near Golden Square, St James, London.
       1743: Globe, in Crown Court, St Ann's, Soho, London.
       1746-47: Queen's Head in Crown Court, Knaves Acre, London.
       1753-70: Crown Court, Soho, London.
       Mathematical instrument maker to the Dukes of Gloucester and Cumberland. He was
       succeeded by James Search.
       made: Slide rules, particularly ... *adding any other factor, division, gauge point, etc., if*
       directed by letter (post paid) from the customer.

1735   *Benjamin Martin* 1735 - 1777
       1736-40: South St, Chichester, Sussex.
       London: 1756-62: Hadleys Quadrant & Visual Glasses near Crane Court, Fleet St.
       1756-59: Two doors from Crane Court, Fleet St. — 1756-82: Resident in Fleet St.
       1760: Four doors east of Crane Court, (later no. 171) Fleet St. — 1761: The New Invented
       Visual Glasses, Fleet St. — 1767-77: 171 Fleet Street.
       Travelling lecturer and instrument maker.
       made: 1767, early slide rules.

1736   *William Mountaine*
       St John's, Southwark, London.
       Mathematics teacher.
       wrote: about John Robertson's inventions, Gunter's sector, slide rule (1778, after
       Robertson's death). Instruments were made by Nairne and Blunt.

1737   ***John Robertson***
        to 1748: Christ's Hospital, London — 1755 - 66: Portsmouth Naval College.
        invented: 1775, improved sliding Gunter with runner.

1740   ***Thomas Hatton*** 1740 - 1774
        1740-69: Preston, Lancashire — 1769-74: London.
        made: 1772, hydrostatic slide rule ● 1772, the joint rule.

1740   ***Joshua Kirby*** 1740 - 1765
        Ipswich, Suffolk, England.
        made: 1761, Architectronic sector.

1740   ***Robert Shirtcliffe*** Gauger.
        London.
        Best known for his book on gauging (1740) where he describes a number of slide rules and
        suggests a new line, the SR line. The rules were made by Coggs and Wyeth.

1742   ***Daniel Voster*** 1742 - 1760
        1742: Cork, Ireland.
        made: 3' rectangular slide rule.

1747   ***Edward Nairne*** 1747 - 1806
        London: Lindsay Row, Chelsea — Golden Spectacles, Reflecting Telescope & Hadley's
        Quadrant, Cornhill opp. Royal Exchange — 1752: Golden Spectacles in Cornhill, opposite
        the Royal Exchange — 1752: Corner of Bartholomew Lane, Threadneedle St. 1753-74:
        Opposite the Royal Exchange in Cornhill — 1772: 20 Cornhill, opposite the Royal Exchange
        — 1772-96: 20 Cornhill.
        Apprenticed to Mathew Loft, he had an international reputation. 1774, partnership with
        Thomas Blunt.
        made: Sector and slide rules.
        1779, supplied James Watt with a slide rule via Mathew Bolton who was visiting London at
        that time.

1747   ***Christopher Stedman*** 1747 - 1774
        London: 1747-50: The Globe on London Bridge — 1758-63: Leadenhall St.
        1774: 24 Leadenhall St.
        made: Slide rules, also resold by James Watt.

1749   ***Edwd. Roberts.*** 1749 - 1784
        London: 1750-56: Grocers Alley in the Old Jewry — 1759-69: Dove Court, Old Jewry.
        1776: 3 Dove Court, Old Jewry.
        Some sources (Delehar) give his dates as 1754 to 1791, see also Edward Roberts at the same
        address from 1790 to 94, this was his son.
        made: c1754-56, excise rules ● c1775, Everard type rule ● 1780, ullaging slide rule in
        boxwood and brass ● Boxwood excise officers slide rule.

1750   ***Heath and Wing*** 1750
        London.
        made: 1750, Leadbetter's improved (Everard) sliding rule.

1750   ***Benjamin Donn***
        Bideford, Devon — 1766: The Library House, King Street, Bristol.
        1774: The Academy, Kingston, near Taunton.
        made: 1764, navigation scale (Gunter's scale), improved 1773, sold by W & S Jones.

1750   *J. Rix* 1750 - 1760
London: Shrewsbury Court, Cripplegate — Shrewsbury Court, in White Cross St, near
Cripple Gate — Salisbury Court.
made: Excise rules x 2 - one spelled "Criple" ● Brewing slide rule.
c1750-60, 4 slide excise rule

1750   *Wood and Lort* 1750 - 1760
Birmingham, England.
Succeeded by Frost and Withnoll, 1767.
made: An improved sliding-rule.

1751   *Tycho Wing* 1751 - 1776
The Strand, London — 1751: Near Exeter Exchange, Strand, London.
Partner to Thomas Heath (1758)
made: Slide rules.

1752   *John Lodge Cowley*
Royal Military Academy, Woolwich, London.
Mathematical teacher, gave William Flower a certificate for his sliding rules (1768).

1752   *J. Try*
Old Bailey No 252, London.
made: Folding rule with logarithmic scale.

1753   *Bradford, Darby & Hulls. (William, Richard and Jonathan)* 1753 - 1754
1753: Chipping Camden, Gloucester, England.
made: 1753, patent money balance and slide rule.

1753   *Richard Rust* 1753 - 1785
London: 1752: At Mr Thomas Rust's, Anchor & Belles, Minories.
1753-78: The Minories — 1776: 125 Minories — 1780: St Catherine's High St.
1781: Corner of St Catherine's Stairs, near the Tower.
made: Slide rules, bought by James Watt for re-sale.

1753   *Joseph Saxspeach* Gauger
Ratcliff, London.
Also sometimes known as Suxpeach.
invented: 1753, The Catholic Organon, or Universal sliding foot rule. First slide rule patent.
Included a telescope.

1755   *John Carlile*
1755: Entry into Bell's Wynd above the Cross, Sign of the Rose, Glasgow, Scotland.
made: Slide rule.

1755   *James Watt*
1755: Apprentice to John Morgan of Finch Lane, London.
1756: returned to Glasgow, making instruments at Glasgow University.
1757-71: In the College, Glasgow — 1759-64: Saltmarket, Glasgow.
1764-71: Trongate, Glasgow.
1775: Joined Mathew Bolton at the Soho factory in Birmingham, also Heathfield.
Watt is known to have bought instruments from Nathanial Hill, T. Jarmain, Charles Lincoln,
James Tomlinson, Richard Rust and Christopher Steadman, which he re-sold possibly with
improvements.
made: 1779, Soho engineering slide rules.

61

1756   *Jesse Ramsden* 1762 - 1800
London: Near ye Little Theatre in ye Hay Market, St James — Next to St James, Piccadilly — Opposite Sackville St, Piccadilly — 1763-66: Strand — 1767-71: Haymarket — 1772-1800: 199 Piccadilly — 1782-1800: (Workshop) 196 Piccadilly.
1771, Best known for his dividing engines, and he worked for many instrument makers including Sissons, Adams, Dollond and Nairne.
made: 1771, dividing engine.

1756   *Edward Troughton*
as John Troughton:
London: 1756: Surrey Street, Strand — The Orrery, 136 Fleet Street.
Apprenticed to his brother John, and later his partner. The firm took over the business of B. Cole at the Orrery in 1782. Firm known as J&E Troughton. In 1826 took William Simms as his partner.
made: 1791, brass 7.5" slide rule.

1758   *T. Hawkins.*
Little known, there was a William Hawkins 1728-50 known to have sold other instruments, the other T(homas) Hawkins were 50 to 100 years later.
made: Boxwood slide rules.

1760   *Thomas Blunt* 1760 - 1823
London: 1760-1822: 22 Cornhill — 1816: 22 Cornhill opposite the Royal Exchange — 1814-20: 136 Minories, London.
1760, Apprenticed to Nairne and then partner to him until 1791. Nairne died in 1806, but the joint name was used until 1822 when Blunt died. He was succeeded by Thomas Harris who continued to use the name Blunt until 1827.
made: Slide rules and other instruments.

1760   *Peter Dollond* 1760 - 1780
Father was James Dollond. He employed many craftsmen.

1760   *A. Fletcher* Mathematical teacher.
Broughton, England.
1762, a book on gauging and surveying, which used Coggeshall slide rules for illustration, and also mentioned Everard rules as still being in use.

1760   *William Flower* Mathematical teacher.
London.
invented: 1768, various improved sliding rules, made and sold by John Bennet.

1761   *Samuel Clark* Mathematical Teacher.
London.
He gave a certificate of accuracy to William Flower slide rules, see also Cowley.

c1760?   *Taston.*
made: Slide rules.

1761   *G.F. Brander*
Augsburg, Germany.
made: Lambert's first German rules. Wood and metal, 4' long.

1761   *A.W. Faber* 1761 - 1996
Stein near Nürnberg, Germany; et al.
Became Faber-Castell in 1906. See Chapter 6.

made: 1894 Improved electrical & mechanical slide rule ● 1894 Faber's Rechenstab●. 1902, The Faber log-log rule ● 1906, log-log slide rule.

1761   *Christopher Jacob* 1761 - 1775
1761: Snow Hill, Birmingham — 1761: Bilstone St, Wolverhampton.
1767: Smallbrook St, Birmingham — 1770: Chapel Row, Birmingham.
1773-75: Gosty Green, Birmingham.
made: Sector ● Slide rule.

1762   *W. Frost & T. Withnold* 1762 - 1775
1767: Birmingham — 1770-75: 14 Litchfield St, Birmingham.
Successors to Wood & Lort.
made: Advertised … *all sorts of rules, sectors, scales etc. as improved by Wood and Lort.*

1764   *John Troughton* 1768 - 1788
London: 1764: Surrey Street, Strand — 1768-71: Crown Court, Fleet St. —
1771-78: 17 Dean St, Fetter Lane — 1778-82: 1 Queen Square, Bartholomew Close—
1782-88: 136 Fleet St.
In partnership with his nephew Edward (1760), and later with his younger brother.
made: Divided and engraved instruments for the trade (wholesaler).
John & Edward produced slide rules.

1764   *Joseph Troughton*
Surrey Street, Strand, London.
Joseph was apprenticed to his brother John and was his partner for a few years, the firm
being known as J&J Troughton. On his death in 1770 his brother Edward (Jnr.) was brought
into the firm and went on to make it particularly famous.

1773   *J. Waddington*
London.
made: Slide rule.

1774   *John Dicas* 1774 - 1797
Liverpool: 1774: 73 Hanover St. — 1777: 25 Cleveland Square — 1781: 71 Duke St.
1787: Duke St. — 1790 - 98: 27 Pool Lane — 1790: 29 Pool Lane.
Dicas was succeeded by his daughters Mary and Ann - married to Benjamin Gammage, set up
the Dicas Patent Hydrometer Manufactory, 1829 to 37, at 11 (1829) and 17 Clarence St (1832
- 34), and 133 Brownlow St, Liverpool (1837).
made: 1780, patent hydrometer and slide rule ● 1790, hydrometer adopted as US standard
● Ivory slide rule - non-logarithmic - with Dicas hydrometer.

1774   *Nairne & Blunt* 1774 - 1793
Fronting the Royal Exchange in Cornhill, London — 1783: 20 Cornhill, opposite the Royal
Exchange, London — See also Nairne, and Blunt, separately.
made: John Robertson's slide rule ● c1790, brass slide rule with 4 slides.

1775   *William Jones*
Soho, Birmingham.
made: c1780, James Watt rules.

1775   *Henry Wood & Thomas Withnoll* 1775 - 1780
29 Great Charles St, Birmingham, England.
Coggeshall's second slide rule is supposed to have been made by a Wood & Withnoll almost
100 years earlier in 1682. What, if any, relationship there was with this firm is not known
made: Slide rule.

63

1776   *Thomas McIntosh* 1776 - 1784
London: Archimedes & Golden Spectacles, opposite Long Acre, four doors from Queen St.
— 7 Great Queen St, Lincoln's Inn Fields — 1776-79: (Workshop), Wild St, Queen St. —
1776: 33 near Exeter Change in the Strand — 1780: 7 Great Queen St, Lincoln's Inn Fields
— 1780: Archimedes & Golden Spectacles, Great Queen St, Lincoln's Inn Fields — 1784:
40 Great Queen St, Lincoln's Inn Fields.
made: 1799, sliding rule ● 1802, sliding Gunter for navigation.

1777   *Walter Field*
made: Excise rules.

c1779   *Barling*
Limehouse Hole, London.
The Barling family were involved with the "Navigation Warehouse" in Limehouse for a
number of years, from approx 1822 to about 1840. See Barling W.H., 1822; Barling
Elizabeth, 1826, and Barling Frederick, 1836. All were associated with the Warehouse.
made: c1779, Soho rule of boxwood and ivory.

c1780   *J.H. Adams*
Liverpool, England.
Note that there were many Adamses in London, nothing further is known of this one.
made: c1780, excise rule.

1780   *William O. Carey* 1789 - 1891
London: 277 Strand — 1786-90: 272 Strand — 1794-1822: 182 Strand.
1800: 182 near Norfolk St, Strand — 1821-90: 181 Strand, London.
See also John Carey. Also spelled Cary (mistakenly).
made: 1813, bankers slide rule ● 1814, scale of chemical equivalents. (Wollastons slide
rule) ● 1815, Silvanus Bevan's engineer's rule ● 1815, slide rule for currency and bullion
use, paper on wood ● c1822, Benjamin Bevan slide rule for accountants & surveyors ●
1857, Roget's slide rule ● 1878, Ewart's cattle gauge.

1780   *William Simms* 1780 - 1871
1780-81: 44 Coleshill St, Birmingham — 1793: Birmingham — 1794: London.
1808-12: Bowman's Buildings, Aldersgate Street, London.
1818-22: 4 Broadway, Blackfriars, London.
William Simms was a son of James Simms and the father of William Simms (Jnr) who became
a partner of Edward Troughton in 1826. See also Troughton & Simms. Two other sons,
James and George, carried on the business in Blackfriars after he died.

1780   *Joseph Stutchbury* 1780 -1826
London: 1791-96: Bird in Hand Court, Cheapside — 1797-1800: Dove Court, Old Jewry
— 1802-26: 3 Dove Court, Old Jewry.
Also known as Stitchbury and Stuchbury.
made: Slide rule and excise officers scale ● 5' jointed wine gauging rod ●
Boxwood wine diagonal ● 1780, Gauging slide rule with four slides ●
6" Everard style rule with 4 slides & imperial markings.

1780   *Mathew Richmond*
According to Andrew McKay, Richmond was a mathematics teacher and writer.
made: A maritime scale (Gunter's?).

1782   *John Carey 1782 - 1831*
London: 1782: Johnson's Court, Fleet St. — 1783: Corner of Arundel Street, Strand.
1783-90: 188 Strand — 1792-1805: 181 Strand — 1820-21: 85/6 St James's Street.
The business moved to St James's Street when 181 Strand was destroyed by fire. It was
subsequently rebuilt and together with 182 was occupied by his brother William. John Snr.
retired in 1831 when the business was taken over by his sons George and John (Jnr.)
made: Primarily globes, later slide rules.

1782   *George Margetts 1782 - 1804*
1770: Old Woodstock, Oxfordshire — London: 1782: Ludgate Hill — 1789: 42 Penton St,
Islington — 1790: Penton St, Parish of St James, Clerkenwell —
1792: 21 King St, Cheapside — 1801: 3 Cheapside — 1804: St Michael's Alley.
Partner with Nicholson (see 1787)
made: Slide rule ● 4' slide rule for navigation ● Calculating machine.

1783   *I(J). Trafford*
Windlebury, Oxford, England.
made: Carpenter's sliding joint rule.

1784   *William & Samuel (W&S) Jones 1784 - 1859*
London: 27 Holborn Hill — 1792-1800: 135 Holborn, next to Furnival's Inn —
1800-60: 30 Holborn, (Holborn Bridge) — 1801-05: 32 Holborn Hill —
35 Lower Holborn (or) Holborn Hill — 1818: 30 Lower Holborn —
1852: 30 Opposite Furnival's Inn, Holborn — Archimedes, 30 Lower Holborn.
William was an apprentice of Benjamin Martin, both were the sons of John Jones.
made: Numerous instruments - wholesalers.
Late C18, "Soho" slide rule, 10.25" ● c1800, wooden slide rule
● 1818, Henderson's double slide rule.

1784   *Thomas Parnell 1784 - 1811*
London: Mariner & Quadrant, 94 Lower East Smithfield — Mariner & Quadrant,
2 Lower East Smithfield — 93 Lower East Smithfield — 94 near the Hermitage Bridge,
Lower East Smithfield — 1784: 11 Cannon St, Ratcliff Highway —
1793: 25 East Smithfield — At Mr Hughes, Nightingale Lane.
made: Slide rule.

1784   *John Rabone & Sons*
Hockley Abbey Works, Birmingham.
There are various Rabone's in Birmingham at a later date, including Thomas Rabone (1830-7)
who made Barometers at 12 Court, Broad Street; Frederick Street, and Great Hampton Street,
Birmingham; there was also Rabone and Mason (1834-7) at 61 St Paul's Square, Birmingham,
who also made Barometers. This last firm later became John Rabone only. Relationships are
not known. Stopped making slide rules by 1963.
made: 1811, Routledge slide rule ● 1850, carpenter's rule. (Sir William Armstrong's rule)
● Carpenter's two foot rule with logarithmic slide ● 1880's, two foot boxwood and ivory
rules for Routledge, Carrett, Hawthorn, Slater and Wilkinson ● 24" and 36" boxwood
contraction rules.

1784   *Alexander Wellington 1784 - 1812*
London: 20 Crown Court, Wardour St. — Crown Court, St Ann's, Soho —
1784: Sherborne Lane, Lombard St. — 1788-1812: 20 Crown Court, Princes St Soho.
1805: 20 Crown Court, Golden Square.

Wellington was succeeded by Laban Cooke, Joined by his son 1814/5. Other sources say Wellington was succeeded by J. Fitch & J. Jones c1818.
made: 1805, J. Routledge's engineer's slide rule.

1787   *Haas & Hurter*
partnership 1790-95
1792: Bull Yard, St Ann's Court, Soho, London.

Jacob Bernard Haas 1789 - 1828
Bull Yard, St Ann's Court, Soho, London.

Johann Heinrich Hurter 1787 - 1799
53 Great Marlborough St. Oxford St, London.
1790: 58 Great Marlborough St. Oxford St, London
made: 5" brass circular slide rule.

1787   *Nicholson* 1787 - 1802
London.
made: 1802?, double length (4') sliding Gunter for navigation.

1788   *Edward Roberts* 1788 - 1796
Grocers Alley, London.
1790-94: 3 Dove Court, Old Jewry, London.
Son of Edward Roberts (see 1749)
made: slide rules, various ● Everard style rule with 4 slides.

1789   *Thomas Barnett* 1789 - 1810
London: 4 Moores Yard, Old Fish St, near Doctors Commons — 1789: 61 Tower Street —
1790-94: 61 Great Tower St. — 1795-96: 6 Tower St. —
1799-1802: 21 East St, Lambeth — 1804: East St, East Place.
made: 6" four slide boxwood proof rule

1790   *Dring & Fage* 1790 - 1940
London: 1790-92: 21 Gracechurch St. — 1790: 4 Albion Place —
1792-96: 6 Tooley Street, London Bridge, Borough —
1796-1804: 248 Tooley Street, London Bridge, Borough. —
1801: 8 Crooked Lane — 1804-1844: 20 Tooley St, London Bridge —.
Excise slide rule. — 1804: 109 Upper East Smithfield —
1843-44: 10 Duke St, Tooley St. — 1845-82: 19&20 Tooley St. —
18??: 320 Tooley St, London Bridge — 1883-1902: 145 Strand, WC
1903-38: 56 Stamford St, SE - Timetable slide rule.
made: Kentish slide rule ● S. Coulson's rule (S, Coulson, Redcar)
● Boxwood gauger's slide rule ● Ivory slide rule ● c1820, spirit rule in ivory
● 1830, brewing or temperature slide for use with saccharometer ●
1839, Thomas Kentish's Compound slide rule ● 1840, Coulson's slide rule (boxwood)
● 1840, timetable or rent rule for storage calculations (box with Ivory slide) ● 1840,
boxwood slide rule for calculating storage charges ● 1830/40, acidometer rule, boxwood
● 1850/60, improved saccharometer in wood and ivory ● 1850/60, ullage and costing rule
● 1850, 4' jointed gauging rod ● 1855, timetable or rent rule ● 1860, boxwood vacuity
rod or out stick ● 1868, excise officers slide rule, Verie pattern ● 1871, Kentish slide rule
● 1876, slide rule for finding the contents of timber, 2' ● 1876, Coulson's slide rule.

• 1880, circular brass slide rule • 1881, Thomson's log-log rule • 1887, Burt's patent measuring and cubing slide rule • 1900, Ewart's cattle gauge • 24" boxwood calculating rule with single slide designed by James Liddell of 3 George St, Falkirk, Scotland; for calculating flooring etc.

1792   *Thomas Webb* 1793 - 1832
Bristol: 1793-98: Earl St — 1820-22: 3 Doors from the Infirmary, Earl St. —
1823-30: 2 Doors from the Infirmary, Earl St. — 1831-32: 3 Doors from the Infirmary, Earl St.
made: Slide rule.

1795   *Lewis & Briggs* 1795 - 1799
52 Bow Lane, Cheapside, London.
also known as Briggs and Lewis.
made: Gauging instruments • Slide rules • Excise slide rule.

1797   *Richard Bakewell* 1797 - 1826
Birmingham: 1797-1820: Loveday St. — 1825-26: 49 Loveday St.
made: Slide rule.

1799   *Thomas Coulsell* 1799 - 1836
London: 42 & 153 Union St, Borough — 1794: Loman's Pond, Borough —
1799-1814: 29 Queen St, Borough — 1815-28: 41 Union Street, Borough, Southwark.
1829-36: 153 Union Street, Borough, Southwark.
May have been related to William Coulsell, final address is Elizabeth Coulsell - maybe his widow at 41 Union Street from 1839-42.
made: Rules in box and ivory.

1799   *Caesar Tagliabue* 1799 - 1839
London: 1795-1800: 294 Holborn — 1807-09: 26 — 1816: 11 Brook St, Holborn.
1822-29: 28 Cross Street, Hatton Garden — 1829-39: 23 Hatton Garden,
(with Casella)
By 1846 the partnership was Casella and Tagliabue, eventually purchased by Negretti and Zambra.
made: Slide rules and other instruments.

1800   *Thomas Cox & Co* 1800 - 1825
Birmingham?
See also F.B. Cox who may have succeeded.
made: Slide rule.

1800   *B. Harrison* 1800 - 1900
11 Princes St, Spitalfields, London.
made: Slide rules.

1801   *Robert Atkins* 1801 - 1806
136 Fenchurch Street, London.
Succeeded by George Atkins, 1806 - 1814, and George Atkins & Co, 1813 - 1822.
made: Slide rule.

1804   *Benjamin Bevan*
Leighton Buzzard, Bushey Heath, England.
made: 1813, sliding rule with divisions to three places of decimals, used Ramsden's dividing machine.

1804   *William Elliott 1804 - 1849*
London: 1817: 26 Wilderness Row, Goswell St. — 1817-27: 21 Great Newport St., St
Martin's Lane — 1830-33: 227 High Holborn — 1835-49: 268 High Holborn.
There were numerous "Elliotts" in the directories, the relationships are not clear, though the
best known was Elliott Bros who succeeded William Elliott and Sons of 1850-53, whose father
was this Elliott.

1805   *H. Lovi 1805 - 1827*
Edinburgh?
made: Slide rule with tables by Isabella Lovi (his mother?).

1805   *Nicholson & Margetts*
succeeded Nicholson.
made: Slide rule

1805   *J. Routledge 1775 - 1829*
1800-10: Leeds, England — 1810-14: Bolton, England. (Thomas Swift & Co, Foundry) —
1814-18: Bolton - his own ironmongers — 1824-29: Warsaw, Poland, where he died.
designed: 1808, Routledge's engineer's improved sliding rule, made by Wellington.

1805   *Thomas Tustian 1805 - 1830*
London: 2 Princes Square, Ratcliffe Highway — 1805: 1 Princes Square, Ratcliffe Highway
— 1836-39: 18 Princes Square, Ratcliffe Highway —
1838-39: 18 Princes Square, St George East.
made: Slide rule

1806   *Charles Augustus Schmalcalder 1806 - 1840*
London: 1806-07: (Residence?) 6 Little Newport St, St Ann's Soho. —
1812: Strand — 1810-26: 82 Strand — 1827-40: 399 Strand.
Inventor and first maker of prismatic compass.
made: Mathematical instruments including slide rules.

1807   *Robert Brettell Bate (R.B. Bate) 1807 - 1847*
1804: residence, Foster Lane, (Clerkenwell?)
London: 1807-28: 17 Poultry — 1824-47: 21 Poultry — 1824-42: 20 Poultry —
1846-47: 33 Royal Exchange — 1847: (Residence) Hampstead.
Optician who made slide rules.
made: Special prison slide rule ● Boxwood slide rule, divided duodecimally ●
Boxwood gaugers slide rule ● Early 1800's, nautical slide rule ● 1800, boxwood head rod
● 1815, boxwood slide rule, for use with Sikes hydrometer ● 1816, Bate handbook,
published 1816; 2nd edn. 1823. Mentions 24" and 36" versions of Thomson Lunar distance
slide rule #1 (is this the Nautical slide rule?) ● 1817, Silvanus Bevan's gaugers sliding rule
● 1818, calculating slide rule, 28" & 56" ● 1818, the Bate rule (exported to France) ●
1824, Bate's ready reckoner (paper on wood) ● 1824, slide rule for treadmill (prison slide
rule?) ● 1824, interest scale ● 1827, Soho slide rule, 28" & 56" ● Single and multiple
slide rules for calculation and conversion. ● Sheep gauge.

1808   *T. Kitchenmun*
made: 1808, Kitchenmun's new authentic sliding rule.

1809   *Stephen Norris Cooper 1809 - 1841*
1809-11: 119 Great Saffron Hill, Holborn, London.
1811-29: 25 Miller St, Manchester — 1834-41: 16 Miller St, Manchester.
made: Slide rule.

1809   **William Coulsell (& Son)** 1809 - 1851
       1816-40: 9 Castle Street, Southwark, London.   (W. Coulsell)
       1845: 41 Union St, Borough High Street, London.
       1849-51: 9 Castle St, Borough, London.
       4 Salmon Lane, Limehouse, London   (W. Coulsell & Son)
       made: Two-fold slide rule.

1810   **Gilbert and Son, W & T Gilbert**
       The Navigation Warehouse, 148 Leadenhall Street, London.
       William Gilbert was the son of John Gilbert (Jnr), later with Thomas Gilbert became M.I.M.
       to the East India Company (1820)
       made: 1828, Roget's slide rule.

1811   **Zachariah Belcher** 1811 - 1829
       1811-18: Bright St. Sheffield, England.
       1822-25: 4 Bright St, Sheffield Moor, Sheffield, England.
       sometime known as Belshaw.
       made: Wood and ivory rules.

1812   **John Newman** 1812 - 1856
       London: 1807: 11 Windmill Row, Camberwell — 1812-16: Lisle St. Leicester Square —
       1817: 7 Lisle St, Leicester Square — 1822: 8 Lisle St, Leicester Square —
       1827-56: 122 Regent Street.
       made: Chemical equivalent slide rule.

1813   **John Stanton** 1813 - 1843
       London: 111 Shoe Lane, Holborn — 1813: 82 Shoe Lane, Holborn —
       1816-40: 73 Shoe Lane, Holborn — 1844-45: (Residence) Rye Lane, Peckham —
       1846-64: Lloyd St, Pentonville — Great Turnstile, Holborn.
       made: Ivory folding 2' rule.

1814   **Peter Mark Roget** Scientist.
       Roget was also famous of the author of the well known "Thesaurus" in use to this day.
       designed: 1815, the log-log scale, and a slide rule to use it. The rule was a long spiral
       pattern rule made for him by Rooker.

1815   **John M. Rooker** 1816 - 1846
       London: 1815: 27 Bridgewater St, Somers Town — 1824: Bridgewater St, Somers Town —
       1826: Guildford Place, Kennington — 1841-46: 1 Little Queen Street, Holborn — 1845: 26
       East St, Lambs Conduit St.
       made: 1815, Roget's spiral scale slide rule ● Boulton & Watt's Soho engineer's slide rule
       ● 1827, Farey's improved Soho slide rule, 25" ● 1840, Woolgar's calculator and pocket-
       book rule ● 1857, Roget's log-log rule, 10" ● Boxwood slide rule, 25".

1815   **M. Collardeau**
       Paris, France.
       made: c1815, Jomard's règle à calcul.

1816   **Hoyau**
       France.
       made: Cylindrical rules. (Boites à calculer).

1817   *Sylvanus Bevan* Gauger.
May have been a relative of Benjamin Bevan.
designed: 1817, improved sliding rule for excise officers, made by R.B. Bate.

1817   *William Hobcraft (Snr).* 1817 - 1856
London: 28 King St, Snow Hill — 28 Cow Lane, Smithfield —
1817: 56 Whitecross St, Fore St. — 1822: 12 Beech St, Barbican —
1825: 13 Barbican — 1834: 18 Barbican — 1837-52: 14 Barbican.
Succeeded by William Hobcraft Jnr, 1845 - 1870
1845: 38 Prince's Street, Leicester Square — 1849: 14 Great Turnstile —
1851: 62 Dean St, Soho — 1855-70: 419 Oxford St.
made: Mathematical rules.

1818   *T(homas) Aston* 1818 - 1850
Birmingham: 1818: Colehill St. — 1825-29: 20 Bartholemew Rd. —
1835: Jenners Row — 1850: 25½ Willis St.
See also Thomas Aston at similar addresses in Birmingham, 1841 to 1862. Both made rules.
made: 2 foot, 2 fold rule with Routledge scales.

1818   *James Fitch & John Jones* 1818 - 1819
Crown Court, Soho, London — 20 Crown St, Princes St, Soho, London.
Successors to the late Arthur Wellington of London.
made: c1818, Routledge rule.

1819   *Mathew Iley* Gauger.
1 Somerset Street, Portman Square, London.
described: Sliding rule in practical applications.

18--   *Isaac Sargent*
France.
made: Slide rule.

18--   *Loftus.*
146 Oxford St — 321 Oxford St, London.
made: Excise slide rule, (146, Oxford St) ● Alcohol slide rule.
● Gauging rule, (321 Oxford St) ● Comparative rule for spirits (boxwood).

18--   *P. Corson.*
made: Excise rules.

1820   *Francis Blakemore Cox* 1828 - 1862
Birmingham: 1828-50: Camden St. — 1835-62: 50 Camden St.
made: c1820, carpenter's rule.

1820   *I. Newman*
made: Wollaston's slide rule of chemical equivalents.

1820   *M & E Emanuel* 1831 - 1879
1820: 1 Bevis Marks, St Mary Axe, London.
1825: 9-10 Bevis Marks, St Mary Axe, London,   M. Emanuel and Co.
also Portsmouth and Portsea   E. & E. Emanuel
1831-1912: 3 The Hard, Portsea, England.
1831-1879: 101 High St, Portsmouth, London.
made: 6" ivory sector (E. & E. Emanuel).

1820   *Cesare Zambra* 1820 - 1841
         1821-40: Saffron Walden, Essex, England.
         1840-41: 23 Brook St, Holborn, London.
         Father of Joseph Warren Zambra (born 1822) who was taken on by Enrico (Henry) Negretti in 1850 to form the well known firm of Negretti and Zambra.

1821   *Lenoir*
         France, succeeded by Collardeau and Gravet
         Etienne Lenoir born 1744.
         First instruments made in 1791, became a company with various partners later.
         made: Lenoir's règle à calcul.
         became Gravet-Lenoir — then Tavernier-Gravet, — Rue Mayet 19, Paris, France. **made:** Boxwood slide rule • Boxwood slide rule with cursor • Boxwood slide rule, Mannheim with cursor • 1851, Mannheim's rules. (Gravet et Lenoir)
         • 1868, Moinot rule. Règle Logarithmique pour la Tacheometric • 1873, Goulier. Règle pour les Levers Tacheometriques • 1876, De Montrichard slide rule. Règle pour le Cubage des Bois • 1878, Cherapashinskii slide rule
         • 1883, Cherapashinskii slide rule • 1886, Lebrun rule. Règle pour les Calculs de Terrassements • 1888, Sanguet rule. Règle pour les Levers Tacheometriques
         • 1892, Bosrammier rule. Règle pour les Levers Tacheometriques
         • 1897, Gallice rule. Règle pour les Calculs Nautiques • 1897, Beghin's règle à calcul.
         • 1903, Leven slide rule • 1907, Règle des Ecoles (Règle Beghin)

1821   *Clouet*
         France.
         made: c1821, Clouet's règle à calcul.

1821   *Joseph Long* 1821 - 1846
         London: 1821-84: 20 Little Tower Street — 1885-1936: 43 East Cheap.
         Joseph Long died in about 1846, but the firm continued under the same name for many years.
         made: 1824, conversion slide rule at change from Winchester to Imperial measure
         • 1829, excise slide rule - see Bibliography • Boxwood slide rule
         • 1850, comparative or reducing slide rule (ivory) • 1870, temperature and pricing rule
         • Ullage and proof rule.

1822   *W.H. Barling* 1822 - 1824
         Limehouse Hole, Limehouse, London.
         The first member of the Barling family involved with the "Navigation Warehouse". See Barling Elizabeth, 1826, and Barling Frederick, 1836.
         made: Various navigational instruments.

1822   *Glazebrook*
         May have been an excise man, corresponded with the Board of Longitude about a slide rule he had invented.

1822   *William Roberts* 1822 - 1848
         Sheffield: 1822: 21 West Bar — 1822-25: 1 Sportsman's Inn Yard, Westbar — 1828-30: Gibralter St. — 1833: Court 23, Gibralter St. — 1833-48: 15 Westbar Green — 1837: 22 Gibralter St. — 1841: 14 Steelhouse Lane.
         Was apprenticed to Edward Roberts.
         made: Boxwood and ivory rules.

1823   *J.W. Woolgar* Mathematician.
         Leeds, England — 1827: 47 Essex Street, Strand, London.

Lewes, Sussex, England.

Woolgar is supposed to have "generalised" the logarithmic scale, also applied it to slide rules to annuities.

designed: Pocket calculator made by John Rooker ● 1828, a 6" plotting and projection scale, made by Elliott.

1824   *George Dollond* 1819 - 1866

London: 1852-66: 59 St Paul's Churchyard — 1859-65: 61 Paternoster Row.

later Birmingham as well.

The firm of "Dollond" started with John Dollond in approx 1750 in Spitalfields, he had 2 sons and 2 daughters who married into the trade, this George "Dollond" was born George Huggins, son of William Huggins who married Susan Dollond, one daughter, and changed his name.

The firm continues to this day as opticians.

made: 1824, with the change to Imperial Measure, George Dollond divided the first standard yard ● 24" timber rule with sprung ends.

1824   *F .Dübler*

Berlin, Germany.

made: Slide rule.

1824   *Thomas Underhill* 1824 - 1827

Liverpool: 1824: 124 Whitechapel — 1825: 6 Coopers Row — 1827: 32 Mersey St.

There was also a Thomas Underhill in Manchester, 1834 - 1881.

made: Excise rule?

1825   *J. Cail* 1825 - 1865

Newcastle upon Tyne: 1825-38: 2 New Bridge St. — 1838-55: 61 Pilgrim St. — 1836-44: (Residence) 44 Northumberland St. — 1844-55: 45 Quayside, 4 doors West of the Customs House — 1855: 8 Grey St. — 1855-65: 21 Grey St.

made: Engineer's slide rule ● 1844, Ewart's cattle gauge.

1825   *Laban Cook(e)* 1825 - 1834

21 Crown Court, (Pulteney St), Soho, London.

Maker to the Board of Excise, succeeded Wellington at Crown Court.

made: Gauging instruments, including ullaging slide rule ● Excise slide rule.

● 2 slide gauging rule with Imperial gauge points ● Everard style rule with 4 slides.

1825   *Philip Myers* 1825 - 1835

Nottingham: 1825-28: Smithy Row — 1830: 1 Okeham Street — 1834-35: Pelham St. — later, London.

made: Ivory sector.

1826   *Elizabeth Barling* 1826

Limehouse Hole, Limehouse, London.

The second member of the Barling family involved with the "Navigation Warehouse". See W.H. Barling, 1822, and Barling Frederick, 1836.

made: Various navigational instruments.

1826   *Troughton & Simms* 1826 - 1922

London: 1839-44: 136 Fleet Street — 1844-1915: 138 Fleet St. — 1866-1915: (Factory) 340 Woolwich Rd, Charlton, London SE.

See also Edward Troughton and William Simms separately.

made: Boxwood slide rule for trigonometrical and astronomical calculations

● Boxwood slide rule for trigonometrical and navigational calculations.

1827   *Gutteridge, Dunning & Son*
London?
made: Slide rule.

1827   *Holtzapffel & Co* 1827 - 1914
1839-65: 64 Charring Cross, opposite King's Mews, London.
1839-65: (Factory) 127 Long Acre, WC London.
made: Slide rule ● c1885, Ganga Ram slide rule (2nd design).

1827   *Joseph Lamb*
London, a watchmaker.
made: 1827, description of a Concentric Circular Proportioner, an instrument for facilitating calculation. Also known as Lamb's circular slide rule.

1828   *Joseph Shuttleworth* 1828 - 1845
Birmingham: 1828: 63 Lichfield St. — 1829-30: 73 Lichfield St. —
1841: 133 Broad St. — 1845: 2 Newhall St, Birmingham.
made: Ivory sector.

1828   *Augustus De Morgan* Mathematician.
London.
De Morgan was Professor of Mathematics at University College London in 1828, he is well known as a mathematical historian and writer on slide rules in the Penny Cyclopedia.

1829   *Thomas Saddington,*
London.
made: Saddington's rules.

1830   *William Chadburn & Co.* 1830 - 1834
Sheffield: 1830: 40 Lady's Bridge, Nursery St. — 1833: 23 Nursery St. —
Albion Works, 27 Nursery Street, Lady's Bridge, Sheffield.
William Chadburn was an optician working from 81 Wicker St in Sheffield from 1816 to 1830. He was succeeded by the company bearing his name who were wholesalers of all types of mathematical instruments which were factory produced. They were succeeded by Chadburn Bros.

1830   *Benjamin Pike*
394 Broadway, New York, USA.
Pike is best known for his catalogue of over 1200 items, however many of the items were imported from well known English makers including Troughton, Wollaston, Sexton etc. Undoubtedly he also made his own instruments as he advertised himself as ... *a mechanic, a practical workman.*

1832   *G.T. Bramwell* 1832
England.
made: c1832, Hawthorn's slide rule.

1833   *William Fage* 1833 - 1855
London: 1833: 59 Long Lane, Bermondsey — 1834-35: 5 King's Row, Walworth — 1836-37: 10 Great Dover Street, Borough — 1838-40: 3 Great Dover Street, Borough — 1841-42: 62 High St, Borough — 1855: 7 Friar St, Blackfriars Road.
See also Dring and Fage.

1833   *Lewis Casella* 1833 - 1838
Edinburgh, Scotland.

May have been the father of Louis Paschal Casella who traded as Casella & Co at 23 Hatton Garden, London from 1848 to 1860, then as Tagliabue & Casella at the same address from 1865 to 1870. From 1875 to 1901 they were at 147 Holborn Bars, London EC. Louis P. Casella died in 1897.

1833   *George Mander* 1833 - 1870
London: 1830-31: 21 Crown Court, Soho — 1833: 21 Crown St, Princes St, Soho — 1839-51: 21 Great Crown Court, Golden Square — 1845-70: 25 Old Compton Street, Soho
Joined with Isaac Aston to form Aston & Mander.
made: Slide rule.

1834   *William Baylis* 1834 - 1873
Liverpool: 1834: 56 Dale St. — 1841: 67 Dale St. 1841-58: 27 Park Lane — 1841-55: 75 Dale St. — 1845-49: 27 South John St. — 1851-53: 2 Prices St. — 1857: 85 Dale St. — 1858-65: 81 Dale St. 1870-73: 67 Park Lane, West Liverpool.
made: Slide rules.

1834   *William Gregory* 1834 - 1836
Clerkenwell, London: 1834: 8 Berry St. — 1835: 9 Hooper St. — 1836: 2 Francis Court, Berkeley St.
made: Mathematical rules.

1834   *T. Bradburn & Sons* 1834 - 1851
1850: 18 Alcester St, Birmingham, England.
Also known as George and Thomas Bradburn.
made: patented joint rule, (Lion trade mark) ●
c1850, Routledge's engineer's folding slide rule.

1834   *Thomas Underhill* 1834 - 1881
Manchester: 1834-38: 40 Water St. — 1841: 70 Bridge Street, Deansgate — 1848: 4 Old Millgate — 1858-64: 2 Corporation St. — 1868: 4 Corporation St. — 1873-81: 53 Princess St.
There was also a Thomas Underhill in Liverpool, 1824 - 1827. The Manchester Underhill is known as Maker & Optician.
made: c1880, excise rule ● 12" closed-frame rule. Labelled Maker & Optician.

1835   *W(illiam) H(enry) Brown* 1835 - 1866
Birmingham: 11 Caroline St. — 1850: Residence, Monument Lane.
made: 2' two fold rule with Hawthorn's scales.

1835   *Smith* Captain, Royal Engineers.
Wrote a book on the use of Gunter's sliding rule.

1835   *Mrs. Janet Taylor* 1835 - 1875
London: 1831-35: 6 East St, Red Lion St. — 1835-45: 103 Minories — 1 Fen Court, Fenchurch St. — 1845: Nautical Academy & Warehouse, 103-4 Minories — 1845-75: 104 Minories — 1846-58: (Residence) 1 Hammet St, Minories 1858: (Residence) Park Place, The Grove, Camberwell.
Nee, Jane Ann IONN, born 1804, Succeeded her husband George Taylor who was active to about 1854, she died in 1870 but the company continued as Janet Taylor & Co.
made: 1834, patent mariner's calculator.

1836   *Frederick Barling* 1836 - 1840
Limehouse Hole, Limehouse, London.

The final member of the Barling family involved with the "Navigation Warehouse". See W.H. Barling, 1822 and Elizabeth Barling, 1826.
made: Various navigational instruments.

1836   *J. Riley*
Oldham, Lancashire.
A Mechanic (instrument maker) who wrote on the use of slide rules for the cotton spinners and other mill workers, may have made slide rules.

1837   *Chadburn Bros.* 1837 - 1884
Sheffield: 1837: Albion Works, 27 Nursery St. — 1841: Shilo Wheel, 44 Stanley St.
1841-84: 26 Nursery St. — 1847-76: Albion Works, Nursery St. —
1857: (Branch) 71 Lord St, Liverpool, England.
Succeeded Chadburn & Co.
made: Wholesalers of mathematical instruments.

1838   *Henry Barrow* 1838 - 1864
1824-26: 18 Crown Court, Soho, London SW.
1849-51: 26 Oxenden St, Haymarket, London.   (as Henry Barrow & Co).
1850-64: 26 Oxenden St   (as Henry Barrow).
Worked as an "out-worker" for Dollond and Troughton, acquired the business of Thomas Robinson in 1842 traded as Robinson and Barrow, was succeeded by Barrow & Owen.
made: Gunner's rule.

1839   *Thomas Kentish*
Best known for his *Treatise on a box of Instruments, and a Slide Rule for the use of Excisemen, Engineers, Seamen & Schools*.
designed: Slide rule, made by Dring and Fage.

1840   *S. Cock*
Store Office, Excise office, Broad St, London.
made: Gauger's slide rule and rod.

1840   *Joseph Bateman* Scientist.
Middlesex, England.
Bateman was an expert on the slide rule. He exhibited his own 8' slide rule at the British Association, but said that ... *navigators still used a Gunter's rule with a pair of compasses*.
designed: 8' slide rule for the "revenue".

1840   *Mme Dondey Dupre* 1840
Paris, France.
made: c1840, Arithmometre.

1840   *Peter Hope* 1840
64 Seel Street, Liverpool, England.
made: c1840, elapsed time slide rule.

1840   *Enrico (Henry) Angelo Ludovico Negretti* 1840 - 1850
London: 1834-35: Hatton Garden — 1839-48: 19 Leather Lane, Holborn —.
1840: 20 Greville St — 1849: 9 Hatton Garden —
1851: (Residence?) 9 Manchester St, Argyle Square.
Apprenticed to Pizzala, later took over the business of Tagliabue, and later still took on Joseph Zambra as his partner to form Negretti and Zambra.

1841   *T(homas) Aston* 1841 - 1862
Birmingham: 1841-56: 17 Jenners Row — 1862: 17,18 Jenners Row.
See also Thomas Aston at similar addresses in Birmingham, 1818 to 1850. Both made rules.
made: 2 foot, 2 fold rule with Routledge scales.

1841   *Mallin & Co.* 1841
1841: 12 Hampton Court, Birmingham.
May have been related to Abr. and Wm. Mallin.
made: c1841, 2 foot, 2 fold Coggeshall rule.

1841   *Montferrier*
France.
made: Arithmomètre.

1842   *James Chesterman* 1829 - 1847
At Mr. Cutts, Division St, Sheffield, England.
made: c1842, Chestermann's patent improved cattle gauge, on a box with a tape measure.

1844   *Aaron Palmer*
USA.
made: computing scale.

1846   *John E. Fuller*
USA.
made: Slide rules when he obtained Palmer's patent
● 1846, Computing Telegraph ● 1860, Telegraph Computer.

1847   *Th. Baumann*
Berlin, Germany.
made: C. Hoffman's slide rules.

1847   *C.T. Dörfel*
Berlin, Germany.
made: C, Hoffman's slide rules.

1847   *C.G. Grunow*
Berlin, Germany.
made: C. Hoffman's slide rules.

1847   *J. Stambach.*
Aarau, Germany.
made: 1847, Eschmann-Wild's Tachymeterschieber.

1847*   *George Lee & Son*
33, The Hard, Portsmouth, England.
Manufacturers of optical and nautical instruments. Contractors to the Admiralty.
sold: Faber slide rules.

1850   *James Tree & Co* 1850 - 1895
1850-51: 22 Little Charlotte St, Blackfriars, London (as James Tree) — 1851-95: 22 Little Charlotte St (as James Tree & Co) — 1895: 7 Lawrence Lane, London EC.
made: 1850, Routledge's engineer's folding slide rule ● 1850, Ewart's improved cattle gauge ● 2' boxwood rule with ivory slide marked with Robert Hawthorn's scales.

c1850   *E. Preston & Sons*
Birmingham, England.
made: 1850, Routledge rule ● 1861, Carrett's improved slide rule
● James Noble's rule ● 24" folding carpenter's rule.

c1850   *Keep Bros*
made: Slide rules.

c1850   *T.B. Winter*
Newcastle upon Tyne, England.
made: Ewart's cattle gauge.

c1850   *E. Driver.*
Manchester, England.
made: Carpenter's rule (scales from 4 - 40).

c1850   *J.H. Haynes*
Birmingham, England.
made: 2' boxwood rule (proof rule?).

1850   *Sml. (Samuel) Albt. (Albert) Smith.* 1850 - 1862
Birmingham: 1850: 368 Coventry Rd. — 1862: Victoria Place, Coventry Rd.
made: Engineer's rule (Routledge) (scales from 1 - 10).

1850   *G. Altmutter.* 1850
1850: Vienna, Austria.
made: Altmutter's Rechenschieber (in cardboard).

1850   *J. Porro.* 1850
1850: Italy.
made: Porro's Scale Logaritmiche Centesimali.

18??   *Stephens & Co.*
made: Slide rule.

18??   *George A. Cramm*
Eastcote, Middlesex, England.
made: Military rule.

18??   *Richard Pasley*
Sheffield, England.
made: c1824, 2' 2 fold rule. Brass slide says "Girt Line" Imperial Gallon.

18??   *Gray & Selby*
Nottingham, England.
made: Sykes hydrometer and proof rules.

1850   *William Mallin* 1850
50 Newington Row, Birmingham.
see also Abr. Mallin and Mallin & Co, they may have been related.
made: Slide rules.

1850   *Abraham Mallin* 1850 - 1862
Birmingham: 1850: 5 Chapel St. — 1862: Vauxhall Rd.

see also Wm. Mallin and Mallin & Co, they may have been related.
made: Slide rules.

1850   *Apps*
London.
made: Hawthorn's slide rule ● Routledge's engineer's slide rule
● 1850, engineer's circular slide rule, 3.725" diameter.

1850   *W. Werner*
Vienna, Austria.
made: 1850, Werner's Rechenschieber, (boxwood)
● 1850, Werner's Tachymeterschieber.

c1850   *S. Walker & Co.*
Address not known,
Note, there were many Walkers in London and Dublin:
London: Edwd Walker, Strand, 1823-39 — Francis Walker, Wapping Wall, 1829-59.
— John & Alexander Walker, 1823-59.
Dublin: Frederick Walker, 1832-50 — George Walker, 1841-48.
made: c1850, carpenter's rule.

1851   *Isaac Aston* 1851 - 1870
25 Old Compton Street, Soho, London.
Joined George Mander to form Aston & Mander, 1870.
made: Hannyngton's slide rule   (Aston & Mander)

1851   *Hachette*
Paris, France.
made: Lalanne's slide rule (made on glass).

1851   *George Thorsted.*
New York, USA.
made: 1851, Nystrom's calculator.

1851   *William Young.*
Philadelphia, USA.
made: 1851, Nystrom's calculator.

1851   *E. Sedlaczek.*
Vienna, Austria.
made: 1851, Regolo Calcolatorio di Schwind ● 1851, Regolo di Sedlaczek per i calcoli
d'interpolazione ● 1851, Regolo Calcolatorio di Oesterle ● 1856, Regolo di Higgison ●
1856, Regolo per iscopi costruttivi e geodeteci del Prof. L.C. Schulz von Starssnicki.

1851   *John Dennett Potter* 1851 - 1880
1851-82: 31 Poultry, London — 1854-82: 11 King St, London — 1882: (Residence) Broad
Green Lodge, Croydon Surrey — 145 Minories, London; (to date?)
Potter succeeded R.B. Bate as chart supplier to the Admiralty, an appointment still held
to date.
made: Mathematical instruments including slide rules.

1851   *George Ransley* 1851 - 1855
10 Barnet St, Hackney Road, London.

made: 2' two fold rule with £.s.d. scales
- 3' boxwood rule with ivory slide for timber calculations.

1851   ***John & Alfred Rooker*** 1847 - 1901
          1849-59: 26 East St, Red Lion Square, London.   (as John Rooker & Son)
          1851: 26 East St, Foundling, London   (as John & Alfred Rooker)
          1880-1901: 156 Euston Road, London.   (as John Rooker & Son)
          made: Slide rules.

1851   ***F. Werner***
          Vienna, Austria.
          made: Boxwood rules.

1851*   ***W.H. Harling***
          47 Finsbury Pavement, London EC.
          made: Slide rules.

1852*   ***J.H. Steward***
          406 Strand, London.
          made: 1878, Boucher calculator • 1906, Hall's slide rule
          • 1906, "R.H.S." tubular rule.

1853   ***Elliott Bros.***
          London: 1853: 56 Strand — Also 1853: 5 Charing Cross — 449 Strand —
          after 1840, Greenwich — 1892: St. Martins Lane.
          Elliott Bros. succeeded William Elliott & Sons who traded at 56 Strand 1850 to 1853.
          made: Improved Burton's 4.5" howitzer slide rule • 1850, boxwood slide rule, No 1252
          • 1892, Pollit's hydraulic rule.

1853*   ***W.F. Stanley***
          Great Turnstile, Holborn, London.
          History: See Allen, C.J.A, 1953, for the history of company - not much on slide rules.
          made: Boxwood slide rule • Sheppard's patent slide rule • Fullers calculator.
          • 1876, Bevan's engineer's rule • 1876, Lala Ganga Ram's slide rule • 1877, Routledge's
          slide rule • 1877, Hawthorn's slide rule • 1878, Boucher's calculator • 1878, Fuller's
          calculator • 1885, Ganga Ram's slide rule (Second type) • 1891, Ganga Ram's slide rule.
          (Third type) • 1905, Anthony's circular slide rule • 1905, Fuller-Bakewell tubular
          calculator • 1905, Stanley-Boucher watch type calculator • 1905, Gravet (also Gravette)
          • 1906, Honeysett's hydraulic slide rule • 1906, Hale's slide rule • 1906, Froude's slide
          rule • 1906, Sheppard's slide rule • 1906, Hudson's slide rule.

1860   ***C.T. Cooper***
          made: Ham's combined ullage and proof rule • c1860, Ham's proof rule
          • 1870, Cylindrical boxwood rule for ballistics calculations • c1880, Ham's ullaging rule.

1861   ***W. Elliot***
          30, Strand, London — 268, High Holborn, London.
          made: Bayley's "One & Two Slide" rules (Strand) • Ivory sector (Holborn).

1862*   ***Dennert & Pape***
          Altona, Hamburg, Germany.
          History: Established to make surveying, cartographic and drafting instruments. First slide
          rules 1872. 1886, DRP 34583 for laminating white celluloid to Mahogany. 1936 pvc only
          slide rules (ASTRALON) marketed under trade name ARISTO. Aristo sales: 500,000 to

schools and approx 300,000 to engineers. From 1975 slide rule sales decreased by approx 50% per year, 1978 closed slide rule production.
made: 1874, Rechenstab von Dennert und Pape ● 1886, Zellhorn (Celluloid) scales ● 1897, Jordan's slide rule ● 1903, Simplex slide rule ● 1908, New Precision slide rule ● 15" engineer's slide rule, patent 173660.

1864   *Lansberg & Wolpers*
Hanover, Germany.
made: E. Sonne's calculating disc, 7" diameter.

1864   *Landsberg & Parisius,*
Hanover, Germany.
made: Sonne's Rechenshiebe.

c1870   *G.L. Cabot*
United States.
made: circular specific gravity calculator.

187?   *Grove*
made: Gauging rule (boxwood).

1872   *R. Webber*
Ashaffenburg, Germany.
made: Webber's Calculating Circle, 5" diameter. Rechenkreissen.

1875   *C. Culmann* 1875
Zurich, Switzerland.
made: 1875: Rechenschieber fur Distanzmessern ● 1875: Pestalozzi's Rechenschieber

1876   *Longman & Co.*
made: Prof. J.D. Everett's Universal Proportion Tables (card).

1877   *Johan Ritter von Puscariu*
Budapest, Hungary.
made: 1877, Puscariu's Stereometer.

1877   *Wiesenthal und Cie.*
Aachen, Germany.
made: 1877, Hermann's Rechenknecht.

1878   *Henri Chatelain*
Paris, France.
made: Improved Boucher calculator.

1878   *Manlove, Alliott, Fryer & Co,*
Nottingham, England.
made: Boucher calculator.

1878   *Albert Nestler*
Lahr, Baden, Germany.
made: 1899, Universal slide rule (also K&E) ● 1903, Rietz's slide rule ● 1905, Universal slide rule ● 1908, Peter's Universal slide rule ● 1908, Peter & Perry slide rule ● 1908, Precision slide rule ● 1909, Wiring calculator slide rule.

1878   *A.G. Thornton*
Manchester, England.
made: Many different types, see Chapter 6.

1880   *Henry Cherry*. 1880
England.
made: c1880, Cherry's calculator. The earliest grid-iron design.

1880   *Aston & Mander*
25 Old Compton Street, Soho, London.
made: 1884, Hannyngton's extended "washboard" slide rule ● Hoare's Reversing slide rule
(boxwood) ● T. Dixon's slide rule ● Builder's slide rule
● C. Hoare's engineer's slide rule (boxwood).

1882   *Morin*
Paris, France.
made: c1882, Charpentier's Calculimetre.

1882   *Boult*
made: Charpentier calculator.

1885   *Tower*
Mr Beauchamp Tower.
designed: Tower's slide instrument.

1886   *Tesdorpf*
Stuttgart, Germany.
made: 1886, Beyerlen's Rechenrad.

1888   *Keuffel & Esser*
Hoboken, New Jersey, USA.
History: made in excess of 10 million slide rules.
made: Celluloid scales ● 1881, Thacher's calculator ● 1891, Patented adjustable slide rule
(Mannheim) ● 1891, Crane's sewer slide rule ● 1892, Charpentier calculator (made to
1927) ● 1895, Colby's slide rule for stadia reductions ● 1897, Cox's duplex slide rule
● 1897, Colby's sewer computer ● 1899, Universal slide rule ● 1900, Mannheim slide
rule ● 1903, Webb's slide rule ● 1903, Cox's stadia rule ● 1903, Favorite slide rule ●
1906, Sperry's slide rule ● 1909, Polyphase slide rule ● 1909, Adjustable duplex slide rule
● 1909, Log-log duplex slide rule ● 1909, Circular calculator ● 1909, Nordell's slide
rule.

1890   *Kern & Cie*
Aarau, Switzerland.
made: 1890, Kern règle à calcul pour la Stadia Topographique ● 1902, Kern règle à calcul.
● 1902, Matthes' Topographic slide rule. (for the US Geographical survey).

1892   *M. Lallemand*
made: 5" slide rule with a magnifying cursor (the first).

1892   *Avmann and Pillmeier*
Cassel, Germany.
made: Scherer's logarithmic calculating table.

1892   *Wingham*
        The Mint, London.
        designed: Mr A, Wingham's Slide Rule for Calculating Blast-Furnace Charges.

1893   *W. Rider & Son*
        London.
        There was a William Rider of Poplar London, 1819 to 1834, at 22 Ashton St and 121 High
        Street, whether they carried on to 1893 is not known.
        made: 1893, Burt's slide rule No. 101.

1894   *E.G. Soltmann*
        119 Fulton St, New York, USA.
        made: W.H. Breithaupt's reaction scale and general slide rule.

1895   *Gebr. Wichmann*
        Berlin, Germany.
        made: Wichmann's Rechenschieber.

1895   *Lanchester*
        made: 1895, Lanchester's radial cursor.

1897   *Theo Alteneder & Sons,*
        945 Ridge Ave, Philadelphia, PA. USA.
        made: Omnimeter.

1897   *Landsberg.*
        Hanover, Germany.
        made: Landsberg's Rechenschiebe (Sonne's design).

1898   *Lemgo*
        Germany.
        made: 1898, Piper's Logarithmische Skale.

1898   *R. Mehmke*
        Stuttgart, Germany.
        made: 1898, Rechenshieber fur Komplexe Grossen.

late 1800's *W & J Burrow*
        London.
        made: proof rule.

late 1800's *Buss*
        48 Hatton Garden, London.
        made: 24" alcohol rule with two slides, boxwood ● 5" alcohol rule, ivory.

late 1800's *Draper*
        B. Draper & Co, Unknown address.
        made: 2' two fold Routledge slide rule, boxwood.

late 1800's *Farrow & Jackson*
        London and Paris.
        made: Proof rule ● 2 slide rule, with Imperial gauge points.

late 1800's **Higgison**
Birmingham.
made: 12" Carpenters slide rule, boxwood.

late 1800's **Sampson**
Sampson, of Aston (Birmingham).
made: Boxwood slide rule.

late 1800's **Stevenson**
P. Stevenson, Edinburgh.
made: 2 sided boxwood slide rule with four scales on each side.

???? **Ackland**
London.
made: slide rule for optical work.

Table 6 : Alphabetical List of Slide Rule Makers, 1600 - 1900.

The date in the right hand column is when it is estimated that the maker started producing slide rules. This date can be used to cross-reference to the chronological list in **Table 5** which has additional details.

Ackland, ?	????
Adams, George (Snr.)	1733
Adams, J.H.	1780
Allen, Elias	1607
Allen, John	1632
Alteneder, Theo & Sons	1897
Altmutter, G.	1850
Apps, ?	1850
Arnold, Thomas	1700
Assheton, Richard	1659
Aston, Isaac	1851
Aston, Thomas	1818 and 1841
Aston & Mander	1880
Atkins, Robert	1801
Atkinson, James (Snr.)	1667
Atkinson, James (Jnr.)	1700
Avmann & Pillmeier	1892
B.E.	1625
Bakewell, Richard	1797
Barling, W.H.	1822
Barling, Elizabeth	1826
Barling, Frederick	1836
Barnett, Thomas	1789
Barrow, Henry	1838
Bate, Robert Brettell	1807

Bateman, Joseph	1840	
Baumann, Th.	1847	
Baylis, William	1834	
Bedford, Hilkiah	1656	
Belcher, Zachariah	1811	
Bennet(t), John	1735	
Bevan, Benjamin	1804	
Bevan, Sylvanus	1817	
Bissaker, Robert	1642	
Bleyghton, John	1630	also Blaton or Bladon
Blow, Edmund	1704	
Blunt	1774	(with Nairne)
Blunt, Thomas	1760	
Boult, ?	1882	
Bradburn, T. & Sons	1834	
Bradford, William	1753	with Darby and Hulls
Bramwell, G.T.	1832	
Brander, G.F.	1761	
Briggs, ?	1795	with Lewis
Brown, John	1648	
Brown, Thomas	1626	
Brown, W.H.	1835	
Browne, Daniel	1624	
Bramwell, G.T.	1832	
Burrow, W & J.	late 1800's	
Buss, ?	late 1800's	
Butterfield, Michael	1674	
Cabot, G.L. & Co	1870	
Cail, J.	1825	
Carey, John	1782	
Carey, William O.	1780	
Carlile, John	1755	
Carver, Isaac	1667	
Casella, Lewis	1833	
Casella & Co	1903	
Chadburn, William	1830	William Chadburn & Co
Chadburn Bros	1837	
Chatelain, Henri	1878	
Cherry, Henry	1880	
Chesterman, James	1842	
(Choule), Robert	1664	also more commonly Jole
Clark, Samuel	1761	
Clouet, ?	1821	
Cock, S	1840	
Coggeshall, Henry	1644	
Coggs, John (Snr.)	1690	
Coggs, John (Jnr.)	1718	with William Wyeth, 1733
Cole, Benjamin	1695	
Cole(s), Christopher	1664	
Collardeau, Maurice	1815	
Collier, William	1707	
Cooke, T.(I.)	1704	
Cook(e), Laban	1825	

Cooper, C.T.	1860	
Cooper, Stephen Norris	1809	
Corson, P.	18--	
Coulsell, Thomas	1799	
Coulsell, William	1809	William Coulsell & Son
Cowley, John, Lodge	1752	
Cox, F.B.	1820	
Cox, Thomas & Co	1800	
Cramm, George A.	18??	
Culmann, C.	1875	
Culpepper, Edmund	1700	
Cunn, Samuel	1714	
Darby, Richard	1753	with Bradford and Hulls
Dargue Bros.	19??	
Davenport, Robert	1647	
De Morgan, Augustus	1828	
Deane, William	1690	
Dennert & Pape	1862	
Dicas, John	1774	
Digges, Leonard	1656??	
Dring	1790	with Fage
Driver, E.	c1850	
Dollond, George	c1850	
Donkin, B.	c1700	
Donn, Benjamin	1750	
Dörfel, C.T.	1847	
Draper, B.	late 1800's	
Dübler, ?	1824	
Dupre, Mme. Dondey	1840	
Eade, Jonathan	1724	
Earle, William	1700	
Elliot, W.	1861	
Elliott, William	1804	
Elliott Bros	1853	
Emanuel, M & E.	1820	
England, John	1703	
Everard, Thomas	1683	
Faber, A.W.	1761	
Fage	1790	with Dring
Fage, Edward	1667	
Fage, William	1833	
Farrow & Jackson	late 1800's	
Field, Walter	1777	
Fitch, J	1818	with J.Jones
Fletcher, A.	1760	
Flower, Christopher	1650	
Flower, William	1760	
Fowler, John	1721	
Frost, W.	1762	with Withnold
Fuller, John E.	1846	
Gevin, ?	1700	with Le Bas

SLIDE RULES

Gilbert, W & T	1810	Gilbert & Son
Glazebrook, ?	1822	
Good John	1706	
Graham, George	1713	
Gray & Selby	18??	
Gregory, William	1834	
Greatorex, Ralph	1646	
Grove, ?	187?	
Grunow, C.G.	1847	
Gutteridge, Dunning & Son	1827	
Hachette, ?	1851	
Haas, Jacob, Bernard	1787	with Hurter
Harling, W.H.	1851	
Harrison, B.	1800	
Hatton, Thomas	1740	
Hawkins, T.	1758	
Hayes, Walter	1648	
Haynes, J.H.	1850	
Heath, Thomas	1719	
Heath	1750	with Wing
Higgison	late 1800's	
Hill, Thomas	1734	
Hobcraft, William (Snr.)	1817	
Holtzapffel & Co	1827	
Hone, Joseph	1663	
Hope, Peter	1840	
Hopton, Arthur	1601	
Hoyau, ?	1816	
Hulls, Jonathan	1753	with Bradford and Darby
Hunt, William	1673	
Hurter, Johann Heinrich	1787	with Haas
Iley, Mathew	1819	
Jacob, Christopher	1761	
Jenner, Anselm	1679	
Jole, Robert	1664	also Choule
Jones, J.	1818	with J. Fitch
Jones, William	1775	
Jones, W & S	1784	William & Samuel
Keep Bros	c1850	
Kentish, Thomas	1839	
Kern & Cie	1890	
Keuffel & Esser	1888	
Kirby, Joshua	1740	
Kitchenmun, T.	1808	
Lanchester, ?	1895	
Lallemand, M.	1892	
Lamb, Joseph	1827	
Landsberg & Parisius	1864	
Lansberg & Wolpers	1864	
Landsberg	1897	

Lea, Philip	1666	
Leadbetter, Charles	1715	
Le Bas, ?	1700	with Gevin
Lee, George & Son	1847	
Lemgo	1898	
Lenoir, ?	1821	
Lewis, ?	1795	with Briggs
Leybourne, William	1647	
Loft, Matthew	1724	
Loftus	18--	
Long, Joseph	1821	
Longman & Co	1876	
Lort, ?	1750	with Wood
Lovi, H.	1805	
Mallin & Co	1841	
Mallin, Abraham	1850	
Mallin, William	1850	
Mander, George	1833	
Manlove, Alliott Fryer & Co	1878	
Margetts, George	1782	
Margetts	1805	Nicholson and Margetts succeeded Nicholson
Martin, Benjamin	1735	
McIntosh Thomas	1776	
Mehmke, R.	1898	
Milburne, William	1632	
Montferrier, ?	1841	
Morin, ?	1882	
Mountaine, William	1736	
Moxon, Joseph	1648	
Myers, Philip	1825	
Nairne, Edward	1747	
Nairne	1774	(with Blunt)
Negretti, Enrico	1840	
Nestler, Albert	1878	
Newman, I.	1820	
Newman, John	1812	
Newton, Isaac	1663	
Nicholson, ?	1787	
Nicholson	1805	Nicholson and Margetts succeeded Nicholson
Packwood, Christopher	1652	
Parker, Benjamin	1725	
Parnell, Thomas	1784	
Partridge, Seth	1658	
Pasley, Richard	18--	
Pike, Benjamin	1830	
Porro, J.	1850	
Potter, James Dennett	1851	
Powell, Rees	1722	
Preston, E & Sons	c1850	
Price, Charles	1697	
Puller, E.	1900	

Rabone, John & Sons	1784	
Ramsden, Jesse	1756	
Ransley, George	1851	
Richmond, Mathew	1780	
Rider, W. & Son	1893	
Riley, J	1836	
Rix, J.	1750	
Roberts, Edward (Snr.)	1749	
Roberts, Edward (Jnr.)	1788	
Roberts, William	1822	
Robertson, John	1737	
Roget, Peter Mark	1814	
Rooker, John M.	1815	
Rooker, John & Alfred	1851	
Routledge, J.	1805	
Rowley, John	1697	
Rust, Richard	1753	
Saddington, Thomas	1829	
Sampson	late 1800's	
Sargent, Isaac	18--	
Sarjant, Thomas	1713	also Sariant
Saxspeach, Joseph	1753	also Suxpeach
Schmalcalder, Charles	1806	
Scientific Publishing	1898	
Scott, Benjamin	1712	
Sedlaczek, E.	1851	
Selden, Joseph	1694	
Seller, John	1654	
Sennex, John	1702	
Shirtcliffe, Robert	1740	
Shuttleworth, Joseph	1828	
Simms, William	1780	
Smith, Samuel Albert	1850	
Smith, ?	1835	
Soltmann, E.G.	1894	
Stambach, J.	1847	
Stanley, W.F.	1853	
Stanton, John	1813	
Stedman, Christopher	1747	
Stephens & Co.	18??	
Stevenson, P.	late 1800's	
Steward, J.H.	1852	
Stokes, Gabriel	1715	
Stutchbury, Joseph	1780	
Sutton, Baptist	1636	
Sutton, Henry	1649	
Suxpeach, Joseph	1753	also Saxspeach
Tagliabue, Caesar	1799	
Taston, ?	c1760?	
Tavernier,	1821	see Lenoir
Taylor, Mrs. Janet	1835	
Taylor, Robert	1717	
Tesdorpf, ?	1886	

Thompson, Anthony	1638	
Thomson, John	'1629	
Thornton, A.G.	1878	
Thorsted, G.	1851	
Tower, B.	1885	
Trafford, I(J).	1783	
Tree, James & Co	1850	
Troughton, Edward	1756	
Troughton, John	1764	
Troughton, Joseph	1764	
Troughton & Simms	1826	
Try, J.	1752	
Tustian, Thomas	1805	
Tuttell, Thomas	1695	
Underhill, Thomas	1834	
von Puscariu, Johan Ritter	1877	
Voster, Daniel	1742	
Waddington, J.	1773	
Wake(r)ly, Andrew	1631	
Walgrave, William	1668	
Walker, S & Co.	1850	
Warner, John	1684	
Watt, James	1755	
Webb, Thomas	1792	
Webber, R.	1872	
Wells, Joseph	1667	
Wellington, Alexander	1784	
Werner, F.	1851	
Werner, W.	1850	
Whiblin, John	1669	
Whitwell, Charles,	1600	
Wichmann, Gebr	1895	
Wiesenthal u. Cie	1877	
Wing, Tycho	1751	with Heath, 1750
Wingate, Edmund	1624	
Wingfield, John	1669	
Wingham, A.	1892	
Winter, T.B.	c1850	
Withnold, T.	1762	with Frost
Withnoll, Thomas	1775	with Wood
Wood, ?	1750	with Lort
Wood, Henry	1775	with Withnoll
Woolgar, J.W.	1823	
Worgan, John	1682	
Wright, Thomas	1718	
Wyeth, William	1719	with John Coggs, 1733
Wynne, Henry	1654	
Yeff, Robert	1693	
Young, William	1851	
Zambra, Cesare	1820	

Chapter 5.
Slide Rule Types 1850 - 1998.

The slide rule types produced in the years after 1850 were those that ultimately became standards across the manufacturing industry, which at the beginning of this period was still in its infancy. The year 1850 is notable for a number of reasons. It was the start of a flurry of "inventiveness" in scale layouts, slide rule designs and technology improvements in England, France and Germany. It was also the year of Amédée Mannheim's first designs, various updates of which resulted in the first of the "modern" slide rule designs which found favour and went on to become the world-wide industry standard. Finally, 1850 was the beginning of the period that completed the transition from individual custom designs to the manufacture of standard designs in factories.

In England there were a large number of designs from the highly inventive Victorians, all aimed at increasing the effectiveness and accuracy possible with this tool that was just beginning to gain world-wide popularity. The many designs described by Cajori in 1909 are an indication of the increased awareness of slide rule technology at this time. Many of these designs fell by the wayside for any number of reasons, not least of which were inconvenience in use and expense in manufacture. Hwever a number of these designs continued in production for many years. Numbers of extended scale rules were invented, and some of these continued to be manufactured for a hundred years.

International Adoption of Scale Design Standards.

Between 1850 and 1900, as other countries emerged as slide rule manufacturers, slide rule designs increasingly moved across national boundaries. Thus the development of slide rule standards. For example, it is possible to trace the movement of the **Mannheim** design through various countries. The Mannheim, Tavernier-Gravet or Gravet slide rule was featured in manufacturers' catalogues all over the world. As France was England's nearest neighbour, there was a technological closeness between the two countries that resulted in a number of designs becoming common on both sides of the Channel. Almost inevitably, England was the first country to produce Mannheim rules in quantity, starting around 1860, in its new instrument-making factories.

Figure 17 : Mannheim Slide Rule

Germany was probably the next European country to produce Mannheim rules, starting in about 1870, with a number of famous manufacturers - Faber, Nestler and Dennert & Pape - advertising the design from that date.

Cherapashinskii in Russia designed a Mannheim rule in about 1883. This was manufactured in France by Tavernier-Gravet, who had made the first Mannheim rules. The Cherapashinskii design, with elements of the Beghin design, became the Règle des écoles standard in France in about 1907.

The Mannheim was seen in the United States from about 1890, initially as imports from Dennert & Pape that were labelled with the Keuffel & Esser name. K&E went on to become the largest American manufacturer of slide rules, a Mannheim being one of its earliest designs.

The beginnings of the Japanese slide rule manufacturing industry followed a visit to Europe by Dr. R. Hirota and Government Officer T. Kondoue in 1894 to look at various aspects of industry as a whole. They took back information about Mannheim slide rules seen during their visit, and they investigated alternative materials, particularly bamboo, as a vehicle for slide rule manufacture. The Japanese slide rule industry started with the formation of Hemmi Keisanjaku (Sun) in about 1895, and the earliest Sun patents are for a bamboo Mannheim type rule.

A number of other notable scale layouts were produced within this time span, and deserve special mention:

— Beghin's Règle des écoles in France was the earliest with a displaced scale.

— Log-log scales re-developed by Thomson in 1881, and again by Perry in 1901 from Roget's basic concept, became a major sub-class available from all manufacturers.

— Special differential trigonometrical scales were first developed by H. Boardman in 1934; these were the last major scale development.

In parallel with the development of the *layouts* of scales on a slide rule, we also see the development of scale *arrangements* in this period. These arrangements, with the benefit of the considerable practical experience that was then available, inevitably become standards.

The **Rietz** design deserves special consideration, as it became the most commonly produced design in the world. In 1902, approximately 50 years after the Mannheim design appeared, Max Rietz produced his arrangement of scales on a slide rule that was first manufactured by Dennert & Pape. The standard was developed in a number of ways, and there can have been very few manufacturers that did not feature at least one size of Rietz rule in their catalogue. Indeed, some manufacturers listed at least four sizes - 5, 7, 10 and 20 inch - in wood/celluloid as well as all-plastic models.

The final standard arrangement of scales was the **Darmstadt**. It was designed in 1934 in Darmstadt by Professor Dr. Walther, who recognised the need for a standard arrangement that would incorporate log-log scales in a practical way. Although the Darmstadt never reached the popularity of the Rietz layout, most manufacturers produced a version, though not necessarily under that name.

Technology Improvements.

In addition to the standardisation of scale layouts and arrangements, the period between 1850 and 1900 also produced specific technology improvements. These can be considered in two ways: firstly, improvements that added to or enhanced the functionality of a slide rule, and secondly, improvements in manufacturing technology that allowed repeatable quantity production.

In the first category, two major inventions at the end of the 19th century brought the functionality of a slide rule to the standard that would continue for the rest of its life.

Firstly, we have the invention of celluloid laminated to wood, by Dennert and Pape in Altona, Hamburg in 1886. This was a highly significant invention, as it made slide rules easier and clearer to read. The use of celluloid also enabled greater repeatability in the accuracy of the slide rule scales by the use of hot pressing and similar methods, so that production in large quantities was also possible.

Secondly, we have the invention of the true duplex slide rule by Cox in the United States in 1891. Slide rules with multiple slides and scales had been produced since Everard's design of 1683; however

Cox made the decisive step of aligning the indices on the two sides so that results from one side could be transferred to the other by means of a cursor with similarly aligned hairlines. This innovation allowed slide rules with large numbers of scales (upwards of 30) to be produced.

The major improvement in manufacturing technology was the dividing engine. Strangely, its use in the manufacture of slide rules was not general until after World War I, although it was common in other areas of instrument making (sextants etc). **Table 7** lists some Dividing Engine designers and makers, and some important milestones in their use and accuracy.

Dividing Technology

One can observe the evolution of dividing engines from the making of repeatable gears for the manufacture of chronometers to the making of navigation instruments such as astrolabes, sextants, etc., to finally the manufacture of instruments that incorporate the demanding standards necessary for the physical study of light via gratings.

All over the Western world, inventors and technologists were occupied with the problems of accurately and repeatably dividing circles and straight lines, examples being cited from the United States, Russia, Switzerland as well as Great Britain. The history of dividing technology can be seen to move forward from

> — early hand division using simple and crude copying from standard "masters" via callipers etc., to
>
> — the earliest forms of horological gear cutters, vital for accurate and repeatable clock mechanisms used for navigation, thence
>
> — through early automated dividing and ruling engines which were still inaccurate, to
>
> — the most sophisticated ruling engines capable of producing diffraction gratings with 25,000 lines to the inch - 1 micron.   [N.B: 2.5 microns = 0.0001".]

The history of the slide rule mirrors the same kind of evolution over a similar period. The earliest slide rules were obviously hand divided; when the machinery allowed some degree of automation, the sophistication and accuracy of the slide rule increased. It is not known what machinery was used by which manufacturer at any point - unfortunately!

Once again, absolute accuracy is a difficult problem but it is not vital to slide rule manufacture, as has been discussed above. However, the most important function provided by a dividing engine is repeatability and the ability to manufacture a number of devices simultaneously.

Table 7 : Dividing and Ruling Engines
Their Makers and Some Milestones in Accuracy.

The following table lists dividing engine makers, the devices they produced, some users and some milestones in accuracy and technique. Despite their availability, ruling engines were not generally used for the manufacture of slide rules until the latter half of the 19th century. Previously, manufacture was from hand-divided masters. After this time dividing engines became more common, though it is not known who was the first manufacturer to use them.

An asterisk* indicates the inventions or designs which moved the whole technology of dividing scales most significantly forward.

Date	Inventor	Comment/Notes
late C16		Hand dividing enabled accuracy of a few minutes of arc to be achieved.
1660-70	Hooke	Invented the first gear cutting engine.
1677	Moxon*	First publication of *Mechanick Exercises*, which described the previously secret techniques of hand dividing for instruments.
C17 - C18		Hand division continued as the norm, using some form of calipers or beam compass.
1715	J. Rowley	Rowley worked between 1697 and 1728, during which time he is known to have sold dividing engines, designer and manufacturer unknown.
1726	I. Kalmykov	Dividing engines used for astrolabes in Russia.
1739	H. Hindley	First dividing machine for dividing a circle, seen by Smeaton c1741.
1751	Diderot	Early horological gear cutter, this is the earliest type of dividing engine for this type of application.
1767	J. Bird*	Defined a method of dividing astronomical instruments.
1775		Use of all-metal un-graduated callipers for transfer scales from a master pattern.
1777	J. Ramsden	Produced an engine for dividing mathematical instruments.
1779	J. Ramsden*	Dividing straight lines on mathematical instruments.
1787	Fortin	Produced a linear dividing engine.
1788	J. Stancliffe	Made his dividing engine while still an apprentice, used for dividing a sextant for Capt. Gower.
c1789	S. Rehe (Rhee)	Rehe made a dividing engine sometime between 1782 - 1794.
c1790	J. Allen	Dividing machine used for a sextant for Stancliffe. Described in the *Philosophical Magazine* c1811.
1793	Troughton	Made a "copy" of Ramsden's machine.
1793?	Ross	Made a dividing engine.
late C18		Use of a diamond cutter to mark scales produced finer graduations.
1800	Maudsley	A 7 foot brass screw is 1/16 inch inaccurate compared with a standard measure.

93

Date	Inventor	Comment/Notes
1811	T. Jones	Patent dividing engine, the Sectograph, sold for 2 guineas.
1814	H. Kater	Dividing astronomical circles & other instruments.
1826-27	Donkin*	Produced a linear dividing engine.
1830	A. Ross	Made a dividing engine.
c1830	A. Lear	Known to have sold dividing engines, designer unknown.
1839	Saxton	Produced a medal ruling engine.
1840	Norbert	Designed a ruling Engine.
1844	Donkin	Produced a linear dividing machine.
1846	W. Simms	Produced a self-acting circular dividing engine, an improvement to Troughton's, where he was a partner.
1850	J. Brown	Designed 3 dividing engines during the 1850's.
1859	M. Fremont	Produced a dividing engine in the USA, possibly the first.
1865	Theury	Produced a linear dividing engine (Société Genevoise) in Switzerland.
1870	Fasoldt	Produced a ruling engine.
1873	Theury	Produced a circular dividing engine (cf. 1865).
1879	Saegmuller	Produced a dividing machine in the USA.
1880	S. Darling	*Art of manufacturing mathematical Instruments*. A seminal US Patent No 233,488.
1890	Rogers	Designed a lathe with screw correcter.
1912	A. Hilger	Accepted an order from Lt. Col. Campbell (later Lord Blythwood) for a ruling engine to be used at the National Physical Laboratory. This was used continuously from 1920 to the 1950's.
1951	J. Strong	New Johns Hopkins ruling engine.
1956	Moore	Produced a dividing head for A.A. Gage & Co.
1974	D. Horne	*Dividing Ruling & Mask Making* by Adam Hilger. A definitive book for dividing.
1981	C. Evans	Design & construction of a large grating ruling engine.

Note.   Maudsley's 7 foot brass screw of 1800 (described above) is extremely inaccurate compared to modern production methods; however an inaccuracy of 1/16" in 7' equates to 0.01%. This, as absolute accuracy, is more than adequate for a 12 inch slide rule; the most important factor is the repeatability between, for example, twelve such ruled devices. Assuming that slide and stock are ruled at the same time, this level of accuracy is also sufficiently accurate for a slide rule. However, should the twelve rules be scribed individually by a machine with this level of accuracy, then it is highly likely that they would not have interchangeable components due to the inconsistencies between such individually scribed devices.

Classification of Rules

Slide rules have always been available in a number of formats. The common ones of rectilinear and circular (with circular and spiral scales) were supplemented by circular watch type calculators, and the various long scale variants that were produced in this period. We can now look at all these formats in more detail and attempt some form of classification and standard description.

Linear Slide Rules.

From 1850 all major manufacturers produced a range of rectilinear slide rules in various forms and sizes as part of their portfolio. There are a number of standard scale layouts as well as arrangements, and it is thus convenient to subdivide layouts further into standard (or trigonometrical) slide rules and log-log ones. The scale arrangement is then particular to a manufacturer, though as previously described most would have included Rietz and Darmstadt as well as their own specific variants.

The tremendous range of makers, sizes and types of slide rules in this format make them the most likely type to be seen and found by collectors, and also the most difficult to classify in any logical way. An immediate category for classification is the physical format:

Closed (or solid) frame. Here the stock of the rule is in one piece, albeit sometimes with a slit in the base to enable the movement of the slide to be controlled in some way. The benefits of modern materials and manufacturing techniques mean that not all rules, particularly more recent designs, have the stock so split. There are various windows and slots in the back of these rules for access to scales on the back of the slide.

Open frame. Here the stock is made up of two pieces between which the slide can move. The ends of the two pieces of the stock are clamped firmly, sometimes permanently, or in some cases by devices which allow adjustment. These "bridges" are variously plastic or metal. This is the format of all duplex slide rules. The duplex rule is a special case of open frame rule which has scales on both sides.

Size. A further subdivision by size can be used; however, as slide rules were available in so many sizes, this is not practical. The common sizes were 5 inch, 8 inch, 10 inch (the most common), as well as 15 inch, 20 inch, 24 inch, and 40 inch. Both longer and shorter were also known. Some 3 inch rules were made, probably as a gimmick, and rules many feet long for special purposes are described in the literature. Most designs were available in at least 5 inch, 10 inch, and 20 inch lengths, so unless there was something special or unusual about the rule dimensions, one might assume that a rule being described in the following descriptions was a 10 inch version. There are also the very large - several foot - teaching and demonstration rules.

Large numbers of special scales and scale arrangements as well as variations of the standard scales were available on rectilinear slide rules. A convenient sub-classification of linear slide rules is those with standard/trigonometric scales (Mannheim and Rietz standards), and those with log-log scales (Darmstadt standard). It should be noted that these are independent of format (open or closed frame).

The Mannheim standard.

The earliest standard, manufactured from about 1855, is the Mannheim, which has the following scales, and was the first standard rule to be supplied with a cursor.

— On the top edge of the slide is a scale running from 1 to 100, sometimes known as a two-cycle logarithmic scale. This is a squares scale, designated "B".

— Above this on the bottom edge of the gap in the stock is an identical scale, designated "A".

— On the bottom edge of the slide is a scale running from 1 to 10. This is the units scale used for most calculations, sometimes called a single-cycle logarithmic scale, designated "C".

— Below this on the top edge of the bottom of the gap is an identical scale, designated "D".

Until approximately 1930, large numbers of slide rules had no more than these basic four scales. However the Mannheim selection of scales on one face was found to be somewhat limited in scope and was soon supplemented by a number of other standard scales which can be considered a development of the Mannheim.

The Rietz standard.

This uses the standard four Mannheim scales supplemented with additional scales organised as follows:

— Above "A" is a scale from 1 to 1000 (sometimes known as a three-cycle logarithmic scale) for cubes and cube roots, designated "K".

— Also above "A" is a linear scale for the mantissa of common logarithms, designated "L".

— Below "D" are scales for sines, designated "S"; tangents, designated "T"; and the sines and tangents of small angles, designated "ST".

— In the centre of the slide, between "B" and "C", is a reciprocals scale designated "CI". This addition dates from about 1925.

The Rietz name is seen spelled in a number of ways; Rietz as per the inventor; Reitz, sometimes believed to be the English spelling; and in one Japanese case, Ritz, either a spelling mistake or an attempt to suggest quality - we will never know. In about 1950 the standard changed to make all scales available from the front face for direct reading. Thereafter a number of manufacturers produced further variations on this arrangement which were still sold under the Rietz name. Probably the most comprehensive variant is the Nestler Rietz-Duplex, which despite the name bore little relationship to the inventor's original layout.

A common variant, still called Rietz, moved the trigonometrical scales to the reverse of the slide. On some early slide rules, in both open and closed frame construction, the slide has to be reversed and used either the right way up or upside down on the front of the rule. Later versions incorporate slots or other apertures in the reverse of the rule, with a suitable index to enable the trigonometrical ratios to be read without removing the slide.

The Darmstadt standard.

The requirement for slide rules incorporating log-log scales led to the Darmstadt rules where the standard fit of scales was further re-arranged as follows:

— Above "A" are "K", "S" and "T" scales.

— Below "D" are "L" and a scale for $\sqrt{(1-x^2)}$, designated "P".

— The centre of the slide has a "CI" scale.

— The reverse of the slide is marked with three log-log scales, "LL1", "LL2" and "LL3".

As with the Rietz nomenclature, there are variations on the Darmstadt arrangement. There are also Darmstadt arrangements which are called something else, perhaps the most unusual being the Unique Brighton slide rule.

The log-log, logo-log or logometric scales produced such a considerable improvement in the functionality of slide rules that the sub-set or super-set of log-log slide rules is separately considered.

The following descriptions are of representative slide rules from a number of manufacturers, within the various classifications. A more complete listing is given in **Chapter 6.**

Standard Linear Slide Rules.

Aristo.

A German manufacturer with factories in Germany and Austria that produced slide rules from the earliest days, originally under the Dennert & Pape name, which still features in its literature. The majority of Aristo rules are closed frame construction. Commonly seen are the Simplex, Scholar and Studio range of rules, and a Darmstadt rule is mentioned in its literature. Some modern rules are:

-Scholar 0903, made in Germany, has an "L" scale below "D".

-Scholar 0903, made in Austria, has an "L" scale above "A".

-Scholar 0903 VS has a reversible cursor to access the scales on the back of the rule.

-Scholar 0903 LL has a reversible slide to enable the log-log scales to be used on the front.

-Scholar 0903 VS2 is a duplex rule.

-Simplex 0911 comes in various styles.

A.W. Blundell / Blundell Harling / BRL.

A well-known manufacturer of British rules that went through a number of name changes. It made slide rules for sale by other companies in the 1960s and later, notably for W.H. Smith. A soundly made range of rules, generally in plastic, both open and closed frame, in three ranges:

Academy -	Inexpensive plastic rules.
Omega -	The standard range in 5 inch and 10 inch versions, both Reitz and Darmstadt layouts.
Janus -	T50 and T51 duplex rules.

Blundell Harling ceased general manufacture of slide rules in about 1985, but still advertises a Speed, Time & Distance (S.T.D.) calculator and a NATO S.T.D. slide rule, and will make special-purpose slide rules in rectilinear format as well as circular.

John Davis and Sons.

Available until the early 1940's, when it ceased manufacture. Perhaps better known for some of its more unusual specialist rules (see later), it nevertheless had a wide range of normal rules to offer. Of note:

Plutocrat -	Standard Rietz.
Monocrat -	As normal, with additional sine scale above "A", logs below "D" and a tan scale on the slide.
Dafield -	As Monocrat with a split scale divided into 20 for shillings and pence work, as well as hundredweights and quarters.
Glider -	Standard Rietz, notable for its special construction which included a waved spring steel band in the groove of the solid frame construction that gave exceptionally smooth movement.

SLIDE RULES

A.W. Faber / Faber-Castell.

Manufactured in Germany primarily, although there were also factories in Switzerland, these are probably the best-quality slide rule available, though undoubtedly some would debate this. A.W. Faber and the later Faber-Castell rules are generally found in excellent condition, attesting to their robustness and longevity. The company supplied beautiful rules made from Swiss pearwood with celluloid facings as its top-of-the-range, as well as plastic solid frame and duplex rules. The company provided a very wide variety to suit many different applications. Its 1958 catalogue listed ten different slide rule arrangements in three sizes (5 inch, 10 inch and 20 inch) and four styles (wood/celluloid and plastic; with and without Addiator) totalling approximately 30 types. Other slide rules within various collections not mentioned in the 1958 catalogue indicate approximately 100 types of slide rule from this manufacturer, covering all applications from school and college use to specific engineering problems.

To overcome the inability of a slide rule to perform addition and subtraction, Faber-Castell (and some other manufacturers) put a simple adding machine on the back of some rules. This "Addiator" was supplied on 5 inch and 10 inch rules. Later, from the early 1970's, to fight off the calculator for as long as possible, the company produced a design that incorporated an electronic calculator. This in some cases appears sensible, as the slide rule has log-log scales which are not available on the four function calculator. In other cases the slide rule is rendered totally redundant!

Reitz and Darmstadt rules were available.

Keuffel and Esser.

An American company that sold numerous slide rules in England. In the 1930's it had over a dozen different types in a variety of sizes available in its catalogue. It started manufacturing in the late 1800's and ceased in 1976, its last slide rule being donated to the Smithsonian Institution in Washington. The range available was always comprehensive and imaginative; of note are:

- Polyphase- Normal "A", "B", "C" and "D", with a reversed "C" above "A", and a cubes scale below "D".
- Universal - Supplied with two slides.

K&E, together with a number of other American manufacturers, chose not to use the Rietz name on their rules, though similar layouts with different names did feature in their catalogue.

Nestler.

A German company with a wide following in England, Nestler started making slide rules in the traditional wood and celluloid solid frame construction and ultimately moved to plastic open frame construction. Among the earlier slide rules are:

Nestle -   "A" and "B" are normal, "C" is reversed, "D" is a cube scale. Above "A" and below "D" are the elements of a 20 inch scale.

Hanauer -   "A", "B" and "D" are normal, "C" is reversed, above "A" is a square scale, cubes are below "D".

Rietz -   Standard Rietz pattern.

Precision -   "A", "B", and "C" are normal, "D" split into the two halves of a 20 inch scale. Above "A" is a cubes scale, the slide has "L", under the slide are "S", "T", and "ST" scales.

Universal - As Reitz, with square and log scales transposed. Also used the edge of the rule for a cubes scale. Specially arranged for surveying.

Fix -   Normal, but with the "A" scale displaced by $\pi/4$.

Baur -   A 25 cm scale along the front edge was repeated three times, 0-25, 25-50, & 50-75 to provide a convoluted but accurate way of finding cubes and roots.

Anido -   Normal, with a double log scale for logs <1 and >1.

Multiplex -   Normal with a strange "B" scale, in which the left half was reversed, the right-hand half normal.

Sun-Hemmi.

A Japanese company particularly known for its bamboo and celluloid rules, as well as more traditional plastic solid frame devices.

Tavernier-Gravet.

A premier French company that manufactured the original Mannheim rules in Paris in the 1850's, and later exported them to England and elsewhere. They appear to have been in business for some considerable time. The Tavernier-Gravet or Gravet name is sometimes used with the standard Mannheim layout. Of note:

Beghin -   Normal "C" and "D", "A" and "B" displaced by half the length of the rule, thus allowing considerably reduced movement of the slide in the performance of calculations. Reversed "C" in the centre of the slide; reverse face of the slide had sines, tans and squares. The edge of the rule had logs and a scale of gauge points.

Long -   Like Nestler Precision, but with all graduations reversed on the slide. Reverse of slide had cubes.

A.G Thornton / British Thornton PIC.

The PIC trade name which has been associated with Thornton since about 1900, has always been synonymous with quality and innovation. The A.G. Thornton catalogue was always well filled with a wide range of rules of all types for general and special applications. The Boardman differential trigonometric scales feature on these rules. The British Thornton PIC range of 10 inch open frame plastic slide rules were amongst the most popular rules available more recently in the United Kingdom, the range being restricted to:

- AA 010 Comprehensive (a duplex rule).

- AD 050 Log-log.

- AD 060 Standard.

- AD 070 Modern Maths.

There were numerous variants in name and type number. The log-log rule was also available as the P271, the comprehensive as the P221, and there was a P281 Standard Student. There was also a P251.

The earlier A.G. Thornton PIC wood and celluloid rules in solid frame construction were well known for their robustness and were available in the standard range of sizes under a large variety of type numbers.

Unique.

Unique was a British manufacturer from about 1923 to 1975 that made a very large range of types and styles in its very distinctive design. The Unique was definitely cheap and cheerful. Its early design of wooden solid frame with unique paper scales under celluloid was familiar to generations of British school children and students, as well as many engineers and scientists. The Technical Supply Company manufactured slide rules for Unique that are indistinguishable from those manufactured by Unique itself. Unfortunately the construction was not particularly robust and many Unique rules seen for sale are spoiled - portions of the scales have lifted or are missing, and often the rule has warped, making the movement extremely stiff.

There are many different types of construction within the same Unique rule type: spring steel back, thick wooden with slits, thin wooden with slits, plastic open frame and plastic closed frame.

Unique manufactured Universal One, Universal I, Universal II and Brighton, as well as numerous special types in both closed frame wood and open frame plastic. Advertising literature of the day shows some 25 types, the Universal I being designated the bestseller. The open frame plastic included the Study 500, 700 and 900. These are not often seen. Unique slide rules are described in considerable detail in the various editions of Snodgrass's *Teach Yourself the Slide Rule*.

Slide rule sellers.

The following brands are included as a sample of suppliers who did not necessarily make their own rules but had rules made for them by a variety of manufacturers who provided rules with the appropriate logo and type number for these retailers.

Boots.

Simple plastic solid frame slide rules, sold under the Ringplan trade-name. They were probably manufactured by Faber-Castell in Germany, as ascertained by comparing the instruction leaflets. Japanese companies also manufactured rules for Boots.

Helix.

A range of inexpensive plastic slide rules in 5 inch and 10 inch sizes, generally simplex and open frame construction were made in Japan.

W.H. Smith.

The range of rules was made by Blundell. They came in 5 inch and 10 inch versions. Simplified Reitz, Log-log Trig, and a Master Duplex A 505 slide rule are Blundell Academy rules.

Slide rules from Jakar, Royal and T.S. and a host of other manufacturers and sellers can also be found. These are generally of a common design, plastic, and of both open and closed frame design, with scales in the Rietz style. A number have advertising material on them.

Log-log Slide Rules.

A number of slide rule manufacturers covered in the previous section produced log-log slide rules. Indeed it would have been very unusual if any of them did not include at least one, and often more than one, in its catalogue.

The number of log-log scales (designated the "E" or "LL" scales) supplied varies. The maximum number is eight: four log-log, and four anti-log-log, with values from plus 0.0001 to 1000 and minus the same values. Two, four and six log-log scales per rule were also available, giving a more restricted range. These rules could form a separate study or collecting specialisation, there being many variations in positioning and layout of the scales.

The Darmstadt arrangement of scales is a log-log variety.

A brief description of some of the more unusual rules follows:

Dargue Brothers.

A small English firm in Halifax that made slide rules for a few years prior to World War II. Of special note is its Simplon SP10 log-log slide rule, available in the 1930s, which was unique in offering a slide in a dovetail, rather than a groove. Later rules were of the normal tongue and groove construction.

Davis.

- Log-log. -Log-log scales were supplied on a separate slide, with both plus and minus "E" scales.

- Jackson-Davis double slide - A second slide was held with clips and there was a special cursor for reading it. A third slide was also supplied, one being the positive "E" values, the other the negative.

Faber.

- Log-log -Two versions, an earlier one which had two log-log scales on the edge of the rule, and a later version which had the two scales on the face, above the "A" and below the "D" scales. This latter also had some electrical scales in the groove of the rule.

- 1/92 Log-Log -also known as the Engineer's Log-Log, in wood/celluloid with the 52/92 all-plastic Students Log-Log.

Keuffel & Esser.

Numerous types, including:

- Log-log Duplex -Many log-log scales on both sides of the rule.

- Deci-Lon -The most comprehensive selection ever offered, supplied in a very smart leather holster complete with belt loops and other bits for hanging the rule on the person.

- LL-Trig Duplex   } Both are extensions of the log-log duplex, with

- LL-Decitrig Duplex   } either angular or decimal notation of degrees.

Thornton.

- Perry log-log. -   Like the Faber rule, but read in conjunction with the "B" rather than the "D" scale.

- PIC -   Various versions, usually including electrical scales.

British Thornton PIC Comprehensive - good selection of scales.

Unique.

The Unique log-log rule was featured in its catalogue throughout the life of the company, and is probably where the majority of slide rule users in the United Kingdom will have first met this useful set of scales. There were also log-log scales on the Brighton and Dualistic range of rules.

The Yokota rule.

Supplied by J. Davis, and a number of other manufacturers, Yokota had the earliest patents for a scale based on the value of *e*. It was unique in that the scales were 10 inch and not 25 cm, so that the normal inches scale could be used to obtain ordinary logarithms.

Log-log slide rules were also available from a number of the other manufacturers including Faber-Castell, Blundell and Aristo; also retailers including W.H. Smith and Boots.

Circular Slide Rules.

There is a fascinating quote from De Morgan in 1842, referring to the attempts in France in the early 1800's to get acceptance of a type of circular slide rule:

> *12 or 15 years ago an instrument maker in Paris laid down a logarithmic scale on the rim of the box and the lid of a common circular snuff box, ... but either calculators disliked snuff, or snuff-takers calculation, for the scheme was not found to answer.*

Despite being the shape of Oughtred's earliest design, circular slide rules have never found the favour they perhaps deserve, particularly as they are capable of providing accuracy usually obtainable only in much longer rules: a 3 inch diameter slide rule has a 10 inch (approximately) scale length. The fact that for a given accuracy they can be made smaller and more portable or pocketable is their main advantage. One of the reasons given for their lack of popularity is that, even though they do not require as many movements of the scales in a multiple function calculation, considerable manipulation is still required to keep the scales in position so as to be able to read the numbers.

Circular slide rules, while having the benefit of a continuous scale (which negates the problems of having the slide at the wrong end of the linear scale), nevertheless still require considerable movement to perform a lengthy calculations involving a number of multiplication and division sums. Also, on cheap circular rules, parallax between scales and between scales and cursors is a considerable problem. Parallax is created when there are two discs on top of each other. It is the distance between the two scales resulting from the thickness of the upper disc. There is a further gap between the scales and the cursor line. Each of these distances allows different readings to be obtained, depending upon the position of the eye relative to the scale or hairline. Slide rules with only one disc and a cursor are considerably better, as the cursors are generally flexible and can be pressed down onto the scale to avoid parallax. Where two discs are used on higher quality rules, the edge of the inner or smaller disc often will be chamfered so as to present a sharp edge against the lower disc; alternatively the upper disc may be let into the bottom disc.

Circular slide rules can be classified into two main sub-groups divided as to the function of their format. We have the traditional multi-disc/multi-cursor circular device and the watch type calculator, the latter being the "Rolls-Royce" of the type and available in a number of quite splendid designs.

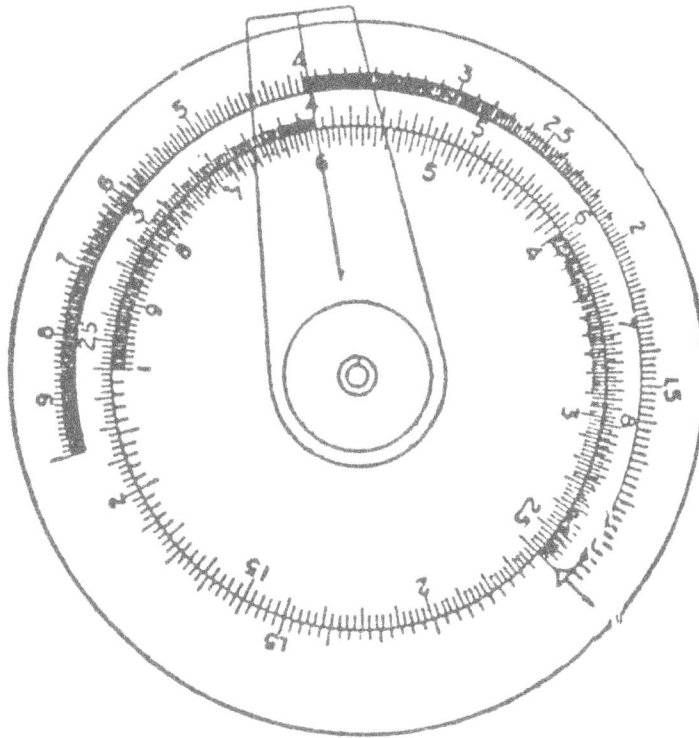

Figure 18 : Picolet Circular Slide Rule

Multi-disc/multi-cursor circular rules.

Most later circular slide rules were produced for promotional purposes, being embellished with advertising material. They were made of plastic, metal (brass and aluminium), or card, with the plastic device being the most common. The formats follow a number of alternatives:

- A base disc with a smaller disc on one side.
- A base disc with a smaller disc and a cursor on one side.
- A base disc with a smaller disc on both sides.
- A base disc with a smaller disc and a cursor on both sides.
- A base disc with a cursor or cursors on one side.
- A base disc with a cursor or cursors on both sides.

Where multiple cursors are used, one of the cursors replaces the small disc and is used to provide the datum point in a calculation.

The simplest circular slide rule has only two scales, the equivalent of the "C" and "D" scales on a linear slide rule. This means that they are only useable for multiplication and division. Nevertheless many circular slide rules were produced with a multiplicity of scales, including some extremely esoteric ones covering specialist applications in many fields.

The following are a cross-section of circular types.

The Picolet Circular Slide Rule.

This was made in the United States c1915-45, and took the form of two discs and a cursor. In addition to the standard two scales, it had a two-thirds length scale outside the main scale on the base disc which was used for performing calculations of cubes and cube roots. The smaller disc also had its scale reversed. See **Figure 18**.

Charpentier Calculator.

The Charpentier calculator featured in K&E catalogues from about the turn of the 20th century, and was a traditional metal disc calculator with multiple cursors.

ALRO Calculators.

These are quite delightful and sophisticated rules, available in a number of sizes. One variant was self-contained within a "box", the lid of which could be used as a stand as well as providing protection for the scales. Another was a more traditional multi-disc circular slide rule. ALRO calculators were often supplied with advertising material.

Small Calculator.

The SMALL calculator was a high-quality circular calculator with scales on the surface of a thick disc as well as on its circumference.

The Gilson Slide Rule.

The Gilson Company of the USA produced a considerable range of circular slide rules with multiple scales as well as some with spiral scales. The Gilson Binary had the most comprehensive set of scales on any circular slide rule. It consisted of an aluminium disc approximately 6 inches in diameter, with two cursors on one side and a single cursor on the other. Some 30 scales were provided, including log-log and fractional scales and scales for screw cutting. It was available over the period 1930 to about 1950. Gilson also produced various other types of rules in a number of sizes.

The Griffin Slide Rule.

This was an example of a cheap plastic two disc, one cursor single-sided rule. Some examples were marked with promotional material from companies all over the world. The two scales provided allow multiplication and division, and standard conversion points were also annotated. This rule exhibited parallax problems to a marked extent, which meant that any calculations performed on it were extremely suspect.

<u>*The Concise Slide Rule.*</u>

Concise was a Japanese company that manufactured a range considered by some to be the ultimate in circular designs. They advertised their rules as a "... *give-away item for friends and business associates imprinted with the name of your company or trademark during the course of manufacture.*"

The range is described in **Chapter 6**.

The Type E was an extremely small rule, in plastic, with two discs and one cursor. The inner disc was chamfered and surprising accuracy was possible with the 2 inch disc (approximately equivalent to a 6 inch linear rule). One example seen is annotated with the logo of an electronics supply company. It also has a quite incredible selection of Japanese to imperial and metric conversion tables built into a separate linear slide within the body. These provide conversion for liquid, linear, area and volumetric measures.

<u>*Unique Dial Calculators.*</u>

Two devices featured in the Unique catalogue over many years were the:
- 10 inch calculator -   Two disc, one cursor single-sided circular slide rule.
- 50 inch calculator -   Single disc with double cursor, capable of considerable accuracy with its spiral scales.

The Unique Dial Calculators are covered in some detail in Snodgrass's *Teach Yourself the Slide Rule*. Despite the considerable detail in the descriptions, however, it has never been established whether these were actually ever manufactured (See **Figure 19**).

Long Scale Disc Calculators.

Long scale disc calculators are found in the preceding circular rules formats, the Long scale of the Fowler types being a good example of the spiral scale at its most effective.

The long scale disc calculator was an attempt to provide greater accuracy on a flat format by using spiral scales in the same way that helical scales were used on tubular devices. A number of the circular slide rules covered earlier in the book have spiral scales, and as such can be considered as part of this section. However, even the most generously proportioned disc calculators cannot match the accuracy of the helical scales on the Fuller Calculator. Even so, some spiral scale disc/circular slide rules had scales of a similar length to an Otis-King calculator.

It should not be forgotten that Oughtred's original design was a long scale calculator with a spiral scale, but it was not until some 250 years later that some measure of popularity for circular designs returned. Models available from the beginning of the 20th century include:

<u>*Fearnley's Universal Calculator.*</u>

Patented at the turn of the 20th century it was not a common design and details are not available.

<u>*Scheuermann's Calculating Instrument.*</u>

Another design from the early part of the 20th century, about which few details are available.

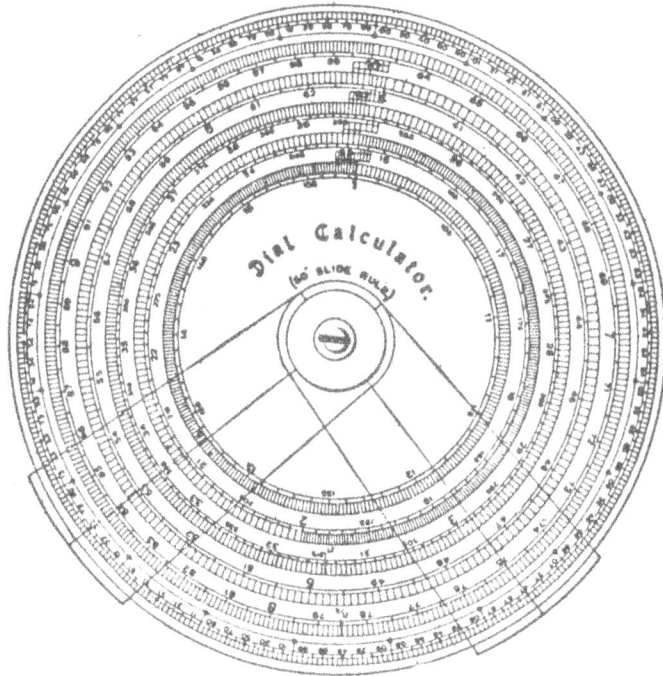

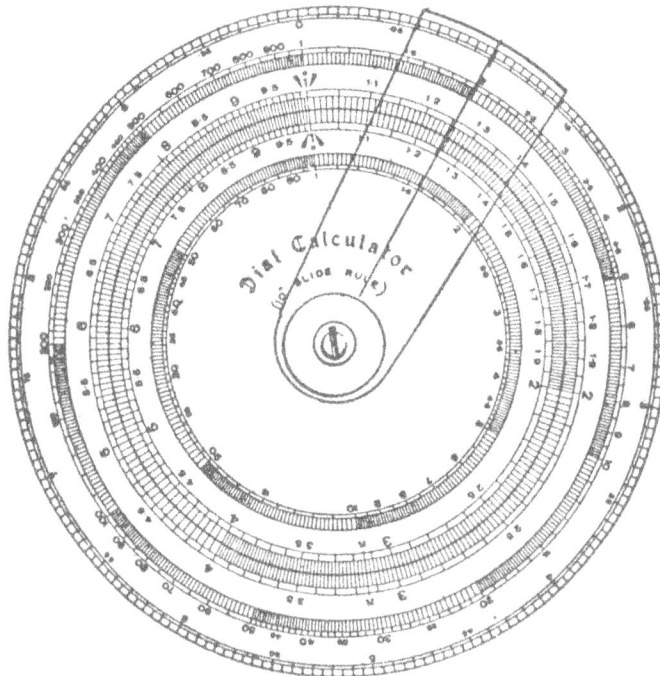

Figure 19 : Unique Dial Calculators

Watch type calculators.

These are extremely attractive instruments, made as wrist and pocket-watch styles. The wrist-watch types generally were produced by watch manufacturers such as TAG Heuer, Breitling, Seiko, Citizen and so on, while the pocket-watch styles were made by a large number of specialist manufacturers. They featured in many other manufacturers' and retailers' catalogues. These devices have their origins in the Boucher calculator of 1876 (see **Figure 20**) and other practical implementations such as the Charpentier calculator and Curtiss's Calculating Disc. These worked on the principle of a disc within a fixed ring, both carrying a scale as with a normal slide rule. The translation into the watch type was a natural progression, and the majority look like a stem-winding watch complete with winder which, via gears, moves either a second disc or the cursor. Rather like some stopwatches, some examples have a second stem, that was used to move the other item, i.e. the main stem was used to move the cursor, the second to move an inner disc. The Halden Calculex, while taking the form of a watch, was the only one without a stem winder but instead used a thumb-nut through the face of the glass to move the second disc.

Both styles of watch calculator were at the top end of the price range. They got relatively more expensive as increased automation and greater use of plastic reduced prices of linear slide rules.

The following include pocket-watch types which were more common in England than elsewhere, as well as some American designs which were available in a number of countries.

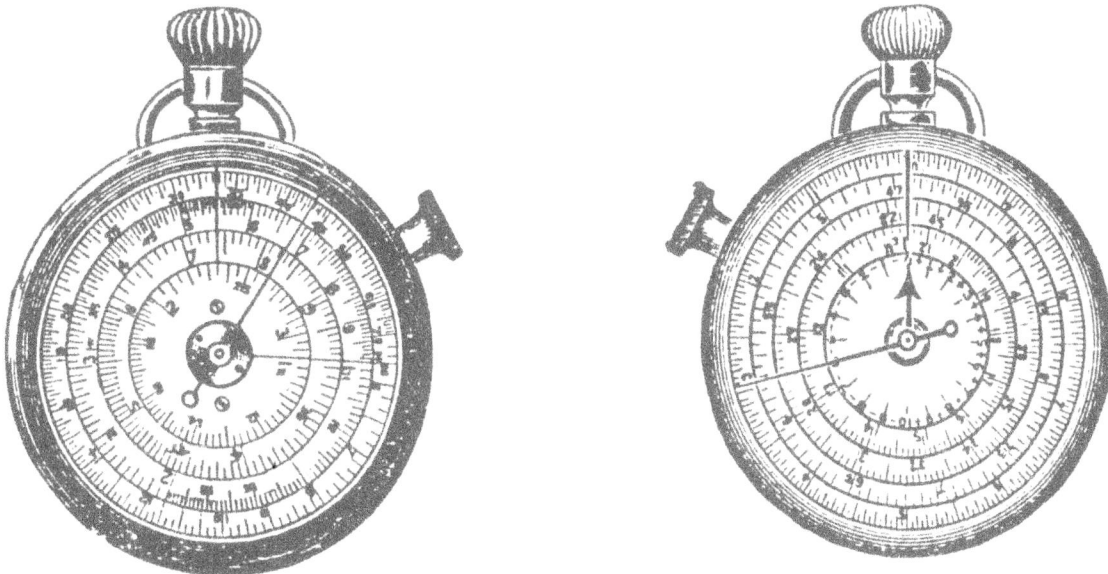

Figure 20 : Watch Type Calculators

Fowler Calculators.

The Fowler Calculator Company of Manchester, England was a major manufacturer of watch type calculators. Various types were available with different scales for different users. All were of the watch form, with front and back dials turned within fixed rings. All types had scales giving square roots, logarithms, as well as sines and tangents of angles.

Fowler calculators of all types were made in three sizes, as follows:

• The smallest (approximately 2½ inches in diameter) for carrying in the waistcoat pocket, was fitted with either one or two dials.

• The medium size, of which the 12-10 and Universal are examples, had single dials approximately 3 inches in diameter.

• The largest size, the Magnum, had a single dial approximately 4½ inches in diameter.

Fowler Long Scale (Type RX) Calculator.

This was the most popular, comprehensive and accurate calculator, with a good range of scales on front and back dials.

Fowler Type H calculator.

A double dial calculator having the back dial of the type RX arranged as a front dial, with a dial giving cubes and cube roots fitted at the back. Alternatively a scale of Imperial to Metric conversions could be fitted as a back dial (like the Artillery calculator).

Fowler Artillery Calculator.

Constructed with either single or "Slide Rule Type" dials. For the solution of artillery range-finding problems. An imperial to metric conversion dial was fitted as standard at the back.

Fowler Circular Slide Rule.

An instrument designed as a replacement for rectilinear rules in a more portable and accurate instrument, where no "end switching" was necessary. It consisted of a dial rotating inside an outer fixed annulus. The back dial gave reciprocals, and cubes and cube roots of numbers; alternatively, an imperial to metric conversion scale could be fitted. Also supplied as a single dial instrument.

Fowler Single Dial, Vest Pocket Calculator.

Intended to fit a vest (waistcoat) pocket, this was fitted with either the front or back dial of the Long scale instrument or with the front dial of the Circular Slide Rule. Available as:

• Type MD - front dial of the type RX (useful for those who wish to perform multiplication and division only).

• Type H (SD) - back dial of the type RX.

• Type CSR (SD) - circular slide rule dial.

Fowler Universal Calculator.

This was a single dial instrument in a Bakelite case which was easy to read and capable of solving all arithmetical calculations. It had a long scale extending over three circles with a total length of 20 inches and gave direct readings of cubes and cube roots and general results to three or four significant figures.

Fowler 12-10 Calculator.

This was approximately 3½ inches in diameter, intended primarily for architects, quantity surveyors, builders, timber merchants and craftsmen, i.e., those who would be dealing with feet and inches, shillings and pence.

Fowler Magnum Calculator.

This was approx 4½ inches in diameter. It was intended for office use or for *"... those with weak sight"*! The long scale was 50 inches, while the normal short scale was 13½ inches. It enabled all of the scales of the two dial instruments to be included in one dial, with the scales carrying additional intermediate graduations. The Long Scale was divided into 100 figured dimensions, each of which was further sub-divided by 10. Gave results to four and five significant figures.

Prices (c1945).

Small vest pocket size, double dial	-	37/6d
Vest pocket size, single dial	-	31/- (also 33/-)
Universal and 12-10	-	37/6d (also 33/-)
Magnum	-	55/- (also 49/6d

Note that there are many inconsistencies in the prices quoted throughout the Fowler Electrical and Mechanical Engineering Pocket Books, which give extensive details about these delightful devices.

Cases.

The vest pocket and small vest pocket sizes were supplied in a nickel case, the Universal and 12-10, and the Magnum in a leather case. The Universal was sometimes supplied in a Bakelite case, and some versions had a Bakelite back to the instrument.

The Halden Calculex.

The Halden Calculex vied for popularity with the Fowler calculators, and was also intended as a vest pocket calculator. It was more akin to the traditional slide rule, but in circular form, and this may have found favour with certain users. It consisted of an outer metal ring carrying a fixed scale, within which was a moveable dial turned by thumb-nuts. The whole was protected by glass faces which could also be separately rotated, and which carried hairline cursors. As the glass faces were extremely close to the face of the dial, the effects of parallax were avoided; as the various elements were directly controlled, this device was more robust than other watch type slide rules because there were no gears. Scales were provided on both sides to enable all normal calculations to be performed. Various models were produced, including a desk-sized model where the thumb-nuts were provided with additional rings to assist with turning the elements (see **Figure 21**).

The Fowler and Halden devices are well described in the Newnes' *Slide Rule Manual.*

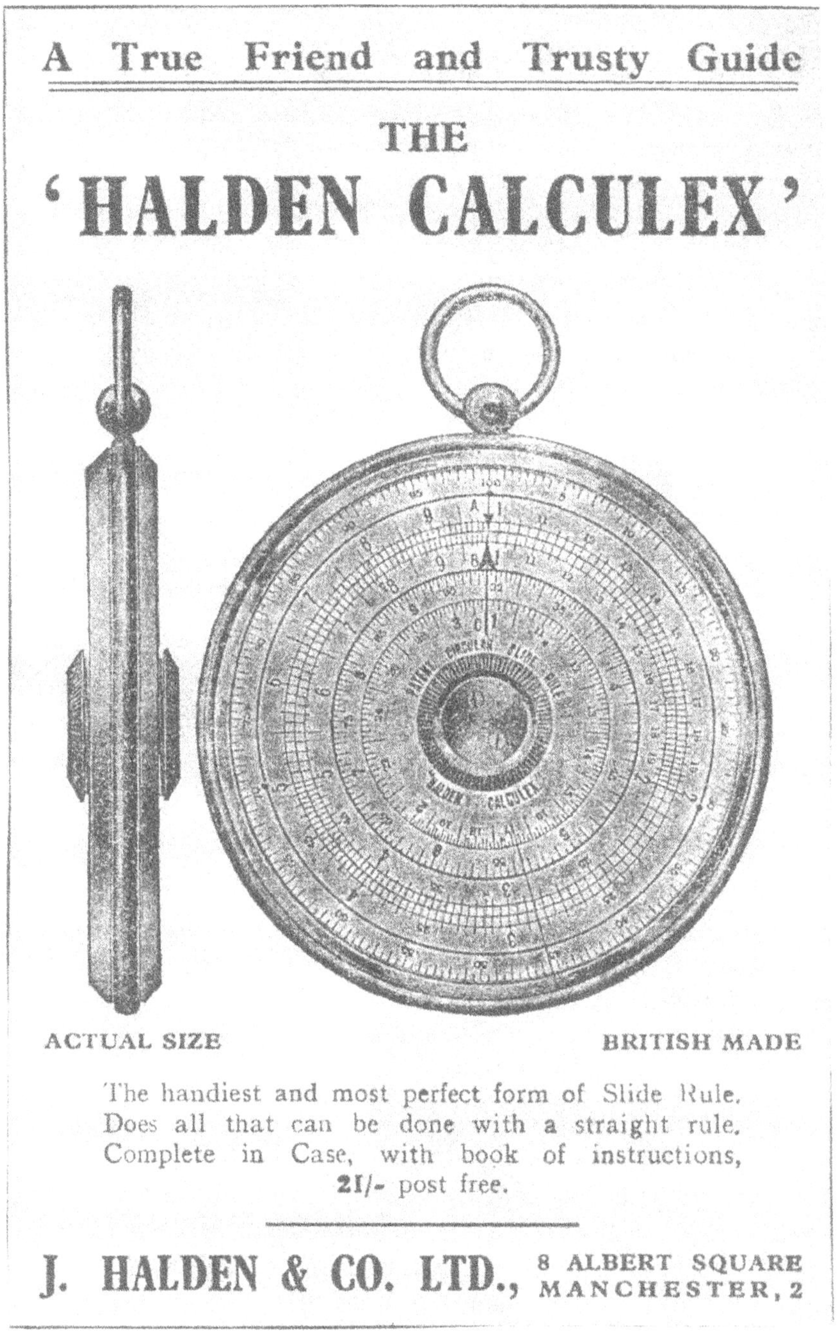

Figure 21 : Halden's Calculex

The Boucher Calculator.

The Boucher calculator was the original design for watch calculators. It was made by several manufacturers in a number of countries. The device illustrated in **Figure 20** was manufactured by W.F. Stanley of London. This has two dials, the back one being fixed, the front one being free to turn on a central axle under the direction of the "winder". Two fine needle pointers were locked together and the action of the side stem moved one over each dial. A fixed index / pointer which was part of the case provided a datum point. Numerous scales were provided enabling squares, logarithms, and sines as well as normal multiplication and division. Some of the scales were in two parts, so increasing the accuracy.

Manufacturers who produced Boucher calculators included Keuffel & Esser in the United States, Stanley and Manlove, Alliot & Fryer in Great Britain and Henri Chatelain in France. The German company Schacht & Westerich produced a version; Wichmann sold versions of the Calculex and the Schacht & Westerich design.

Keuffel & Esser.

Keuffel & Esser supplied many different types of slide rule for the British market, including a number of watch type slide rules. Among these were:

Sperry's Pocket Calculator.

The original watch calculator patented by Sperry, and very similar to the later K&E calculator. Early versions used one knob and two push buttons, later versions also used the concentric thumb-nuts for moving the elements. The scales were differently arranged into three sections giving a scale length of 12½ inches.

The K & E calculator.

A basic two sided calculator with two dials, only one of which revolved. There was an index line engraved on the glass. The two pointers and the dial were moved by concentric thumb-nuts on one spindle. The two pointers moved together by means of a milled nut, the central dial under the action of a thumb-nut. The scales were in the form of two turn spirals for accuracy.

Other manufacturers of watch type calculators included Dyson (who produced a textile calculator), and a number of unknown continental makers who produced unusual calculators. There are also various Russian pocket-watch calculators available.

Long Scale Slide Rules.

Tubular Slide Rules.

Probably the most common form of long scale slide rule, the tubular slide rule is one of the most effective ways of achieving extremely impressive scale lengths in a workable overall size. Most had helical scales, this being particularly practical as a method of putting a scale many feet long onto a short(ish) tube. However, the obvious limitation of this format was that it was not suitable for the addition of the many other scales which were found to be so useful on other styles of slide rule. For many years the Fuller calculator was the choice as the slide rule/calculator where the requirement was for greatest accuracy (five figures) without the need to resort to logarithms or other methods of calculation. Neville Shute's autobiography has an engrossing section which describes a team of men working out the stresses in the framework of the R100 airship with Fuller calculators. It is worth reading if for no other reason than to see how technology in calculating has evolved over the last 70 years.

Tubular devices have always had a fervent minority following. There were not that many types produced; however those that succeeded did so for many years and were made in large quantities. The Otis-King was still on sale when slide rules ceased to be produced, and the Fuller was sold for over 80 years; over 150,000 Otis-King and over 80,000 Fuller rules were sold.

It is interesting to compare the price of the tubular calculator with that of other slide rules at the various times. In the early 1920s, following the introduction of the Otis-King, Gallenkamp sold that calculator for £1 1s/0d; a Rietz slide rule cost £1 2s/6d; a Castell precision 10 inch rule 15s/9d; the 20 inch model £2 15/0d; a Boucher's watch calculator 18s/9d; and the Fuller tubular calculator £5 10s/0d. Initially it can be seen that the price differential was not much more expensive than a linear rule.

In 1938, the Otis-King calculator sold for 22s/6d when a good-quality linear wood/celluloid slide rule was available for approximately 6s/0d. In 1960 the Otis-King sold for about £5 and a wood/celluloid slide rule cost approximately 30s. Cheap slide rules were then available for about 13s from Unique.

These later price differences were the probable reason why the rule was only popular with a minority of users.

The Otis-King calculator.

This device was a well-produced tubular slide rule which was sold in a number of forms, the most common being:

- The type K - sold with only the scales necessary for multiplication and division.
- The type L - sold with a logarithms scale as well.

There were also various models with monetary scales, though it would appear that these only were available for only a limited period, at the start of production in about 1922-23.

All types were approximately 1¼ inches in diameter and 6 inches long when closed, opening out to approximately 10 inches at full extension. The helical scales were 66 inches long, and were on non-stretchable velum paper; the various sliding elements were made of metal. The whole was supplied in an attractive leather wallet, altogether a very professional piece of equipment.

Fuller Calculating Rule

This device was available from 1878 from a number of firms (see **Chapter 6**), the best known being W.F Stanley, A.G. Thornton, and Keuffel & Esser. This was altogether a more massive piece of equipment than the majority of other tubular calculators. It was available in three main forms (though there are two versions of No 2):

- No 1 model   -single scale (multiplication etc.) only.
- No 2 model   -with sine and log scales as well.
- Special   -Fuller-Bakewell for tacheometry.

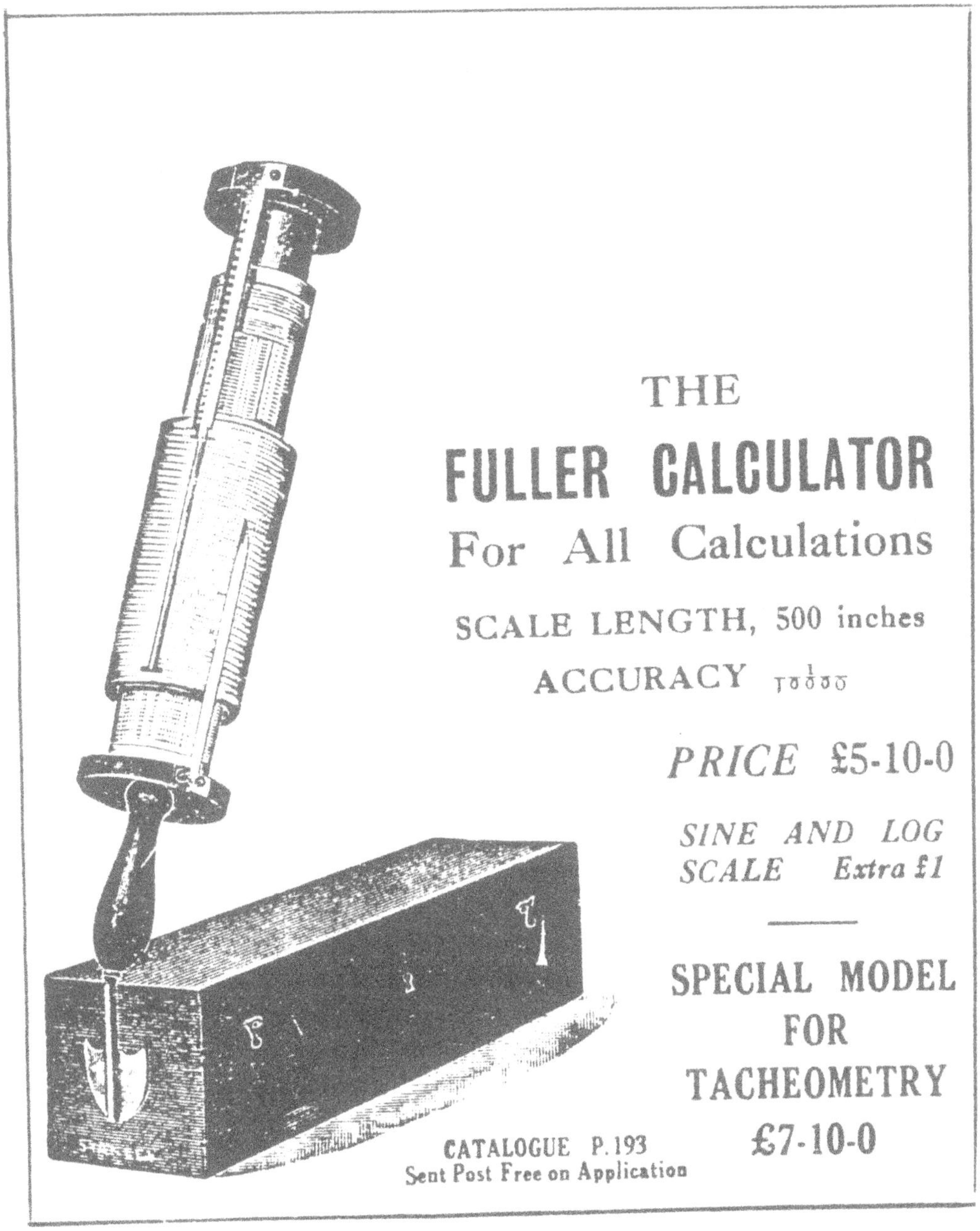

THE

FULLER CALCULATOR

For All Calculations

SCALE LENGTH, 500 inches

ACCURACY $\frac{1}{10000}$

PRICE £5-10-0

SINE AND LOG SCALE *Extra £1*

SPECIAL MODEL FOR TACHEOMETRY £7-10-0

CATALOGUE P.193
Sent Post Free on Application

Figure 22 : Fuller's Calculator

The Fuller Calculating Rule (see **Figure 22**) was approximately 12 inches long and 3 inches in diameter, made of wood, more recently with Perspex slide and indices. The rolling pin shaped rule was supplied in a wooden box, and with its equivalent 500 inch (nearly 42 feet) scale length was capable of four and five figure results. At approx £5/10s/0d. in 1938, it was very expensive and it is doubtful whether many private individuals had the means to afford one. An example seen in Australia was used by the New South Wales Department of Lands until the mid-1960s.

The RHS Calculating Rule.

This is similar to the Otis-King, with a sliding cursor on a tube. Invented by Professor R.H. Smith in 1899, it was made by J.H, Steward of London and was available at the same time as the O-K and Fuller. Effective scale length was 50 inches on a device 9½ inches long and 3/4 inches diameter. An earlier version is shorter and thinner with only a 40 inch scale length. It is not known why it was not at least as popular as the Otis-King.

Grid-iron Slide Rules.

The Victorian genius for producing incredible devices is nowhere better illustrated than in the other form of long scale rule, the grid-iron rule. There are a number of variants. All break the total scale into a number of shorter lengths, the grid-iron examples setting out the shorter scales into some form of grid. This form of rule was first proposed by Dr. J.D. Everett in 1866, and his ideas were taken up by Hannyngton with his Extended Slide Rule. Cherry in 1880 was able to show that the scales did not need repeating, provided that two fixed index points were marked in addition to the natural indices of the scale. The ideas were developed to provide extended scale length by breaking the scale down into sections, and then mounting these sections so as to allow calculations to be performed conveniently, or at least semi-conveniently!

In 1903 Anderson produced a multi-scale slide rule in rectilinear form, perhaps one of the most convenient forms. In the United Kingdom, Unique offered a multi-scale slide rule in two sizes - the Five-Ten with 10 inch scale accuracy on a 5 inch rule, and the Ten-Twenty with 20 inch scale accuracy on a 10 inch rule - but these were not particularly convenient to use. They are not considered further in this section as they are not true grid-iron rules within the definition of scales in some form of grid.

Cherry's Calculator.

This is one of the earliest forms of grid-iron rule. The format is a flat sheet of card or celluloid, ruled with the lines of logarithmic scales, and a transparent overlay sheet with an identical set of scales to those on the lower sheet. Multiplication of one number by another is performed by placing an index on the overlay sheet against a factor on the bottom sheet. The position of whichever index appears under the transparent sheet is noted, and this point is then moved to the other factor, whereupon the result is found under whichever index lies on the bottom sheet.

A very simple "Do-it-Yourself" variant of the Cherry calculator can be made by ruling a number of logarithmic scales and photocopying onto overhead projector film to provide the overlay. Alternatively, a photocopier can also be used to enlarge an illustration of such a calculator and then to copy onto OHP film for the overlay.

Proell's Pocket Calculator.

This is a variant of Cherry's calculator, the overlay in this case being an inverted copy of the base sheet. Usage is exactly the same as for Cherry's calculator.

Hannyngton's Extended Slide rule.

Major-General Hannyngton produced a slide rule where the scales were broken into ten sections which were screwed, as with a wash board, onto a flat base. A cursor, which was a miniature wash board in its own right, could be moved around within the area of the main scales to perform the calculations required. Three sizes of Hannyngton slide rule were produced, 10 foot, 20 foot and 40 foot equivalent scale length, the largest of which is 32 inches by 8¼ inches and has 11 bars. The others are physically smaller with fewer bars.

Thacher's Calculating Instrument.

This consisted of a cylinder 4 inches in diameter and 18 inches long, which could be both rotated and moved longitudinally within an open framework consisting of twenty triangular bars. These bars were connected at their ends to rings, which could be rotated in standards fixed to a baseboard. The scale on the cylinder consisted of 40 sectional lengths, with the right-hand half of each section repeated on the left-hand half of the next section. This resulted in there being two complete scales following each other round the cylinder. One edge of the triangular bars contained scales which exactly corresponded to one of the scales on the cylinder; another edge of the triangular bars contained a scale of square roots.

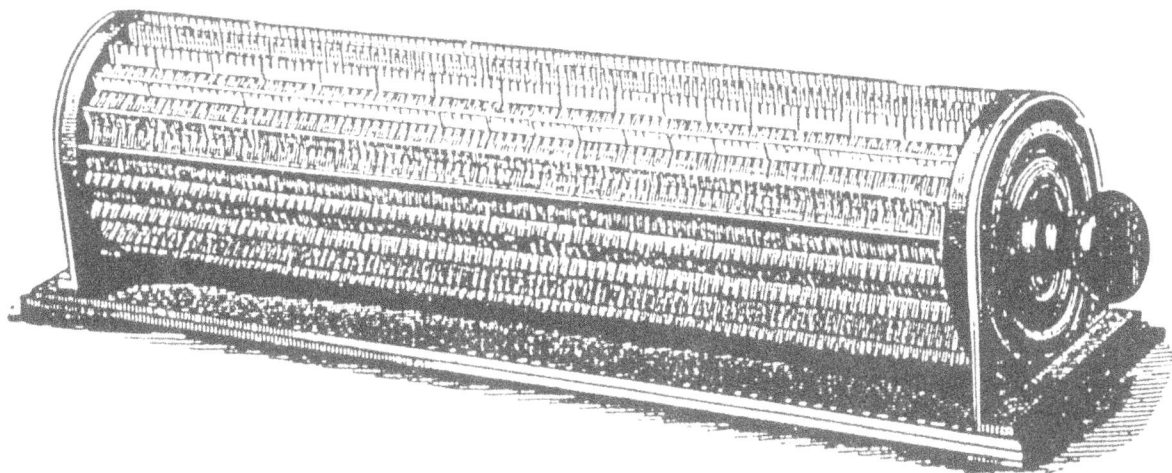

Figure 23 : Thacher's Calculator

By rotating the cylinder and moving it longitudinally, any graduation on the cylinder could be brought into agreement with any other graduation on the triangular bars. The whole could be rotated in the standards so as to bring any section of scale into view. Results were read as with any other linear rule, and for convenience and accuracy a magnifier was provided on a moveable bar. This rule (see **Figure 23**) had an effective scale length of 360 inches (30 feet) and was capable of four or five figure accuracy.

The Nestler Cylindrical Calculator.

This was similar to the Thacher calculator, but with the implementation reversed. The cylinder rotated within fixed supports, while the slide was a cage of bars able to rotate and move longitudinally relative to the cylinder. The scales were as in the Thacher instrument, i.e. repeated. Clips supplied to attach to the slide bars were useful for in-between values and for locating constants. Even more massive than the Thacher instrument, this device came in three sizes, the largest having a 21 inch long cylinder some 6 inches in diameter, with an equivalent scale length of 492 inches (41 feet).

The LOGA Rechenwalze.

Very similar to the Nestler cylinder (which it probably preceded), it was available in 7 different scale lengths including 1.2, 2.4, 7.5, 10, 15 and 24 metre lengths. These delightful Swiss tubular calculators were made by the firm Daemen-Schmid, which later became Loga and made standard slide rules as well (see **Figure 25**).

The Cooper Slide Rule.

This was a more sophisticated version of the Cherry calculator, supplied by W.F. Stanley of London. The logarithmic scale, which was 100 inches long, was arranged as twenty parallel lines engraved on a baseboard of white celluloid about 8 inches by 7 inches. On this baseboard was a frame capable of vertical movement. Within this frame was an inner frame capable of horizontal movement, which carried a transparent celluloid sheet marked with four index lines - one near each corner. Any of these index lines could be brought to bear on any of the graduations on the baseboard by moving the two frames. A moveable index, to record intermediate values, was supplied in the form of a large-headed steel pin clipped to the inclined surface of a metal block with a rubber covering on its base. The rubber covering ensured that the index would not move once placed on the celluloid cover.

This rule (see **Figure 24**) was particularly effective for percentage calculations as well as all the normal uses of a rule, and with an additional scale of logarithms it could also be used for all powers, roots and so on.

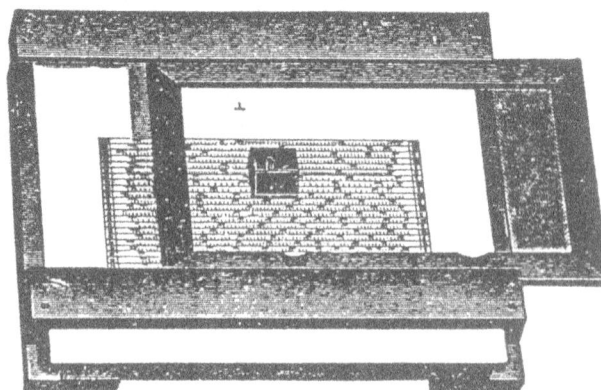

Figure 24 : Cooper's Calculator

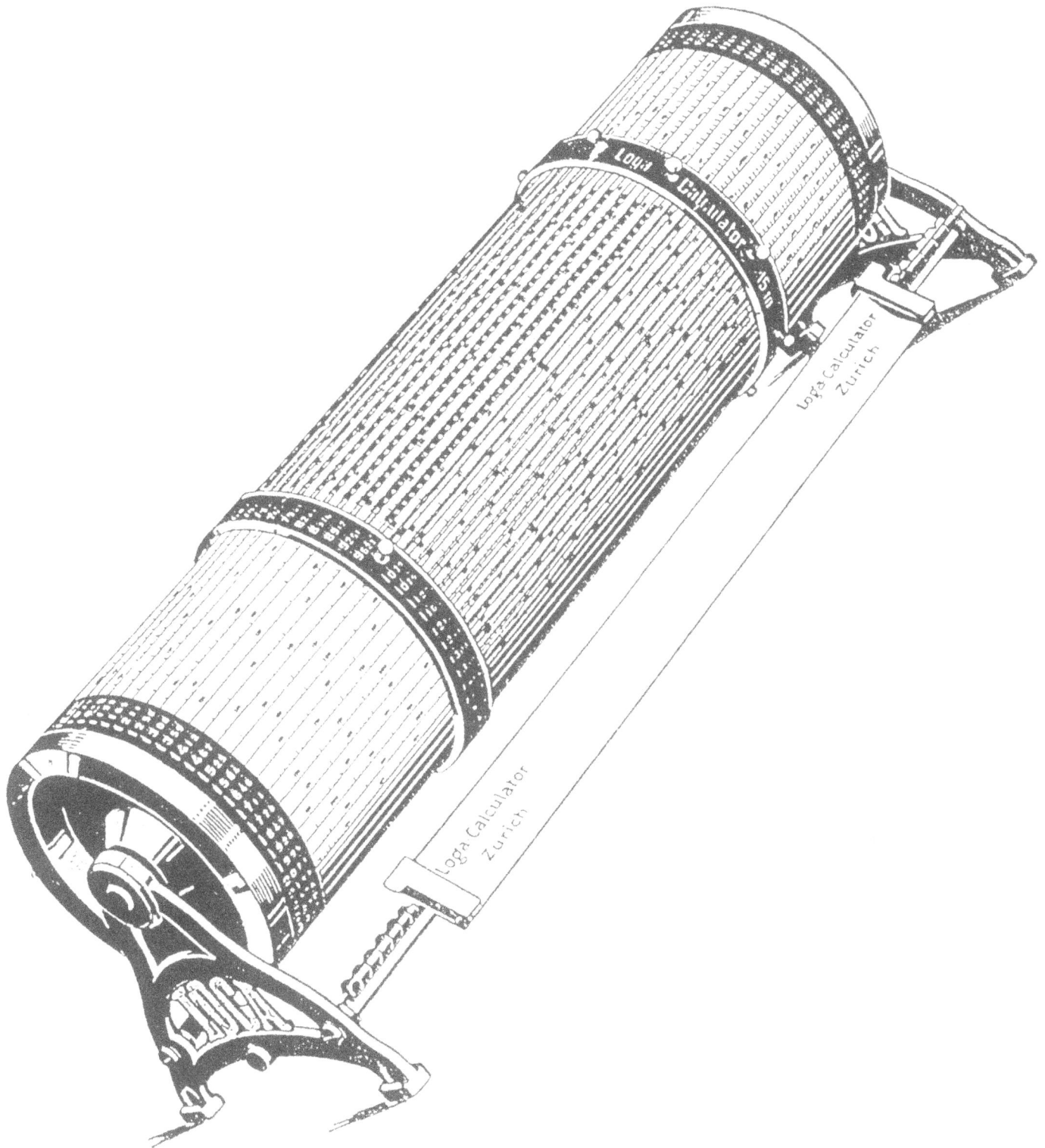

Figure 25 : The Loga Calculator

Specific Function Slide Rules.

The enormous number of different specific areas of technology requiring their own particular variations of mathematical expression and calculation resulted in a multitude of specific or special function slide rules. They easily provide one of the largest areas for continued research, and a rich area for the collector. Specific function slide rules were produced in all the standard formats and in all the usual materials: circular and linear, card, plastic, and sophisticated wood/celluloid. They cover almost every possible area of interest.

Amongst the simplest, there were simple conversion calculators, often made out of card and given out as promotional items by major companies. The many forms of decimal convertor rules produced when the United Kingdom converted to decimal currency in 1972, are an example. The same type is still being produced today to simplify the conversion of all the other items that are slowly being decimalized in the United Kingdom. One of the most sophisticated conversion calculators (also probably the most ghoulish and unusual) is the Radiac slide rule, in rectilinear and circular forms, which give figures for human survival rates in a nuclear blast area.

In between are numerous different rules which cover the gamut of applications. These rules are discussed below in no particular order, but have been collected into similar areas of specialisation.

Engine Power calculators.

Many forms exist. A typical design was the one designed by C.N. Pickworth, which had two slides in edge contact within a broad stock, and which calculated the numerous functions of size, horse power, efficiency etc. for steam, gas and oil engines.

Gunnery/Artillery slide rules.

It is not surprising that this branch of science features its own sub-section of slide rules, the earliest "modern" slide rule being the invention of Mannheim. Of particular note due to some special function are:

The Davis-Stokes Field Gunnery Slide Rule.

This device featured 19 different slides which could be fitted into a stock whose two sides were separated by a flexible celluloid strip. The various slides enabled numerous calculations for performance of both "encounter" and "entrenched" field gunnery. The one version seen in the literature was specifically for the 18 pounder quick-firing field gun. Whether there were versions for other weapons is not known, but it is highly probable. This rule, supplied by the John Davis company, was also equipped with a 1 in 20,000 scale to assist in map reading, and was obviously a complete tool for the artillery officer using it.

The Macleod Field Artillery Slide Rule.

Supplied by J.H. Steward, this was an example of a more general purpose artillery slide rule which was geared to the solution of the various triangular problems which feature in this science.

118

A.G. Thornton and *Blundell-Harling* produced general purpose artillery slide rules. The Blundell versions had two cursors; why is not known. Specialist examples for specific artillery pieces were also produced by Elliott Bros, Nicholson and other manufacturers.

Piecework calculators.

Numerous versions of these devices were produced, and indeed many normal slide rules were supplied with scales for performing piecework calculations - a considerable boon to the wages clerk who had to calculate the wages for a factory workforce on a weekly basis.

Two devices of particular note (as the largest slide rules on record), were supplied by John Davis and Co. of Derby, and in all probability are no longer in existence. They are:

The Smith-Davis Piecework Balance Calculator.

This had two scales 11 foot long marked from 1 penny to £20, so arranged that they could be used for money and time calculation. The scales were placed on the rims of two similar wheels so that the divided edges came together. These wheels were mounted on axles, pressed together under spring pressure, and were able to move as one. To enable the scales to be suitably arranged, a treadle gear freed the wheels whereupon they could be moved individually to set the amount of the balance to the weekly wage. The wheels were then locked together by releasing the treadle, and the two could be turned together, enabling individual wages to be read off directly.

The Smith-Davis Premium Calculator.

A smaller version with scales a mere 4 ft 6 in long.

Electrical slide rules.

Perhaps the most common form of specialist slide rule, these were available in numerous forms. Many standard slide rules were supplied with some electrical conversion capability, even if it was only the conversion of electrical power from kilowatts to horsepower by means of additional markings on the cursor. John Davis, Nestler, Unique, Faber, Faber-Castell, Keuffel & Esser, as well as Thornton, all had versions of electrical rules, the major differences being the placement of the scales and the method of reading them. In a number of cases these scales were in the groove or well of a solid frame slide rule, and so required some form of extension to the slide to enable them to be read.

Wireless slide Rules.

These were effectively a sub-set of the electrical slide rule. The John Davis company supplied the Davis-Martin Wireless slide rule, which was equipped with special scales of inductance and capacitance to enable specific calculation of wavelength.

Navigation slide rules.

This is another function covered by diverse special rules, many being multi-function to enable plotting to be carried out as well. They would make a fascinating research and/or collecting specialisation, and as many of the rules were produced in extremely large numbers for the armed forces, there is no shortage of types or numbers available.

Unique Navigational Slide Rule.

This is the simplest level of navigational slide rule and was introduced in the early days of World War II. It was in the normal Unique format of wood with paper scales covered in a see-through material. It carried twin sine and tangent scales marked in degrees and minutes, one on the stock and one on the slide in each case, in addition to the standard "A", "B", "C" and "D" scales.

Various manufacturers manufactured splendid wood/celluloid rules in an 8 inch format for navigation. These were available in a variety of languages as well as formats.

The Dalton Navigation Computer.

Known to generations of pilots in the world's air forces as well as to private pilots, this was patented by Philip Dalton in 1937. It was available from a number of suppliers in several formats. The following are typical:

- *Mark IVA.* Basically a two-sided hand-held device. One side was a very basic two disc circular slide rule with no cursor, incorporating a number of scales specifically to do with aircraft navigation. The other side had a simple rotatable celluloid screen which was used as a plotting table. Under this could be moved a linear slide which had scales for wind drift and aircraft speed. This device enabled all the basic navigational functions to be performed conveniently and accurately.

- *Mark IIID.* A much more sophisticated device, in which the linear slide was replaced with an endless belt which could be brought under the plotting window. The circular slide rule, with its special functions, was on a hinged plate covering the plotting disc, and the whole could be strapped to the thigh for ease of access while in flight.

- *Type E6B* is the United States equivalent device available from numerous suppliers.

Commercial Slide Rules.

The definition of a commercial rule is as broad as possible, including any slide rule that allows calculation involving money or costing exercises. Most manufacturers listed at least one such rule in their catalogues. In the main, they had specific scales for monetary calculations (e.g., The Unique Monetary rule, Nestler Commercial etc.); for use in the United Kingdom, of course, these monetary scales were in the pre-decimal £ s d scales. Some types were specifically supplied with scales for the calculation of profit, discounts, or other percentages used in the business community. Some particularly clever rules were as follows:

Farmar's Profit-calculating Rule.

Available from J. Casartelli & Sons of Manchester, as well as Farmar's themselves, this ingenious rule had part of the stock replaced with a ten section set of money scales, 3/4 inch in diameter, which could be rotated and brought into contact with the slide.

Nestler Commercial Rule.

This had scales enabling percentages, interest and discounts in 1/16% increments to be worked out, the "A", "B", and "D" scales being normal. The reverse of the slide had reciprocals for the monetary scales and scales which related to metric equivalents of money.

Faber-Castell Business Rule 1/22.

This had some fairly unusual scales, including non-decimal monetary scales in two parts.

Unique.

Unique produced two rules which can be classified as commercial:
- Unique Commercial rule: fitted with "H", "K", "M/N", "R", "C/D", "U", "V" scales.
- Unique Monetary rule: fitted with "M2", "M1/B", "C/D", "M3" scales.

These offered a powerful combination of scales for their respective functions; even so there was always a perception that commercial calculations were not the province of a slide rule.

Load Computers.

The science of loading ships and aircraft was well served by a number of slide rules specifically graduated for the loads that they carried. These rules were graduated in different quantities depending on the applications, an example being a particularly nice pair of specialised plastic circular double-sided slide rules with crescent-shaped cursors presented in a high-quality leather wallet by the American Hydromath Company. Each of these rules pertained to a different type of ship, the *Fort* or the *Liberty*, and enabled the user to calculate the amount of cargo of different weights which could be stowed in the numerous holds, or between the decks.

Alcohol Rules.

Many of the earliest slide rules, particularly Everard's design, were, of course, gauging rules; these have been covered in some detail earlier. Later versions were produced to simplify the tax calculations relating to the strength of alcohol, as well as for price setting.

Farmar's Spirit Rule.

A range of wooden rules. The No.1 size was a 20 inch rule for working out the strength of alcohol, and the price to be charged for diluted spirits.

Other specific function rules.

The range is enormous. Seen in catalogues are rules for:
- Merchants, another name for commercial slide rules.
- Surveyors.
- Chemists (Nestler, Unique and Keuffel & Esser).
- Stadio, for use in calculations for the building trade.
- Construction industry.
- Rate fixers (Faber-Castell).

Also found are very specific application rules such as:
- *Cranes* Sewer Slide Rule.
- *Allan* Friction Head Slide Rule.
- Cattle Rule, for calculating the fat on an animal. It is a 10 inch wooden rule with two slides adjacent to each other.
- The Radiac Slide Rule. It is a plastic 10 inch slide rule, with two well-separated slides in an approximately 4 inch wide stock. It is also available in circular form, with a transposable second disc allowing calculations for land or sea. Both types calculate the length of time a human being can survive in an atomic blast area. There is also a circular Nuclear Bomb Effects Computer which calculates damage radii.
- Photographic exposure calculators. There are many versions in a large selection of different units (Weston, H&D, Scheiner etc.).

Data Slides.

Literally hundreds of special calculators and advertising/promotional products (converters etc.) were produced to cover specific products and special applications. These are not covered in this review of slide rules, but a number are listed in **Chapter 6**. They are still available from a number of firms in Europe and the USA, such as Blundell Harling, IWA, Perrygraph and others.

Table 8 : Slide Rule Designs and Their Makers, 1901 - 1998.

DATE:	SLIDE RULE:	MAKER:
1901	Muller's Hydraulischer Rechenschieber	Austria.
1901	Riebel's Geodetischer Rechenschieber	Austria.
1901	Schweth's Rechenschieber (probably earlier)	
1901	Slide rule for electrical calculations	A.E. Colgate, New York.
1901	Thacher-Schofield engineer's slide rule	Eugene Dietzgen Co, Chicago.
1901	Pierre Weiss, règle à calcul	
1901	Barth's slide rule for lathe settings for maximum output	Bethlehem Steel Co, USA.
	Barth's rule for strength of spur gears	Bethlehem Steel Co, USA.
	Barth's speed slide rule	Bethlehem Steel Co, USA.
	Barth's time slide rule	Bethlehem Steel Co, USA.
1902	Honeysett's hydraulic slide rule	W.F. Stanley, London.
c1902	Hale's slide rule for indicator diagrams	W.F. Stanley, London.
c1902	Froude's displacement rule	W.F. Stanley, London.
1902	Sheppard's cubing and squaring slide rule	W.F. Stanley, London.
1902	Simplex slide rule	John Davis & Son, Derby.
1902	The Smith-Davis premium calculator	John Davis & Son, Derby.
1902	The Smith-Davis piecework balance calculator	John Davis & Son, Derby.
1902	The Davis log-log rule	John Davis & Son, Derby.
1902	The Faber log-log Rule	A.W. Faber, Germany.
1902	Hall's nautical slide rule	J.H. Steward, London.
c1902	Hudson's horse-power scale	W.F. Stanley, London.
c1902	Hudson's shaft, beam and girder scale	W.F. Stanley, London.
c1902	Hudson's pump scale	W.F. Stanley, London.
c1902	Hudson's photo-exposure scale	W.F. Stanley, London.
1902	The Jackson-Davis double slide rule	John Davis & Son, Derby.
1902	Cox's high pressure fluid discharge computer	
1902	Furle's Rechenblatter	Mayer & Muller, Berlin.
1902	Matthe's US Geographical Survey topographic slide rule	Kern & Cie, Switzerland.
1902	The Perry log-log rule	A.G. Thornton, Manchester.
1902	Proell's Rechentafel	R. Proell, Dresden.
1903	Leven, règle pour les reports de bourse	Tavernier-Gravet, Paris.
1903	Rietz's Rechenschieber	A. Nestler, Lahr.
1903	Simplex Taschenrechenschieber	Dennert & Pape, Altona.
1903	Frank's Einscala-Rechenschieber	A. Martz, Stuttgart.

DATE:	SLIDE RULE:	MAKER:
1903	Hazen-Williams' hydraulic slide rule	G.G. Ledder, Boston.
1903	Knowles' calculating scale	
1903	Webb's stadia slide rule (cylindrical)	Keuffel & Esser, New York.
1903	Goodchild mathematical chart	Keuffel & Esser, New York.
1904	Baines' slide rule for solving equations	
1904	Mougnie, règle pour le calcul des conduites d'eau ...	
1904	Goldschmidt's rule with two runners	D. Goldschmidt, New York.
1904	Wurth-Micha, règle pour le calcul des distributions de vapeur	Wurth-Micha, Liege.
1904	M. Schnitzel's Logarithmischer Kubizierungsmass-stab	
1905	Derivry, carte à calcul	
1905	Toulon, règle pour les calculs de terrassements.	
1905	F.J. Vaes, Echelles Binaires	Holland.
1905	Anthony's improved circular hydraulic calculator	W.F. Stanley, London.
1905	Fuller-Bakewell calculating slide scale	W.F. Stanley, London.
1905	Stanley-Boucher calculator	W.F. Stanley, London.
1905	Gravet's tacheometer slide rule	W.F. Stanley, London.
1905	Moehlenbruck's instrument	
1905	Union slide rule (Mannheim)	Eugene Dietzgen, Chicago.
1905	Mack improved slide rule (Mannheim)	Eugene Dietzgen, Chicago.
1905	Masera's Rechenschieber	Zurich.
1905	Bell's stadia slide rule	Arthur L. Bell
1905	Malassis, cercles logarithmiques	
1905	Viaris, ruban calculateur	
c1905	Chadwick's improved slide rule	J. Chadwick, London.
1906	Sperry's pocket calculator (watch type)	Keuffel & Esser, New York.
c1906	Burnham's circular slide rule	Warner & Trasey, USA.
1906	The "R.H.S." calcultor (tubular)	J.H. Steward, London.
1906	Cox's K&E stadia slide rules	Keuffel & Esser, New York.
1906	Halden's Calculex (watch-type circular)	J. Halden & Co, Manchester.
1906	W.H. Harling's calculating rule	W.H. Harling, London.
1906	W.H. Harling's calculating rule. Rietz pattern.	W.H. Harling, London.
1906	Mechanical engineer pocket calculator (circular)	W.H. Harling, London.
1906	Miggett's cost estimating slide rule	W.L. Miggett, USA.
1906	Miggett's cost estimating slide rule (in conical form)	W.L. Miggett, USA.
1906	Rosenthal's multiplex slide rule	Eugene Dietzgen, Chicago.
1906	Nickel's runnerless slide rule	F.F. Nickel, New Jersey.

DATE:	SLIDE RULE:	MAKER:
1906	Robert's slide rule for wiring calculations	E.P. Roberts, Pittsburg.
1906	Niehan's Metalltachymeterschieber	
1906	Fearnley's universal calculator	
1907	Règle des Ecoles (règle Beghin)	Tavernier-Gravet, Paris.
1907	Slide Rule for calculating sag in wires	R. Wood, New York.
1907	Tilsley's slide rule for strength of gear teeth	H. Tilsley.
1907	Younger's logarithmic chart for computing cost (replacement for Miggett)	
1907	"Midget" slide rule	Kolesch & Co, New York.
1907	"Vest Pocket" slide rule	Kolesch & Co, New York.
1907	"Triplex" slide rule	Kolesch & Co, New York.
1907	The L.E. Knott slide rule	L.E. Knott Apparatus Co, Boston.
1908	Peter's Universal Rechenschieber	A. Nestler, Lahr.
1908	"Peter & Perry" rule	A. Nestler, Lahr.
1908	Precision slide rule	A. Nestler, Lahr.
1908	Pickworth's power computer	C.N. Pickworth, Manchester.
1908	Niethammer's Prazisions-Schulstab	Germany.
1908	Neuer Prazisions-Rechenstab	Dennert & Pape, Altona.
1908	Griffin and Son's card slide rule	J.J. Griffin, London.
1908	Daemen-Schmid's Rechenwalze	Orlikon
1908	Koppe's Barometerhohenrechenschieber	
1908	Vogler's Barometerhohenrechenschieber	
1908	Hammer's Barometerhohenrechenschieber	
1908	J.L. Hall's structural slide rule, (column slide rule, beam slide rule and Mannheim slide rule in one)	Slide Rule Club, New York.
1908	Edge's weight computer for structural shapes	Edge Computer Sales, New York.
1908	Perry's new slide rule	A.G. Thornton, Manchester.
1908	Wynne-Roberts' pocket hydraulic calculator	R.O. Wynne-Roberts.
1908	Eichhorn's trigonometric slide rule	M.J. Eichhorn, Chicago.
1908	Celluloid slide rules, with spare logo logarithmic slide as arranged by Dunlop & Jackson	John Davis & Son, Derby.
1909	The K. & E. Polyphase slide rule	Keuffel & Esser, New York.
1909	The K. & E. adjustable duplex slide rule	Keuffel & Esser, New York.
1909	The K. & E. adjustable log-log duplex slide rule	Keuffel & Esser, New York.
1909	The K. & E. circular calculator (watch type)	Keuffel & Esser, New York.
1909	Nestler's slide rule for wiring calculations	A. Nestler, Lahr.
1909	Woodworth's slide rule for electrical wireman	Lewis Institute, Chicago.

DATE:	SLIDE RULE:	MAKER:
1909	Woodworth's slide rule for calculations with volts, amperes, ohms and watts	Lewis Institute, Chicago.
1909	Nordell's slide rule for computing run-off to sewers	Keuffel & Esser, New York.
1909	Schumacher's Rechenschieber mit Teilung in gleiche Intervalle.	Munich.
1933	Differential Trig/log-log slide rule by Boardman	A.G. Thornton, Manchester.
1935	Darmstadt arrangement of scales	Faber-Castell, Germany.

Chapter 6.

Slide Rule Manufacturers 1850 - 1998

The multitude of instrument makers plying their trade from the 17th to the first half of the 19th century shook down into several companies that manufactured a wide range of instruments including slide rules, as well as several companies whose sole business was the manufacture and supply of slide rules. As time passed, a number of these were taken over by, or joined forces with, other companies in the same line in a process of rationalisation which resulted in fewer but larger companies. A number of manufacturers also started in business during this period. Most of these manufacturers lasted only as long as the slide rule, though one or two soldiered on for a number of years, making other instruments after they stopped slide rule production before they too went out of business.

Those companies devoted exclusively to slide rules were the first to go out of business when electronic calculators started to eat into their business in the 1970's. There were some brave attempts to keep the slide rule alive. The most ironical was that of Faber-Castell, which produced a range of slide rules with electronic calculators on the reverse, to replace the Addiator which featured on earlier models. Faber-Castell weathered the demise of the slide rule, as they were also major manufacturers of drawing and other mathematical instruments.

In their catalogues a number of well known wholesale and retail companies featured slide rules whose manufacturer was not always obvious. Blundell Harling and Unique in the UK, and several Japanese manufacturers, supplied numbers of slide rules to these companies. Looking at the retailers' catalogues, as well as at the slide rules themselves, it is clear that different suppliers were used, probably to save money. **Table 9** (p.128) names the manufacturer wherever possible, but complete accuracy is difficult.

The last years of slide rule production also featured the greatest increase in sales of standard slide rule types, the most common being the Mannheim and Rietz, and later, Darmstadt and various log-log designs. The Rietz design was probably the most common slide rule produced, the great majority of manufacturers having at least one Rietz model in their range. The notable exception was Keuffel & Esser in the USA, which never had a Rietz design designated as such in their catalogues, but used different names for slide rules with Rietz scale layouts.

There were a number of "specials" which were manufactured and sold by only one company. These were somewhat idiosyncratic, and examples include the Dargue Bilateral and a number of military or artillery designs which were unique in having two cursors and a number of specific gauge points. Even these can be classified into a number of types, which are covered in detail in **Chapter 5** and are included in this chapter.

Table 9: Descriptions of manufacturers and retailers.

The manufacturers and retailers listed in **Table 9** are those that were active during the 20th century until the end of large scale manufacturing. At the end of the alphabetical listing of companies are "Miscellaneous" and "Other Manufacturers" about which very little is known. One or two companies are still making small quantities of specialist rules and some are still making slide charts for advertising purposes. If a company was formed from a union of firms, the names and updated addresses are shown where appropriate. Companies are listed alphabetically. The *Date* when the firm was established is given if available. The *Address* shows where the company operated and, if possible, when.

Cross-referencing between manufacturers is shown in **BOLD,** and not all slide rules have been repeated within both or all references to a particular manufacturer. There is some repetition, and this often illustrates how different sellers gave different names or numbers to the same item. Different *Factories* used by the companies are listed if known. The short *History* gives an idea of how the company may have changed during its life. The lack of detail in this area is indicative of just how little is known of some of the companies!

Table 9: Descriptions of slide rules.

Slide rules are listed under the rule's manufacturer or retailer. Thus, a Boots slide rule will be listed under Boots even though it may have been made by Faber-Castell or Blundell. Where possible a cross-reference to the specific maker of a particular slide rule is provided under the *Comment* heading, or there may be a general statement within the *History* portion, e.g., Wolters-Noordhoff slide rules were generally supplied by Fuji.

Each slide rule is covered by a one or two line description. The one line description covers simplex slide rules where there are scales either on one side only, or where the slide may also have scales on the reverse. Duplex slide rules have a two line description, the reverse scales being on the second line.

A key describing the contents of each field used in **Table 9** is outlined below. Slide rule scales and gauge points are described in **Appendix 5.**

Table 9: Key to Field Names and Their Contents

Field:	Contents:		
1	Type Name/Type Number		
2	Length or Diameter in cms of scales; " gives dimension in inches, ' gives dimension in feet.		
	Class of rule:	B	= Beaker
		C	= Circular
		G	= Grid-iron
		L	= Linear
		P	= Pencil
		S	= Special function
		T	= Tubular
		U	= Others (specify)
		W	= Watch type calculator
3	Construction:	CF	= Closed frame
		CFD	= Closed frame duplex
		Dup	= Duplex
		FM	= Folded metal
		FP	= Folded plastic

	MD	= Multi-disc circular
	MC	= Multi cursor circular
	OF	= Open frame
	Oth	= Other (specify)

4   Material:

A	= Aluminium
Ba	= Bakelite
B	= Bamboo
C	= Celluloid
M	= Metal
P	= Paper/cardboard
Pl	= Plastic
W	= Wood
W/C	= Wood/celluloid
W/p	= Wood/painted

5   Cursor:

m	= Metal knife edge
m/c	= Metal/celluloid
m/g	= Metal/glass
mgm	= Metal/glass magnifying
m/p	= Metal/plastic
pl	= Plastic
plf	= Plastic folded
plm	= Plastic magnifying

6   Scales: See Table 4 for listing

/	= stock/slide interface
//	= reverse of slide
/ \	= edge of slide rule
\\	= in well of rule i.e. under the slide

Also used for description or other details.

In some examples, a price in "money of the day" is given.

7   Comment:

1.	Numeric covers variants.
a.	Alphabetical is for another comment, or, gives year(s) of model availability.

An example from the Aristo range is as follows:

1	**2**	**3**	**4**	**5**	**6**	**7**
Duplex Mod 968 Studio	25L	Dup	Pl	pl	LL01,LL02,LL03,A/B,L,C/LL3,LL2,LL1 A/B,C/D	

This shows that the Aristo Duplex Model 968 Duplex slide rule is a 25cm linear slide rule in plastic with a plastic cursor, with LL01,LL02,LL03,A/B,L,C/LL3,LL2,LL1 scales on the front and A/B,C/D scales on the reverse.

Table 10 : Slide Rule Manufacturers and Retailers, 1901 - 1998.

ABAC (GB)
Address: Association of British Aero Clubs, England
History: Responsible for the specification of several Flight Computers/Calculators e.g. Navigation Computer Mk II, which would then have been made by a number of firms.

ABBOT BROWN (GB)
Address: Abbot Brown Ltd., Beaminster, Dorset, England.
History: Not known, but made a series of specialist slide rules including the Bender Optical Rule (used for the calculation of spectacle lenses and frames).

Slide Rules:

1	2	3	4	5	6	7
Bender Optical Slide Rule	6"L					

ABCO (USA)
Address: Not known.
Factories: Not known.
History: 1957, Sold a version or versions of the **OTIS-KING** calculator under the name ABCO Cylindrical calculator.

Slide Rules: Other than **OTIS-KING**, none know

ACU / ACU-MATH / ACU-RULE (USA)
Address: Acu-Rule Manufacturing Co, St Louis Mo. Mt. Olive, Illinois, USA
Factories: Not known.
History: c 1948, sold the **ACU-MATH** slide rule range, in existence to at least 1965. The Acu-math Duplex slide rules may have been made by **STERLING PLASTICS**.

Slide Rules:

1	2	3	4	5	6	7
No 12	5"L	CF	Pl	pl	K,A/B,CI,C/D,L	
No 30 Log-log decimal trig	10"L	Dup	Pl	pl	L,LL1,DF/CF,CIF,CI,C/D,LL3.LL2 LL0,LL00,K,A/B,C/D,S,ST,T	
No 150 Log-log						
Trig and Log-Log						1965
No 400	10"L	OF	Pl	pl	K,L,A/B,CI,C/D,S,T	
No 400 Student Mannheim	10"L	OF	Pl	pl	as 400, different name.	
No 511 Mannheim Professional	10"L	?	?	?	K,L,A/B,CI,C/D,S,T	
No 600	?L	?	?	?		
No 900 Professional Mannheim Trig	10"L	Dup	Pl	pl	K,A/B,T,ST,S/D,DI,L equivs,DF/CF,CIF,CI,C/D,equivs	
No 1240	5"L	CF?	Pl	pl	K,A/B,CI,C,D,L	

ACUDIAL (Japan)
Address: See **FULLERTON** and **DARGUE**.
History: A range of circular slide rules made by **FULLERTON** and sold by **DARGUE**.

AERO (USA)
Address: Aeroproducts Research Inc, Los Angeles, California, USA.
History: May have been part of **AGA** later, in which case may have made and sold **RELIABILITY & CONFIDENCE** slide rules c1960. Also sold Flight Computers c 1969 to 1976.

AGA (USA)
Address: Aero Geo Astro Corporation, USA
History: c1960, made and sold **RELIABILITY & CONFIDENCE** slide rules. See also **AERO** and **ASTRO**.

AGNEW (GB?)

Address: R.F. Agnew, address unknown.
History: known from a Weight Calculator.

AHREND (Holland)

Address: (Early) Wed. J. Ahrend en Zoon (the widow J. Ahrend and Son); (Later and to date) J. Ahrend & Son, Holland.
History: Well known Dutch office supplies retailer who are still in business to date. They may also have made some of their own slide rules (Holland Slide Rule Co?) in the early period. Known to sell **ARISTO** with the Ahrend label, also **HEMMI** similarly labelled.

Slide Rules:

1	2	3	4	5	6	7
Hollandia Rekenstok	L					early.
Ahrend Rekenstok	L				may have been the same as above	
Elektro 15	L					

AIRTOUR (GB)

Address: Airtour Flight Equipment Ltd, Elstree, Herts, England.
History: Sold Flight Computers, made by others.

Slide Rules:

1	2	3	4	5	6	7
Pooley's CRP-1 Flight Computer	U	n/a	Pl	n/a	Plastic Daltons type Flight Computer	
Rule LPC-1	3"C	MD	Pl	pl	For calculating direction of approach	

ALCO (Japan)

Address: Japan
History: Sold a 5" slide rule with an Addiator on the reverse. Note this Addiator is identical to a **KINGSON** Addiator with slide rule (Hong-Kong).

ALCOTT (USA)

Address: Alcott Calculator Company, P.O. Box 547396, Orlando, Florida, USA.
Phone: +1 407 295 5457; Fax: +1 407 295 4464, E-mail: calcott@ix.netcom.com
History: Formed 1952, and still in business today producing slide charts and wheel charts etc., with appropriate advertising logos and so on. A recent list from the World-Wide Web is shown below:

Slide Rules:

1	2	3	4	5	6	7
The JOBBER "4"	n/a				Handheld dimensional electronic calculator for building sizing.	
The Steel Weight Finder					Steel weight slide chart.	
Structural Steel Detail Guide					Details of 240 steel sections.	
Piping Selector 150/300					Wheel chart with various fittings dimensions.	
Piping Selector "SB"					Slide chart with make up dimensions.	
Aluminium Weight Finder					Slide chart with weights of various structural shapes in 10 alloys.	
Steel and Aluminium Coil Length & Weight Calculator					As described in its name.	
Mortgage Payment/Monthly Payment Calculator					As described in its name.	
Profit Calculator					As described.	
Selling Price Calculator					As described.	
The Date Forecaster					As described.	
The US to Metric and the Metric to English Conversion Calculators					As described.	
International Travel Converter					As described.	
Blackjack Calculator					As described.	
Mileage Calculator					As described.	
Fat Calorie/Food Label Guide					As described.	

ALDERMAN
Address: Not known.
History: Not known.

Slide Rules:

1	2	3	4	5	6	7
-	6"L	CF?	I/M	n/a	A/B,C/D; ivory/metal	

ALOE (USA)
Address: A.S. Aloe Co. USA. - or, A.L. Aloe Company, 414-416 North Broadway Street, City?, USA
History: c1894, A.L. Aloe sold a Colby's Stadia Rule; c1927, A.S.Aloe sold the **OTIS-KING** Calculator in the USA (Models K & L). Are they the same Aloe Company?

Slide Rules:

1	2	3	4	5	6	7
Colby Stadia rule	50"L	T	W	n/a	Log scale/Arc scale; One slides in a slot in the base, a 'T'	c1894
Otis King, Models K & L	T				see **CARBIC.**	c1927

ALLEGHENY (USA)
Address: Allegheny Plastics Inc, Coraopolis, PA. 15108, USA.
History: Made a range of Air Navigation Computers and Gunnery calculators for the US Armed Forces, also specialist calculators.

Slide Rules:

1	2	3	4	5	6	7
Computer Attack, type CPU-73A/P	9"C	MD	Pl	pl	Gunnery scales	c1980
Duer Spring Calculator						c1947

ALRO (Holland)
Address: ALRO, Development Company of Patents Ltd. The Hague, Holland.
Factories: The design and piece part manufacture were carried out by various Dutch firms, the parts were then assembled in ALROs small factory in what used to be a stables.
History: Horses-head logo. Established 1935, manufactured various clever circular slide rules in a self contained box which also doubles as a desk stand. They still continue in business (but not making slide rules) as part of US parent DataCard following a number of takeovers. They made some 175,000 slide rules during the period of slide rule production.

Slide Rules:

1	2	3	4	5	6	7
Type 10.R	6C				General technical scales	
Type 55.Nmmi	6C				Multiplication & division.	
Type 50.Ng	6C				Graphic artists, with conversion for print information.	
Type 50.Nvm	6C				Multiplication & Division, used in aviation.	
Type 101. Commercial	6C				Multiplication & division with a spiral scale.	
Type 100.R	13C				Constructional engineers.	
Type 200.R	13C				Mechanical engineers.	
Type 300.D	13C				Darmstadt.	
Type 400.D	13C				Similar to Darmstadt, plus constants.	
Type 500.N	13C				Basic scales.	
Type 600.E	13C				Electrotechnology scales.	
Type Chemistry scales	13C					
Type Beton	13C				Reinforced concrete.	
Type Piret	13C				Reduction calculations for wire-drawing.	
Type Brouhon	13C				Timber calculations.	
Type Ha-Tex	13C				Textile calculations, with UK currency and yards.	
Type GoA	13C				System Van Roon, sea and air navigation.	
Type 1010. Commercial	13C				Multiplication & division with 2.5m spiral scale.	
Type small KLM disc	13C				No metal box, conversion tables for flying.	
Type KLM disc	18C				scales for course and wind speed.	
Type Bio-Rhythm disc					scales according to Dr. Fliess.	
Type Clinical	18C				Scales according to Dr. A. Lips.	
Type Military					Circular slide rule for field artillery.	

Type 746, Radio calculation disc,					manufactured by "De Muiderkring" publishing.	
Type Chemical Disc					manufactured by "NV Chemie Producten"	
Audio Slide Rule	L	CF	P	n/a	scales for audio calculations made by ALRO Print for Philips.	1987?

Also a range of specially marked discs with Company Logos or special order scales, Phillips and the Dutch Post Office are examples.

Quantities made:	6 cm dia	25,000 standard, plus 25,000 special
	13 cm dia	100,000 standard, plus 20,000 special
	18 cm dia	5,000

ALTENDER (USA)
Address: Theo. Altender & Co. Philadelphia, USA.
History: c1897 made the Omnimeter, see also **DIETZGEN**.

ALVIN (Japan?)
Address: Alvin, Japan.
History: Known from a #1151 Elite Multilog, double log duplex decitrig with electrical scales, otherwise not known.

AMERICAN BLUEPRINT (USA)
Address: American Blueprint Co Inc, New York, USA.
History: See **US BLUEPRINT**.

Slide Rules:

1	2	3	4	5	6	7
Mannheim polyphase 2025	25L	?	W/C	?	A/B,CI,C/D,K//S,L,T	1948
Military slide rule	25L	?	W/C	?	Special scales.	

AMERICAN SLIDE CHART CORP (USA)
Address: American Slide Chart Corp, 25W550 Geneva Road, P.O. Box 111, Wheaton. Illinois 60187, USA.
Factories: Not known
History: Associated with **SLIDE CHARTS (BRITAIN) LIMITED**. Known examples are made from cardboard and plastic, and have been commissioned by a companies for advertising or semi-technical purposes.

Slide Rules:

1	2	3	4	5	6	7
Computar CCTV Range calculator	10.5C	MD	Pl	pl	2D,1C; specialist CCTV optical scales	1993
Modem Selection Guide	U	CF	C	n/a	made for Frederick Electronics; U=9"x4"	1984

ANTICA (Italy)
Address: See Vittorio **MARTINI**
History: The Logo for slide rules made by Vittorio **MARTINI** of Italy.

APPOULLOT (France)
Address: Léon Appoullot, Saint Brieuc, Seine et Oise, France. On slide rule: Saint-Brice-sous-Foret, Seine et Oise, France.
History: Patent granted to Leon Appoullot of Paris France.
Made a spiral scale circular slide rule embodying some clever features, patented in France and Germany in 1921 and 1922 respectively.

Slide Rules:

1	2	3	4	5	6	7
T1,						
T2,	C	MD	M	m	C,L,B,T,CI,	
T3,	C	MD	M	m	C,L,B,T,CI, bigger or smaller than T2?	
T4,						

ARICI (Italy)

Address: Arici, Italy.

History: Made a number of military multi-purpose slide rules of considerable complexity.

Slide Rules:

1	2	3	4	5	6	7
Arici 55	U	MD/L	M		3 circular and one linear rule on one plate. Artillery calculations.	

ARISTO (Germany)

Address: Aristo Werke, Altona, Hamburg. 1960, 10 Juliusstrasse, 2000 Hamburg 50.

History: Established as **DENNERT & PAPE** on 1.7.1862, a name they kept to 1924. From 1925 to 1936 they were **DUPA**, after 1936 they used the name **ARISTO** for their new range of plastic slide rules manufactured from 1936 to 1978.

UK agency held by Technical Sales (**see T.S**) of London to 1948 when Aristo UK was set up.

Aristo (D&P) were the first to patent the lamination of celluloid (Zelhorn) to wood which became the standard for all other manufacturers, providing clear and bright scales which were easier to read and engrave.

Factories: Hamburg; H or 2. Geretsried; G or 3. Wörgl; W or 4. (The first figure or letter in the date code stands for the factory). Also a factory at Bludenz in Austria, dates unknown.

Slide Rules:

1	2	3	4	5	6	7
1961: Aristo Circular Computer						
1962: System Darmstadt Slide Rule						
1964: Aristo Studio slide rule						

From the Lawes Rabjohns Technical Catalogue - 1952

Mod 945	25L	CF	Pl	pl	DF/CF,CI,C/D,K//S,T,L folded at 3.6?
Mod 1045	50L				as 945
Mod 99 Rietz Pattern	25L	CF	Pl	pl	K,A/B,CI,C/D,L//S,T,L
Mod 109 Rietz	50L				as 99
Mod 914 Electrical	25L	CF	Pl	pl	LL,A/B,CI,C/D,LL//S,T,L\\V,D
Mod 967 Darmstadt	25L	CF	Pl	pl	K,A/B,(1/x²),CI,C/D,P,S,T??
Mod 1067 Darmstadt	50L				as 967
Duplex Mod 966 Studio	25L	Dup	Pl	pl	T,ST,DF/CF,CIF,CI,C/D,P,S A/B,C/D??
Duplex Mod 968 Studio	25L	Dup	Pl	pl	LL01,LL02,LL03,A/B,L,C/LL3,LL2,LL1 A/B,C/D??
Duplex Mod 902P Students	25L	Dup	Pl	pl	L,A/B,CI,C/D,S,T

where C,D refer to the Mod 967 superscript x²: K,A/B,$(1/x^2)$,CI,C/D,P,S,T??

Pocket slide rules

Mod 812 Simplex	12L	CF	Pl	pl	A/B,C/D//S,L,T
Mod 845 Simplex	12L	CF	Pl	pl	A/B,C/D//S,L,T
Mod 89 Rietz Pattern	12L	CF	Pl	pl	K,A/B,CI,C/D,L//S,T,L
Mod 814 Electrical	12L	CF	Pl	pl	LL,A/B,CI,C/D,LL//S,T,L\\V,D 5" Plastic
Mod 867 Darmstadt	12L	CF	Pl	pl	K,A/B,(1/x²),CI,C/D,P,S,T??

All can be found with magnifying cursor.

From the Stanley "A" catalogue - 1958.

Simplex 911	25L	CF	Pl	pl	Fundamental & Squares.
Scholar 0903	25L	CF	Pl	pl	Fund, Recip, Squares, Cubes, Mant, Trig.
ScholarLL 0903LL	25L	CF	Pl	pl	Fund, Recip, Squares, Cubes, Mant, Trig, LL.
Commerce 0905	25L	CF	Pl	pl	F, Fold, R, Interest, Conversion marks.
Rietz 89	12L	CF	Pl	pl	F, R, Sq, Cub, Mant, Trig.
Rietz 99	25L	CF	Pl	pl	F, R, Sq, Cub, Mant, Trig.
Rietz 109	50L	CF	Pl	pl	F, R, Sq, Cub, Mant, Trig.
Multi Rietz 829	12L	Dup	Pl	pl	F, Fold, R, Sq, Cub, Mant, Trig.
Multi Rietz 909	25L	Dup	Pl	pl	F, Fold, R, Sq, Cub, Mant, Trig.
Darmstadt 867U	12L	CF	Pl	pl	F, R, Sq, Cub, M, Trig, L-L.
Darmstadt 967U	25L	CF	Pl	pl	F, R, Sq, Cub, M, Trig, L-L.
Darmstadt 1067U	50L	CF	Pl	pl	F, R, Sq, Cub, M, Trig, L-L.
Studio 868	12L	Dup	Pl	pl	F, Fold, R, Sq, Cub, Man, Trig, L-L.
Studio 968	25L	Dup	Pl	pl	F, Fold, R, Sq, Cub, Man, Trig, L-L.
Studio 1068	50L	Dup	Pl	pl	F, Fold, R, Sq, Cub, Man, Trig, L-L.
Multilog 970	25L	Dup	Pl	pl	F, Fold, R, Sq, Cub, Man, Trig, L-L.

Hyperbolog 971	25L	Dup	Pl	pl	F, Fold, R, Sq, Cub, Man, Trig, L-L, plus Hyperbolic.	
Electro 814	12L	CF	Pl	pl	F, R, Sq, Cub, M, Trig, L-L.	
Electro 914	25L	CF	Pl	pl	F, R, Sq, Cub, M, Trig, L-L.	

also:

Junior No0901	25L	CF	Pl	pl		
Scholar No0903	25L	CF	Pl	pl		
Scholar No0903VS	25L	CF	Pl	pl	reversible slide	>1958
Scholar No0903VS2	25L	CF	Pl	pl	reversible slide	>1958
Scholar No0903LL	25L	CF	Pl	pl	reversible slide	
TriLog Nr0908	25L	Dup	Pl	pl		1968?
Simplex No0911	25L	CF	Pl	pl		
Elektro Nr 914	25L	CF	Pl	pl	inc 'K' on edge	>1958
Studio No0968	25L	Dup	Pl	pl/m		1968
0903	25L	CF	Pl	pl		
CommerzIII Nr623	4.75C	MD	Pl	pl		1968
Hyperbolog	25L	CF	Pl	pl		
Multi-Rietz	25L	CF	Pl	pl		
Multi-log	25L	CF	Pl	pl		
Darmstadt	25L	CF	Pl	pl		
40104						

Comment:

1.	Numerous variants with square and rounded Bridges, with and without the ARISTO name on the Bridge; made in Germany and Austria; Studio with metal and plastic edged cursors. Early rules with cardboard cases, later with plastic tubes, grey and red/white. Dating Aristo rules is based on the word ARISTOFLEX with the numbers 0 to 9 for the second and third figures of the date code. The first figure gives the factory - see above.

<u>1955 Aristo catalogue</u> shows the following slide rules, generally as 1960, but has a different Logo - a circle with a segment delineated. Slide rules are listed as follows, details as 1960 listing,

89, Rietz	15L	CF	Pl	pl		c.
99, Rietz	30L	CF	Pl	pl		c.
109, Rietz	55L	CF	Pl	pl		c.
9/150, Rietz Demonstration	180L	CF	Pl	pl		
814, Electro	15L	CF	Pl	pl		
914, Electro	30L	CF	Pl	pl		
867U, Darmstadt	15L	CF	Pl	pl		c.
967U, Darmstadt	30L	CF	Pl	pl		
1067U, Darmstadt	55L	CF	Pl	pl		
67/150, Darmstadt Demonstration	180L	CF	Pl	pl		
829, MultiRietz	15L	Dup	Pl	pl		
929, MultiRietz	30L	Dup	Pl	pl		
868, Studio	15L	Dup	Pl	pl		
968, Studio	30L	Dup	Pl	pl		
68/150, Studio Demonstration	180L	Dup	Pl	pl		
970, MultiLog	30L	Dup	Pl	pl		
971, HyperboLog	30L	Dup	Pl	pl		
0903, Scholar	30L	CF	Pl	pl		
3/150, Scholar Demonstration	180L	CF	Pl	pl		
0903 LL, Scholar LL	30L	CF	Pl	pl		
3LL/150, Scholar LL Demonstration	180L	CF	Pl	pl		
0905, Commerce	30L	CF	Pl	pl		
5/150, Commerce Demonstration	180L	CF	Pl	pl		
845, Merchant	15L	CF	Pl	pl		
945, Merchant	30L	CF	Pl	pl		
1045, Merchant	55L	CF	Pl	pl		
965, Commerce II	30L	CF	Pl	pl		
810, Puck	12L	CF	Pl	pl	A/B,CI,C/D//S,L,T	
811, Puck	12L	CF	Pl	pl	A/B,CI,C/D	
812, Simplex	15L	CF	Pl	pl	A/B,C/D//S,L,T	
89 NZ Rietz	15L	CF	Pl	pl	as Aristo 89, with scales R5, R10, and R20 on back of slide	
807, Flex	15L	CF	Pl	pl	as 867U but thinner and omitting scales on back of slide	
958, Surveyor	30L	Dup	Pl	pl		c.
939, Ferro Concrete	30L	Dup	Pl	pl	Goetsch Pattern	
916, Weight of Materials	30L	CF	Pl	pl	Length/Width,Thickness/Weight; Stockhusen Pattern	

918, Hydraulic	30L	CF	Pl	pl	Vikari Pattern
947, Materials Testing	30L	CF	Pl	pl	
10048, Temperature Radiation	30L	CF	Pl	pl	Czerny Pattern
10052, Damping Slide Rule	15L	CF	Pl	pl	For Telecommunications Technicians
602, Circular Computer	6C	MD	Pl	n/a	C/D, with advertising material
670, Motorists Computer	2"C	MD	Pl	n/a	metric measures
671, Motorists Computer	2"C	MD	Pl	n/a	imperial measure
615, Aviator circular computer	14C	MD	Pl	pl	dual faced
620, Circular Computer	9C	MD	Pl	pl	C/D
621, Circular Computer for Businessmen	9C	MD	Pl	pl	multi-scale

<u>1956 Aristo catalogue</u> shows the following slide rules, generally as 1955 and 1960, but has the circular Logo, and some different slide rules as follows, details as 1960 listing,

615, Aviat					
616, Aviat G					
1068, Studio	50L				
911, Simplex	25L	CF	Pl	pl	A/B,C/D
(811, Puck is not in this catalogue)					

<u>1960 Aristo catalogue:</u>

0911 Simplex	25L	OF	Pl	pl	cm,A/B,C/D,ins	
11/150 Simplex Demonstration	150L	OF	Pl	pl	cm,A/B,C/D,ins	
0901 Junior	25L	OF	Pl	pl	DF/CF,CIF,CI,C/D,A	
1/150 Junior Demonstration	150L	OF	Pl	pl	DF/CF,CIF,CI,C/D,A	
101 Junior Projection	20L	OF	Pl	n/a	DF/CF,CIF,CI,C/D,A	
0903 Scholar	25L	OF	Pl	pl	L,K,A/B,CI,C/D,S,ST,T//cm,ins; Rietz system.	
3/100 Scholar Demonstration	100L	OF	Pl	pl	L,K,A/B,CI,C/D,S,ST,T//cm,ins; Rietz system.	
3/150 Scholar Demonstration	150L	OF	Pl	pl	L,K,A/B,CI,C/D,S,ST,T//cm,ins; Rietz system.	
0903LL Scholar Log-Log (Darmstadt)	25L	OF	Pl	pl	L,K,A/B,CI,C/D,S,ST,T//cm,-/S,LL2,LL3/-ins; Rev slide	
3LL/150 Scholar Log-Log Demo. (Darm)	150L	OF	Pl	pl	L,K,A/B,CI,C/D,S,ST,T//cm,-/S,LL2,LL3/-ins; Rev slide	
103LL Scholar Log-Log Proj. (Darm)	20L	OF	Pl	n/a	L,K,A/B,CI,C/D,S,ST,T//cm,-/S,LL2,LL3/-ins; Rev slide	
0903VS Scholar VS	25L	Dup	Pl	pl	L,K,A/B,CI,C/D,ST,T cm,DF/CF,C/D,ins; Reversible cursor.	
3VS/150 Scholar VS Demonstration	150L	Dup	Pl	pl	L,K,A/B,CI,C/D,ST,T cm,DF/CF,C/D,ins	
0903VS-2 Scholar Log-Log as Scholar VS, but with a Double sided cursor.						
0908 Trilog	25L	Dup	Pl	pl	T1,T2,A/B,BI,CI,C/D,P,S,ST LL1,LL2,LL3,DF/CF,CIF,C/D,L,K	
8/150 Trilog Demonstration	150L	Dup	Pl	pl	T1,T2,A/B,BI,CI,C/D,P,S,ST LL1,LL2,LL3,DF/CF,CIF,C/D,L,K	
108 Trilog Projection	20L	Dup	Pl	n/a	T1,T2,A/B,BI,CI,C/D,P,S,ST LL1,LL2,LL3,DF/CF,CIF,C/D,L,K	
0905 School Commerce	25L	OF	Pl	pl	%,KZ/T2,P,T1/Z,M,£,s-d//cm.ins	a.
5/150 School Commerce Demonstration	150L	OF	Pl	pl	%,KZ/T2,P,T1/Z,M,£,s-d//cm.ins	
845, Commerce 1	12.5L	CF	Pl	pl	%,KZ/T2,P2,P1,T1/Z,M//£,s-d	a,b.
955, Commerce 1	25L	CF	Pl	pl	%,KZ/T2,P2,P1,T1/Z,M,£,s-d//ZZ3,ZZ2,ZZ1,ZZ%,T1	
1055, Commerce 1	50L	CF	Pl	pl	%,KZ/T2,P2,P1,T1/Z,M,£,s-d//ZZ3,ZZ2,ZZ1,ZZ%,T1	
965, Commerce II	25L	CF	Pl	pl	%,KZ/T2,P2,P1,T1/Z,M,£,s-d//ZZ3,ZZ2,ZZ1,ZZ%,T1	
89, Rietz	12.5L	CF	Pl	pl	K,A/B,CI,C/D,L//S,ST,T	c.
99, Rietz	25L	CF	Pl	pl	K,A/B,CI,C/D,L//S,ST,T	c.
109, Rietz	50L	CF	Pl	pl	K,A/B,CI,C/D,L//S,ST,T	c.
9/150, Rietz Demonstration	150L	CF	Pl	pl	K,A/B,CI,C/D,L//S,ST,T	c.
89NZ, Rietz	12.5L	CF	Pl	pl	K,A/B,CI,C/D,L//S,ST,T; mantissa scale and a scale of preferred numbers based on the Renard system.	
829, MultiRietz	12.5L	Dup	Pl	pl	K,DF/CF,CIF,CI,C/D,L P,A/B,T,ST,S,C/D,DI	
0929, MultiRietz	25L	Dup	Pl	pl	K,DF/CF,CIF,CI,C/D,L P,A/B,T,ST,S,C/D,DI	
29/150, MultiRietz Demonstration	150L	Dup	Pl	pl	K,DF/CF,CIF,CI,C/D,L P,A/B,T,ST,S,C/D,DI	
867, Darmstadt	12.5L	OF	Pl	pl	K,A/B,BI,CI,C/D,P,S,T//L,LL1,LL2,LL3	d.
967, Darmstadt	25L	OF	Pl	pl	K,A/B,BI,CI,C/D,P,S,T//L,LL1,LL2,LL3	
1067, Darmstadt	50L	OF	Pl	pl	K,A/B,BI,CI,C/D,P,S,T//L,LL1,LL2,LL3	
67/150, Darmstadt Demonstration	150L	OF	Pl	pl	K,A/B,BI,CI,C/D,P,S,T//L,LL1,LL2,LL3	
868, Studio	12.5L	Dup	Pl	pl	T,ST,DF/CF,CIF,CI,C/D,P,S	

0968, Studio	25L	Dup Pl	pl	LL01,LL02,LL03,A/B,L,K,C/D,LL3.Ll2,LL1 T,ST,DF/CF,CIF,CI,C/D,P,S
1068, Studio	50L	Dup Pl	pl	LL01,LL02,LL03,A/B,L,K,C/D,LL3.Ll2,LL1 T,ST,DF/CF,CIF,CI,C/D,P,S
68/150, Studio Demonstration	150L	Dup Pl	pl	LL01,LL02,LL03,A/B,L,K,C/D,LL3.Ll2,LL1 T,ST,DF/CF,CIF,CI,C/D,P,S
168, Studio Projection	8"L	Dup Pl	n/a	LL01,LL02,LL03,A/B,L,K,C/D,LL3.Ll2,LL1 T,ST,DF/CF,CIF,CI,C/D,P,S
870, MultiLog	12.5L	Dup Pl	pl	LL01,LL02,LL03,DF/CF,CIF,L,CI,C/D,LL3,LL2,LL1 LL00,K,A/B,T,ST,S,C/D,DI,LL0
0970, MultiLog	25L	Dup Pl	pl	LL01,LL02,LL03,DF/CF,CIF,L,CI,C/D,LL3,LL2,LL1 LL00,K,A/B,T,ST,S,C/D,DI,LL0
01070, MultiLog	50L	Dup Pl	pl	LL01,LL02,LL03,DF/CF,CIF,L,CI,C/D,LL3,LL2,LL1 LL00,K,A/B,T,ST,S,C/D,DI,LL0
70/150, MultiLog Demonstration	150L	Dup Pl	pl	LL01,LL02,LL03,DF/CF,CIF,L,CI,C/D,LL3,LL2,LL1 LL00,K,A/B,T,ST,S,C/D,DI,LL0
170, MultiLog Projection	8"L	Dup Pl	n/a	LL01,LL02,LL03,DF/CF,CIF,L,CI,C/D,LL3,LL2,LL1 LL00,K,A/B,T,ST,S,C/D,DI,LL0
0971, HyperboLog	25L	Dup Pl	pl	LL01,LL02,LL03,DF/CF,CIF,L,CI,C/D,LL3,LL2,LL1 Th,K,A/B,T,ST,S,C/D,DI,Sh2,Sh1
815, Electro	12.5L	CF Pl	pl	K,A/B,BI,CI,C/D,VD,Eff//S,ST,L,T
915, Electro	25L	CF Pl	pl	K,A/B,BI,CI,C/D,LL3,LL2,VD,Eff//S,ST,L,T
852, Attenuation	12.5L	CF Pl	pl	U1,U2,A/B,Neper,C/D,dB//mV,V
920, Federfix	L	CF Pl	n/a	Nomographic rule 370mm x 203mm, arc cursor, calcs to DIN 2089
921, TAUmax	10"L	CF Pl	pl	2 slides, scales for spiral springs to DIN 2089
922, Temperature Radiation	25L	OF Pl	pl	Czerny System
939/x, Ferro Concrete	25L	OF Pl	pl	Ferro concrete calculations, $x=8$, 10, 15, 15/35 for various stresses. Extra wide cursor.
956, Air Photo	25L	Dup Pl	pl	aerial photo ratios T,K,A/B,CI,C/D,ST,S
0958, Surveyor	25L	Dup Pl	pl	Y,ST,DF/CF,CIF,CI,C/D,P,S $1\text{-cos},1\text{-cos},\text{A/B},\tan a/2^{-2},\sin.\cos,\cos^2,\text{D,L,K}$
930, Textil	25L	Dup Pl	pl	$\text{L,K,A/B,BI,C/D,Ne}_1,\text{Ne}_2,\text{Ne}_3,a_{1e},\text{T/1"}$ $a_c,a_c,\text{tex},a_m,\text{Nm,Ne}_w,\text{Ne}_K,\text{Ne}_L,N_2,N_1,\text{Ne}_B,\text{Td}$ (1st 5 scales equiv to DF,CF,CI,C/D)
0602, Circular Computer	6C	MD Pl	pl	C/D,%
622, Calculator	8.5C	MD Pl	pl	Constants,C/D,CI,%
623, Commerce III	8.5C	MD Pl	pl	K,A,C/D,DI
630, Chemistry	20C	MD Pl	pl	C/D,Chemi,etc
670, Motorists Computer	4C	MD Pl	n/a	Fuel consumption,Speed,Time,Distance; Metric units
671, Motorists Computer	4C	MD Pl	n/a	Fuel consumption,Speed,Time,Distance; Imperial units
610, Aviat, Air Navigation Computer	10C	MD Pl	pl	mult,div,speed,time,dist,fuel cons true airspeed,true altitude, density altitude, speed of sound etc
613, Aviat, Air Navigation Computer	xx	MD Pl	pl	mult,div,speed,time,dist,fuel cons; 9.6 x 17.2 plotting surface, various speeds available from 60 - 150 mph
614, Aviat-Sport, Air Nav Computer	14C	MD Pl	pl	mult,div,speed,time,dist,fuel cons, simpler version of 615, true airspeed,true altitude, density altitude, speed of sound etc
615, Aviat, Air Navigation Computer	14C	MD Pl	pl	mult,div,speed,time,dist,fuel cons true airspeed,true altitude, density altitude, speed of sound etc
615W, Aviat, Air Nav Comp, Demo	62C	MD Pl	pl	mult,div,speed,time,dist,fuel cons true airspeed,true altitude, density altitude, speed of sound etc true airspeed,true altitude, density altitude, speed of sound etc
616, Aviat G, Air Navigation Computer	xx	MD Pl	pl	mult,div,speed,time,dist,fuel cons; 11.5 x 25 plotting surface, true airspeed,true altitude, density altitude, speed of sound etc
617, Aviat, Air Navigation Computer	xx	MD Pl	pl	mult,div,speed,time,dist,fuel cons; 14 x 25 plotting surface, various speeds available from 60 - 150 mph true airspeed,true altitude, density altitude, speed of sound etc
617W, Aviat, Air Nav Comp, Demo	xx	MD Pl	pl	mult,div,speed,time,dist,fuel cons; 14 x 25 plotting surface, various speeds available from 60 - 150 mph true airspeed,true altitude, density altitude, speed of sound etc
810, Puck	10L	CF Pl	pl	A/B,CI,C/D,K//S,L,T; for advertising, goodwill gifts
815, Electro	L	CF Pl	pl	slide rule in pocket wallet
845, Commerce I	L	CF Pl	pl	slide rule in pocket wallet e.
867, Darmstadt	L	CF Pl	pl	slide rule in pocket wallet
89, Rietz	L	CF Pl	pl	slide rule in pocket wallet e.

1964 Drawing Office Material Manufacturers and Dealers Association catalogue:

89, Rietz	12.5L	CF	Pl	pl	K,A/B,CI,C/D,L//S,ST,T	c.
99, Rietz	25L	CF	Pl	pl	K,A/B,CI,C/D,L//S,ST,T	c.
109, Rietz	50L	CF	Pl	pl	K,A/B,CI,C/D,L//S,ST,T	c.
150, Rietz Demonstration	150L	CF	Pl	pl	K,A/B,CI,C/D,L//S,ST,T	c.
815, Electro	12.5L	CF	Pl	pl	K,A/B,BI,CI,C/D,VD,Eff//S,ST,L,T	
915, Electro	25L	CF	Pl	pl	K,A/B,BI,CI,C/D,LL3,LL2,VD,Eff//S,ST,L,T	
867, Darmstadt	12.5L	OF	Pl	pl	K,A/B,BI,CI,C/D,P,S,T//L,LL1,LL2,LL3	d.
967, Darmstadt	25L	OF	Pl	pl	K,A/B,BI,CI,C/D,P,S,T//L,LL1,LL2,LL3	
1067, Darmstadt	50L	OF	Pl	pl	K,A/B,BI,CI,C/D,P,S,T//L,LL1,LL2,LL3	
67/150, Darmstadt Demonstration	150L	OF	Pl	pl	K,A/B,BI,CI,C/D,P,S,T//L,LL1,LL2,LL3	
868, Studio	12.5L	Dup	Pl	pl	T,ST,DF/CF,CIF,CI,C/D,P,S LL01,LL02,LL03,A/B,L,K,C/D,LL3.Ll2,LL1	
0968, Studio	25L	Dup	Pl	pl	T,ST,DF/CF,CIF,CI,C/D,P,S LL01,LL02,LL03,A/B,L,K,C/D,LL3.Ll2,LL1	
01068, Studio	50L	Dup	Pl	pl	T,ST,DF/CF,CIF,CI,C/D,P,S LL01,LL02,LL03,A/B,L,K,C/D,LL3.Ll2,LL1	
68/150, Studio Demonstration	150L	Dup	Pl	pl	T,ST,DF/CF,CIF,CI,C/D,P,S LL01,LL02,LL03,A/B,L,K,C/D,LL3.Ll2,LL1	
0903 Scholar	25L	OF	Pl	pl	L,K,A/B,CI,C/D,S,ST,T//cm,ins; Rietz system.	
3/100 Scholar Demonstration	100L	OF	Pl	pl	L,K,A/B,CI,C/D,S,ST,T//cm,ins; Rietz system.	
3/150 Scholar Demonstration	150L	OF	Pl	pl	L,K,A/B,CI,C/D,S,ST,T//cm,ins; Rietz system.	
0903LL Scholar Log-Log (Darmstadt)	25L	OF	Pl	pl	L,K,A/B,CI,C/D,S,ST,T//cm,-/S,LL2,LL3/-ins; Rev slide	
3LL/150 Scholar Log-Log Demonstration	150L	OF	Pl	pl	L,K,A/B,CI,C/D,S,ST,T//cm,-/S,LL2,LL3/-ins; Rev slide	
0903VS Scholar VS	25L	Dup	Pl	pl	L,K,A/B,CI,C/D,ST,T cm,DF/CF,C/D,ins; Reversible cursor.	
3VS/150 Scholar VS Demonstration	150L	Dup	Pl	pl	L,K,A/B,CI,C/D,ST,T cm,DF/CF,C/D,ins	
89, Rietz	12.5L	CF	Pl	pl	K,A/B,CI,C/D,L//S,ST,T. In leather case.	
89NZ, Rietz	12.5L	CF	Pl	pl	K,A/B,CI,C/D,L//S,ST,T; extra scales (Renard Numbers) R5, R10 and R20 (DIN232) on rear face.	

Comment:

a.	Scale is folded at 3.6
b.	Available with inch rather than cms scale on edge
c.	Available with trig scales graduated in 400 intervals rather than the normal sexagesimal. model numbers: x/400^g.
d.	Available with trig scales graduated centesimally, code 867 U/400^g.
e.	Also available in loose leaf book binders.

ASA (USA)

Address: ASA, address unknown.

History: c1992, known from a number of E6B Flight Computers.

ASTON & MANDER (GB)

Address: Aston & Mander, 25 Old Compton Street, Soho, London.   61 Old Compton Street, Soho, London.

History: Established 1870 when George Mander and Isaac Aston joined forces, however, Hannyngton in 1890 claims that A&M *have been in business over 100 years* implying a much earlier date of 1790 for their formation.

1833, George Mander, Instrument maker.

1850, Isaac Aston, Instrument maker.

1884: Made Hannyngton's "washboard" extended scale slide rule.

Hannyngton's extended slide rule available in 3 sizes as follows:

Slide Rules:

1	2	3	4	5	6	7
120" rule,	11 bar stock, 8 bar slide				Price £4.-.-	
60" rule,	9 bar stock, 5 bar slide				Price £2.-.-	
30" rule,					Price £1.10.-	
					Square Root slide extra.	
Harrow Mark Reducer slide rule	L	?	W	?	(sold from 61 Old Compton St)	a.
Slide Rule Mk VI	L	??	M	?	Rustfree metal, Military slide rule	1915 b.

Comment:

a.	Believed to have been designed by a schoolmaster at Harrow school to calculaate prorata mark reductions for scaling of examination marks.
b.	Applied Observation of Shot slide rule.

ASTRO (USA)
Address: Astro Corp, address not known.
History: See **AGA**. Sold **RELIABILITY & CONFIDENCE** slide rules c1960.

AUSTIN (GB)
Address: James Austin & Sons, England.
History: 10" linear slide rule, a Mild steel calculator, normal single slide and cursor.

AVIATION TECHNOLOGY (USA)
Address: Not known.
History: Known from a c1976 Plastic disc DME (Distance Measuring Equipment?).

AZZOLA (Italy?)
Address: E. Azzola & Co, not known.
History: Not known.

Slide Rules:

1	2	3	4	5	6	7
(Rietz)	15L	CF	W/C	m/g	cm\K,A/B,CI,C/D,L\1:25//S,ST,T	

BAKER (GB)
Address: Charles Baker, 244 High Holborn, London WC1.
History: 1917 - Advert in Pickworth.
6", 10" and 20" slide rules. also new and secondhand mathematical and surveying instruments.
Sold or "made" Fullers Calculating Slide Scale. Listed in the 1922 *Dictionary of British Scientific Instruments*.

BALL & CO (Germany)
Address: Ball & Co, Aachen, Germany.
History: Not known. The one example may have been made by another manufacturer for Ball.

Slide Rules:

1	2	3	4	5	6	7
System Rietz	10"L	CF	W/C	m/g	K,A/B,CI,C/D,L//S,ST,T	

BARBOTHEU (France)
Address: Barbotheu, 17 Rue Béranger, St. Etienne (Loire) 6-26u, Paris, France.
History: Not known. 17 Rue Béranger is known from an early catalogue, any connection with **MANUFRANCE** of St. Etienne is tenuous, but is left in pending correction or confirmation.

BARNARD (USA)
Address: See **STANLEY RULE & LEVEL CO** amongst others.
History: Designed a Coordinate Calculating Rule, made by **STANLEY RULE & LEVEL CO**.

BASIC (USA)
Address: Basic Slide Rule Co.
History: c1968 produced a Basic Slide rule chart, the Quickchart.

BAUSCH & LOMB (Switzerland?)
Address: Not known.
History: c1973, 3 disc cardboard and plastic ZTS Scale matching calculator.

Slide Rule Manufacturers and Retailers

BAUER (Germany)
Address: Rudolf Bauer & Co., Norderstedt, Germany.
History: Sold slide rules under the trade name **FEDRA**. These were made for them by **GRAPHOPLEX**.

Slide Rules:

1	2	3	4	5	6	7
64/621 Rietz	25L	CF	Pl			
68/641 Darmstadt	25L	CF	Pl			
1614 Schul-Log-Log	25L	CF	Pl			
1600 Schule	25L	CF	Pl			
65/690 Technicus	25L	Dup	Pl			
694A Tschnilog	25L	Dup	Pl			
64 Rietz	25L	CF	Pl		ex Faber-Castell	
65 Technicus	25L	CF	Pl		ex Faber-Castell	
68 Darmstadt	25L	Dup	Pl		ex Aristo	
63 Rietz	12.5L	CF	Pl		ex Faber-Castell	

BELCHER BROS (USA)
Address: Not known.
History: c1860 made Boxwood and Ivory slide rules

BELL (USA)
Address: Arthur L. Bell, USA
History: c1905, Bell's Stadia slide rule, otherwise history not known.

BELL TELEPHONE LABS (USA)
Address: Bell Telephone Labs. USA.
History: Known from a slide rule for reliability and failure rate calculations.

BERRICK BROS (GB)
Address: Berrick Bros Limited,
c1905, 13/14 Camomile Street, London EC3 (From an early instruction for Castell precision rule, mentions London agency at this address).
c1913, 17 Bury Street, London EC3.
1963, 1969, 20/24 Kirby St, London EC1.
History: 1900-69. Agents for **FABER-CASTELL** particularly, and others including **HEMMI, UNIQUE**
A 1963 catalogue shows the following:

Slide Rules:

1	2	3	4	5	6	7
Faber Castell as follows:						
1/60 Basic	10"L	CF	W/C	pl	58/-	
1/87 Rietz	10"L	CF	W/C	pl	65/-	
1/92 Log-Log	10"L	CF	W/C	pl	80/-	
1/98 Electrical Log-Log	10"L	CF	W/C	pl	80/-	
1/54 Darmstadt	10"L	CF	W/C	pl	80/-	
1/22 Business	10"L	CF	W/C	pl	70/-	
1/28 Super Business	10"L	CF	W/C	pl	70/-	
1/38 Stadia	10"L	CF	W/C	pl	75/-	
1/48 Machine time	10"L	CF	W/C	pl	75/-	
3/11 Concrete	10"L	CF	W/C	pl	80/-	
111/22 Business	10"L	CF	Pl	pl	70/-	
111/87 Rietz	10"L	CF	Pl	pl	65/-	
111/98 Electrical Log-Log	10"L	CF	Pl	pl	80/-	
111/54 Darmstadt	10"L	CF	Pl	pl	80/-	
111/66 Demegraph	10"L	CF	Pl	pl	80/-	
2/82 Duplex	10"L	Dup	Pl	pl	87/-	
2/84 Mathema	10"L	Dup	Pl	pl	115/-	
57/22 "Students" Business	10"L	CF	Pl	pl	27/-	a.
57/87 "Students" Rietz	10"L	CF	Pl	pl	27/6d	a.
57/88 "Students" Advanced Rietz	10"L	CF	Pl	pl	30/-	a.
57/89 "Students" Advanced Log-Log	10"L	CF	Pl	pl	33/-	a.

57/92 "Students" Log-Log	10"L	CF	Pl	pl	33/-	a.
52/82 "Students" DUPLO	10"L	Dup	Pl	pl	42/6d	a.
167/87 "Students" Rietz	5"L	CF	Pl	pl	22/-	a.
67/39 Basic	5"L	CF	Pl	pl	A/B,C/D; 25/-	
67/91 Basic	5"L	CF	Pl	pl	A/B,C/D//S,L,T; 28/-	
67/87 Rietz	5"L	CF	Pl	pl	K,A/B,CI,C/D,L//S,ST,T; 35/-	
67/54B Darmstadt	5"L	CF	Pl	pl	P,K,A/B,CI,C/D,L,S,ST,T//LL1,LL2,LL3; 47/6d	
67/98B Electro	5"L	CF	Pl	pl	46/-	
67/22 Business	5"L	CF	Pl	pl	DF/CF,CI,C/D,L//LL1,LL2,C; 34/6d	
62/82 Duplex	5"L	Dup	Pl	pl	A/B,BI,CI,C/D,P,S	
					LL03,LL02,LL1,K,K',L,CF/DF,LL1,LL2,LL3; 54/-	
111/22A Business/Addiator	10"L	CF	Pl	pl	100/-	
111/54A Darmstadt/Addiator	10"L	CF	Pl	pl	112/-	
111/87A Rietz/Addiator	10"L	CF	Pl	pl	96/-	
67/22R Business/Addiator	5"L	CF	Pl	pl	58/6d	
67/54R Darmstadt/Addiator	5"L	CF	Pl	pl	70/-	
67/87R Rietz/Addiator	5"L	CF	Pl	pl	57/-	
67/98R Electro/Addiator	5"L	CF	Pl	pl	70/-	

Hemmi bamboo slide rules as follows:

No 30 Mannheim Basic	4"L	CF	B/C	pl	A/B,CI,C/D//S,L,T; 25/-
No 32 Mannheim Basic	4"L	CF	B/C	plm	as No 30 with magnifying cursor; 32/-
No 34RK Mannheim Basic	5"L	CF	B/C	pl	A/B,CI,C/D//S,L,T; 27/6d
No 74 Rietz	5"L	CF	B/C	pl	K,A/B,CI,C/D,L//S,ST,T; 30/-
No 2634 Rietz with folded scale	5"L	CF	B/C	pl	K,DF/CF,CIF,CI,C/D,A,TI1,TI2,L,SI; 34/-
No 66 Rietz	6"L	CF	B/C	pl	K,A/B,CI,C/D,L//S,ST,T; 42/6d
No 86K Electrical	6"L	CF	B/C	pl	Eff,Volts,A/B,CI,C/D,LL2,LL3,S,L,T/K; 50/-
No 136 Darmstadt	6"L	CF	B/C	pl	L,K,A/B,CI,C/D,LL3,LL2,LL1//S,T; 45/-
No 45 Rietz with folded scale	8"L	CF	B/C	pl	K,DF/CF,CIF,CI,C/D,A,TI1,TI2,L,SI; 20/-
No 2640 Mannheim Basic	8"L	CF	B/C	pl	L,A/B,CI,C/D//S,K,T; 21/-
No 40RK Mannheim Students	10"L	CF	B/C	pl	A/B,CI,C/D,K//S,L,T; 22/6d
No 64 Rietz	10"L	CF	B/C	pl	K,A/B,CI,C/D,L//S,ST,T; 47/6d
No 50W Mannheim Basic	10"L	CF	B/C	pl	A/B,CI,C/D,K//S,L,T; 40/-
No 80K Electrical	10"L	CF	B/C	pl	Eff,Volts,A/B,CI,C/D,LL2,LL3,S,L,T/K; 53/-
No 130 Darmstadt	10"L	CF	B/C	pl	L,K,A/B,CI,C/D,LL3,LL2,LL1//S,T; 57/-
No 2664S Rietz with folded scale	10"L	CF	B/C	pl	K,DF/CF,CIF,CI,C/D,A,TI1,TI2,L,SI; 47/-
No 2690 Stadia	10"L	CF	B/C	pl	L,D,S,Cos,Cos^2,C/D,A/S,ST,T; 48/-
No 153 Duplex Elec Eng	10"L	Dup	B/C	pl	L,K,A/B,CI,D,T,GΦ
					Φ,RΦ,P,Q,Q',C,LL3,LL2,LL1; 96/-
No 250 Duplex General	10"L	Dup	B/C	pl	DF/CF,CIF,CI,C/D,L
					K,A/B,TI2,T1,SI,D/D1; 86/-
No 251 Duplex Mech Eng	10"L	Dup	B/C	pl	L,DF/CF,CIF,CI,C/D,DI
					K,A/B,T,S,ST,C/D,LL3,LL2,LL1; Folded at $\sqrt{10}$; 95/-
No 255 Duplex Elec Eng	10"L	Dup	B/C	pl	L,K,DF/CF,CIF,CI,C/D,LL3,LL2,LL1
					SH1,SH2,TH,A/B,TI,TI2,SI,C/D,X; 105/-
No 257 Duplex Chem Eng	10"L	Dup	B/C	pl	L,A,DF/CF,CIF,CI,C/D,LL1,LL2,LL3
					ch,M1,M2,$^{\circ}$F,$^{\circ}$C,$^{\circ}$K,F(t),kg/cm^2,mmHg,Atm,inHg,Vac,Lb/in^2 gage,tw; 120/-
No 259 Duplex Mech Eng	10"L	Dup	B/C	pl	LL01,LL02,LL03,DF/CF,CIF,CI,C/D,LL1,LL2,LL3
					L,K,A/B,T,S,ST,C/D,LL0,LL00; folded at π.; 105/-
No 275 Duplex Elec Eng	20"L	Dup	B/C	pl	L,K,DF/CF,CIF,CI,C/D,LL3,LL2,LL1
					SH1,SH2,TH,A/B,TI,TI2,SI,C/D,X; 330/-
No 279 Duplex Mech Eng	20"L	Dup	B/C	pl	LL01,LL02,LL03,DF/CF,CIF,CI,C/D,LL1,LL2,LL3
					L,K,A/B,T,S,ST,C/D,LL0,LL00; 330/-
No 154 Duplex Elec Eng	20"L	Dup	B/C	pl	DF,P',P,Q,CF,CI,S@^{0},A/D,K
					Sx,Tx,TΦ^0,Th,Sh,C/D,L,X; 330/-

Unique rules as follows:

U1, Universal 1	10"L	CF	W/C	pl	Universal One shown in illustration; 16/-
U1/2, Universal 1	5"L	CF	W/C	pl	11/6d
U1/1, Universal 1	5"L	CF	W/C	pl	additional low reading log-log scale; 17/-
U2, Universal II	10"L	CF	W/C	pl	16/-
10L/L, Log-Log	10"L	CF	W/C	pl	11/6d
5L/L, Log-Log	5"L	CF	W/C	pl	8/-
10G, Legible	10"L	CF	W/C	pl	11/6d
5G, Legible	5"L	CF	W/C	pl	8/-

10/20, 10/20 Precision	10"L	CF	W/C pl	16/-
C, Commercial	10"L	CF	W/C pl	16/-
E, Electrical	10"L	CF	W/C pl	16/-
N, Navigational	10"L	CF	W/C pl	17/6d
D1 Dualistic	5"L	CF	W/C pl	17/-
D3 Dualistic	10"L	CF	W/C pl	25/-
B, Brighton	10"L	CF	W/C pl	30/-
M1, Monetary	10"L	CF	W/C pl	16/-

Otis King calculators:
Model K 62/6d
Model L 62/6d

Comments;
a.   The "Students" range have all the same features as the 111/xx range but are of simpler design and construction.
b.   Magnifying cursors with cylindrical lenses and snap-on magnifying attachments are available.

BESELER (USA)
Address: Beseler/Arcom.
History: c1965 sold a transparent projection slide rule.

BETHLEHEM (USA)
Address: Bethlehem Steel Co, USA.
History: c1900 produced a range of specialist slide rules for the steel industry to designs by ? Barth.

Slide Rules:

1	2	3	4	5	6	7
Barth's Slide rule for Lathe settings for maximum output						1901
Barth's Slide rule for Strength of Spur Gears						1901
Barth's Speed Slide Rule						1901
Barth's Time Slide Rule						1901

BILLETER (Switzerland)
Address: Julius Billeter, Zurich, Switzerland.
History: There was Billeter, Max und Bohnhorst, Julius; and Schuppisser & Billeter; both of Zurich. It is not known what the relationship between these firms were.

Slide Rules:

1	2	3	4	5	6	7
Circular slide rule	C					
Cylindrical slide rule	T					
"National" Cylindrical slide rule	T					1920
Rechentafel	52L	?	M/G	n/a	metal/glass rule, Swiss pat 43,463.	

BLUNDELL (GB)
Address: 1953, Blundell Rules Limited, Chaul End Lane, Luton, Beds, England.
1956, Blundell Rules Ltd., Regulus Works, Lynch Lane, Weymouth, England.
1968, Blundell-Harling, Regulus Works, Lynch Lane, Weymouth, Dorset. DT4 9DW.
History: Blundell established in 1852 as Felt makers in Luton. Started making slide rules after the Second World War c1947 when they found that they had specialist machinery lying idle. Moved to Weymouth 1956.
Takeover of **W.H. HARLING** 1966 - a possible date for the formation of **BRL**.
Slide rules sold under the name **BLUNDELL**, **BRL** and **BLUNDELL-HARLING**.

Slide Rules:

1	2	3	4	5	6	7
Blundel-Harling comprehensive listing is as follows:						
1947 to 1949 Bakelite Slide Rules:						
G1	10"L	CF	Ba	m/p	K,A/B,R,C/D,M	
G2	10"L	CF	Ba	m/p	K,A/B,R,C/D,M; extensions to scales	
E3	10"L	CF	Ba	m/p	Volt,A/B,Cos,L/D,Dyn	
L4	10"L	CF	Ba	m/p	UL,A/B,R,C/D,LL	

G5	10"L	CF	Ba	m/p	K,A/B,R,C/D,M; extensions to scales
L6	10"L	CF	Ba	m/p	UL,A/B,R,C/D,LL//S,ST,T

1949 to 1958 Astralon Slide Rules:

AG1	10"L	CF	Pl	m/p	K,A/B,R,C/D,M
AG2	10"L	CF	Pl	m/p	K,A/B,R,C/D,M; extensions to scales
AL4	10"L	CF	Pl	m/p	UL,A/B,R,C/D,LL
AG5	10"L	CF	Pl	m/p	K,A/B,R,C/D,M; extensions to scales
AL6	10"L	CF	Pl	m/p	UL,A/B,R,C/D,LL//S,ST,T
AU7	10"L	CF	Pl	m/p	UL,K,A/B,S,ST,T,R,C/D,M,LL
T11D	10"L	CF	Pl	m/p	LL1,LL2,A/B,R,C/D,P,LL3//S,ST,D
T12D	10"L	CF	Pl	m/p	K,LL1,LL2,CF/DF,CRF,CR,C/D,P,LL3,A//S,ST,T,M
D26	10"L	CF	Pl	m/p	M,K,A/B,R,L,C/D,P,S,T//LL1,LL2,LL3
E13	10"L	CF	Pl	m/p	K,LL1,LL2,DF/CF,CR,CD/A,LL3,A,CD//S,ST,T,M

1949 to 1958 Wood slide rule, Astralon Slide

WG5	10"L	CF	W/Pl	m/p	K,A/B,R,C/D,M

1947 to 1949 Bakelite 5" slide rule

P8	5"L	CF	Ba	m/p	A/B,R,C/D

1948 to 1958 Astralon 5" slide rules

P14	5"L	CF	Pl	m/p	K,A/B,R,C/D,M//S,ST,T	c.
P15	5"L	CF	Pl	m/p	A/B,R,C/D//S,ST,T	c.
P16	5"L	CF	Pl	m/p	UL,A/B,Cos,C/D,CC//S,ST,T\\Volt,Eff	c.
P17	5"L	CF	Pl	m/p	UL,A/B,R,C/D,LL//S,ST,T	c.
P19	5"L	CF	Pl	m/p	A/B,R,C/D	c.
P28	5"L	CF	Pl	m/p	M,K,A/B,R,C/D,P,S,T//LL1,LL2,LL3	c.
P31	5"L	CF	Pl	m/p	A/B,R,C/D; scales not lettered	c.

1958 - 1970, Omega series of slide rules in plastic.
400 Series

401 Rietz	10"L	CF	Pl	m/p	ins\K,A/B,CI,C/D,L//S,T,ST
402 Log-Log Trig	10"L	CF	Pl	m/p	ins\LL2,A/B,CI,C/D,LL3//S,T,ST
403 Darmstadt	10"L	CF	Pl	m/p	ins\L,K,A/B,CI,C/D,P,S,T//LL1,LL2,LL3
404 Electro	10"L	CF	Pl	m/p	ins\LL2,A/B,CI,C/D,LL3//S,L,T
405 Basic	10"L	CF	Pl	m/p	ins\A/B,C/D
406 Simplified Rietz	10"L	CF	Pl	m/p	ins\A/B,CI,C/D//S,T,ST
407 Technico/Log-Log	10"L	CF	Pl	m/p	ins\LL1,LL2,DF/CF,CIF,CI,C/D,P,LL3,A//S,ST,T,L
409 Mercantile	10"L	CF	Pl	m/p	ins\C/D,M1,M2,M3,IL,IR; reading from 1/- to £80
410 Technico	10"L	CF	Pl	m/p	ins\A,DF/CF,CI,C/D,K//B,S,L,T

600 Series

601 Rietz	5"L	CF	Pl	m/p	ins\K,A/B,CI,C/D,L//S,T,ST
602 Log-Log Trig	5"L	CF	Pl	m/p	ins\LL2,A/B,CI,C/D,LL3//S,T,ST
603 Darmstadt	5"L	CF	Pl	m/p	ins\L,K,A/B,CI,C/D,P,S,T//LL1,LL2,LL3
604 Electro	5"L	CF	Pl	m/p	ins\LL2,A/B,CI,C/D,LL3//S,L,T
605 Basic	5"L	CF	Pl	m/p	ins\A/B,C/D
606 Simplified Rietz	5"L	CF	Pl	m/p	ins\A/B,CI,C/D//S,T,ST
607 Technico/Log-Log	5"L	CF	Pl	m/p	ins\LL1,LL2,DF/CF,CIF,CI,C/D,P,LL3,A//S,ST,T,L
608 Five/Ten	5"L	CF	Pl	m/p	ins\C1,CD,D1/D2,C1:1,C1:1,C2:1,D,L//S1,S2,T1,T2
610 Technico	5"L	CF	Pl	m/p	ins\A,DF/CF,CI,C/D,K//B,S,L,T

1957 - 1965 Janus series Duplex Slide rules

T50 Multi-Log Duplex	10"L	Dup	Pl	m/p	LL01,LL02,LL03,CF/DF,CIF,CI,C/D,LL1,LL2,LL3 L,P,A/B,T,ST,S/D,DI,K
JV56 Multi-Log Vector Duplex	10"L	Dup	Pl	m/p	LL01,LL02,LL03,DF/CF,CIF,CI,C/D,LL3,LL2,LL1 Sh1,Sh2,Th,A/C,T,ST,S,D/DI,K,L
T51 Multi-Log Duplex Lightweight	10"L	Dup	Pl	m/p	LL01,LL02,LL03,CF/DF,CIF,CI,C/D,LL1,LL2,LL3 L,P,A/B,T,ST,S,D,DI,K
N52 Navigator	10"L	Dup	Pl	m/p	S,A/B,T1,T2,ST,S,C/D,K,L Temp conv,Density & True Alt,Speed & Dist Conv, True Air speed, CI

1958 - 1970, Academy 300 cheap slide rules

300 Simplified Rietz	10"L	OF	Pl	pl	K,A/B,CI,C/D,L
302 Log-Log Trig	10"L	OF	Pl	pl	L,LL2,LL3,A/B,CI,C/D,S,T,ST
303 Rietz	10"L	OF	Pl	pl	L,K,A/B,CI,C/D,S,ST,T
304 Darmstadt	10"L	OF	Pl	pl	L,K,A/B,CI,C/D,P,S,T//LL1,LL2,LL3
305 Electro	10"L	OF	Pl	pl	A,B,C/D,CI,K,LL2,LL3//S,L,T
307 Air Navigation	10"L	OF	Pl	pl	??
308 Accountants	10"l	OF	Pl	pl	C/D,IL,IR,LL1,LL2,LL3,M/T,S
310 Pocket Circular	3"C	MD	Pl	pl	??
311 Simplified Rietz	5"L	OF	Pl	pl	K,A/B,CI,C/D,L

1970 - 1982, Academy 800 Series cheap slide rules, with metric standards and coloured bands for ease of use.

800 Basic	25L	OF	Pl	plm	A/B,C/D
802 Log-Log Trig	25L	OF	Pl	plm	K,LL2,LL3,A/B,L,CI,C/D,S,T,ST
803 Rietz	25L	OF	Pl	plm	L,K,A/B,CI,C/D,S,T,ST
804 Darmstadt	25L	OF	Pl	plm	L,K,A/B,CI,C/D,P,S,T//LL1,LL2,LL3
805 Electrical Log-Log	25L	OF	Pl	plm	Volt,Eff,A/B,CI,C/D,K,LL2,LL3
808 Commercial Log-Log	25L	OF	Pl	plm	S,LL1,DF/CF,IL,C/D,LL3,LL2
810 Navigational	25L	OF	Pl	plm	speed,time,distance
812 Mannheim Rietz	12.5L	OF	Pl	plm	S,T,A/B,CI,C/D,K,L
813 Mannheim Rietz	25L	OF	Pl	plm	S,T,A/B,CI,C/D,K,L

1965 - 1970, Academy Duplex range, inexpensive.

500 Log-Log	10"L	Dup	Pl	pl	L,K,A/B,CI,C/D,LL3,LL2
					T,ST,DF/CF,CIF,C/D,P,S
502 Log-Log	10"L	Dup	Pl	pl	L.K,A/B,CI,C/D,LL3,LL2,LL1
					T,ST,DF/CF,CIF,CI,C/D,P,S
503 Commerce	10"L	Dup	Pl	pl	L,K,A/B,CI,C/D,S,ST,T
					M1,Year,Days,Weeks,Decrease/Increase,C/D,M2,M3
504 Sea Navigation	10"L	Dup	Pl	pl	S1,A/B,T1,T2,ST,S,C/D,K,L
					P,DF/CF,CIF,CI,C/D
505 Log-Log/Full Recip	10"L	Dup	Pl	pl	LL01,LL02,LL03,A/B,L,CI,C/D,LL3,LL2,LL1
					T,ST,DF/CF,CIF,CI,C/D,P,S
506 Log-Log Electro	10"L	Dup	Pl	pl	LL01,LL02,LL03,DF/CF,CIF,CI,C/D,LL3,LL2,LL1
					S1,S2,Th,A/B,T,ST,S,D/D1,K,L

1970 - 1980, Super Duplex, patented end caps, post decimalisation money scales

503 Commerce	10"L	Dup	Pl	plm	L,K,A/B,CI,C/D,LL3,LL2
					M1,M2,M3,M4,Year,Days,Weeks,Decrease/Increase,C/D
504 Sea Navigation	10"L	Dup	Pl	plm	S1,A/B,T1,T2,ST,C/D,K,L
					P,DF/CF,CIF,CI,C/D,LL3,LL2
505 Log-Log/Full Recip	10"L	Dup	Pl	plm	LL01,LL02,LL03,A/B,L,CI,C/D,LL3,LL2,LL1
					K,P,ST,DF/CF,CIF,CI,C/D,T,S
506 Log-Log Electro	10"L	Dup	Pl	plm	LL01,LL02,LL03,DF/CF,CIF,CI,C/D,LL3,LL2,LL1
					Sh1,Sh2,Th,A/B,T,ST,S,D/D1,K,L

1975 to 1982, Academy Duplex Slide Rules

904	10"L	Dup	Pl	plm	A/B,CI,C/D,ST,S,T
					K,L,DF/CF,CIF,C/D,LL2,LL3
905	10"L	Dup	Pl	plm	LL01,LL02,LL03,A/B,L,CI,C/D,LL1,LL2,LL3
					K,P,ST,DF/CF,CIF,CI,C/D,T,S

1975 to 1982, School Education Slide Rules

S143-20	5"L	OF	Pl	pl	K,A/B,CI,C/D,L	
S143-10	10"L	OF	Pl	pl	K,A/B,CI,C/D,L	
S150-20	5"L	OF	Pl	pl	S,T,A/B,CI,C/D,K,L	
S150-10	10"L	OF	Pl	pl	S,T,A/B,CI,C/D,K,L	
Vector Slide Rule for GEC	U	CF	Plm	pl	special scales for Vector calculations, U=12"x6"	1953

From the Lawes Rabjohns Technical Catalogue - 1952
BRL Slide Rules
Improved Type.

A.G.1	25L	CF	Pl	m/g	K,A/B,CI,C/D,L
A.G.2	25L	CF	Pl	m/p	ditto plus scale extns, R&B
A.G.5 Rietz	25L	CF	Pl	m/p	K,A/B,CI,C/D,L//S,T,L
E.18 Electrical	25L	CF	Pl	m/p	A/B,C/D,V,D&M,t°,//S,ST,T

| A.L.4 Log-Log | 25L | CF | Pl | m/p | LL1,A/B,CI,C/D,LL2 |
| A.L.6 Log-Log | 25L | CF | Pl | m/p | LL1,A/B,CI,C/D,LL2//S,ST,T |

Students Slide Rule.

| Model 1 | 25L | CF | Pl | m/g | K,A/B,CI,C/D,L (no inches) |
| Model 2 | 25L | CF | Pl | m/g | LL1,A/B,CI,C/D,LL2 (no inches) |

Wide faced Slide Rules.

T.11 Technicians	25L	CF	Pl	m/g	LL1,LL2,A/B,CI,C/D,P,LL3,
T.12 Technicians	25L	CF	Pl	m/g	K,LL1,LL2,DF/CF,CRF,CI,C/D,P,LL3,A
T.20 Technicians	50L	CF	Pl	m/g	LL1,LL2,A/B,CI,C/D,P,LL3,
T.13 Technicians	25L	CF	Pl	m/g	K,LL1,LL2,DF/CF,CRF,CI,C/D,P,LL3,A,V,D
T.13c Technicians (ex gauge pt)	25L	CF	Pl	m/g	K,LL1,LL2,DF/CF,CRF,CI,C/D,P,LL3,A,V,D

Pocket Slide Rules

P.14 Rietz	12L	CF	Pl	m/g	K,A/B,CI,C/D,L//S,T,Log
P.15	12L	CF	Pl	m/g	A/B,C/D//S,ST,T
P.16 Electrical	12L	CF	Pl	m/g	LL1,A/B,C/D,V,D&M,t°,LL2//S,ST,T
P.17 Log-Log	12L	CF	Pl	m/g	LL1,A/B,CI,C/D,LL2//S,ST,T
P.19	12L	CF	Pl	m/g	A/B,CI,C/D

Duplex Slide Rules

T.50 Janus	25L	Dup	Pl	pl	L,LL1,DF/CF,CIF,CI,C/D,LL3,LL2
					K,LL0,A/B,CI,C/D,LL00,S,ST,T
T.51 Janus	10"L	Dup	Pl	pl	as T50, but with non adjustable end plates, and lighter in weight.

Magnifying cursor available for BRL rules.

From the Stanley "A" catalogue - 1958.
"B.R.L." Slide rules.
Rietz and General Purpose.

Pattern AG1	25L	CF	Pl	pl	Fundamental, Cubing, Mantissa, Reciprocal
Pattern Academy 300	25L	CF	Pl	pl	Fundamental, Cubing, Mantissa, Reciprocal
Pattern AG2	25L	CF	Pl	pl	Fundamental, Cubing, Mantissa, Reciprocal
Pattern P19D	12L	CF	Pl	pl	Fundamental, Reciprocal
Pattern 405	25L	CF	Pl	pl	Fundamental only
Pattern G23	25L	CF	Pl	pl	Fundamental, Cubing, Mantissa, Reciprocal, with C on back
Pattern 627	25L	CF	Pl	pl	A, C, and D, Folded CF and DF
Pattern 606	12L	CF	Pl	pl	Fundamental, Reciprocal, S, T & ST
Pattern 406	25L	CF	Pl	pl	Fundamental, Mantissa, Reciprocal, S, T
Pattern 601	12L	CF	Pl	pl	Fundamental, Cubing, Mantissa, Reciprocal, S, T, & ST
Pattern 401	25L	CF	Pl	pl	Fundamental, Cubing, Mantissa, Reciprocal, S, T, & ST
Pattern Academy 303	25L	CF	Pl	pl	Fundamental, Cubing, Mantissa, Reciprocal, S, T, & ST

Log-log

Pattern AL4	25L	CF	Pl	pl	Fundamental, Reciprocal and Log-Log
pattern 602	12L	CF	Pl	pl	Fundamental, Log-log, Reciprocal, S, T, ST
Pattern 402	25L	CF	Pl	pl	Fundamental, Log-log, Reciprocal, S, T, ST
Pattern Academy 303	25L	CF	Pl	pl	Fundamental, Log-log, Reciprocal, Cubing Mantissa, S, T,
Pattern 603	25L	CF	Pl	pl	Fundamental, Mantissa, Cubing, Recip, L-l, Pythagorean, S, T,
Pattern 403	25L	CF	Pl	pl	Fundamental, Mantissa, Cubing, Recip, L-l, Pythagorean, S, T,
Pattern Academy 304	25L	CF	Pl	pl	Fundamental, Mantissa, Cubing, Recip, L-l, Pythagorean, S, T,
Pattern T110	25L	CF	Pl	pl	Fundamental, Reciprocal, Log-log, Pythagorean, S, T, ST
Pattern T120	25L	CF	Pl	pl	Fundamental, Pythagorean, Folded, Reciprocal.
Pattern T120	25L	CF	Pl	pl	F, P, Fold, R, Mant, L-L, S, ST
Pattern AU7	25L	CF	Pl	pl	F, R, Mant, L-L, S, T, ST.

Electricians

Pattern 604	12L	CF	Pl	pl	Elec C&D, L-L, S, T, ST, VD, D/M%, Resis, Cos
Pattern Academy 305	25L	CF	Pl	pl	Elec C&D, Cube, Recip, Mant, L-L, S, T, VD, G%, M%, 3ΦM,G
Pattern 404	25L	CF	Pl	pl	F, M, L-L, Recip, Cube, S, T, VD, D&M%, D&MΦ%
Pattern E13	25L	CF	Pl	pl	L-L, A, Cubing, C, D, Recip, Folded, Pythag, Mant, S, T, ST

1970/80.

From instruction leaflets it can be deduced that the B-H range when they ceased manufacture of general slide rules in the late 1960's or early 1970's comprised the following:

Academy

800 Simplex	25L	CF	Pl	pl
802 Log-log	25L	CF	Pl	pl
803 Reitz	25L	CF	Pl	pl
804 Darmstadt	25L	CF	Pl	pl
805 Electro	25L	CF	Pl	pl
506 Duplex	25L	Dup	Pl	pl

Omega

401 Reitz	25L	CF	Pl	pl
405	25L	CF	Pl	pl
Darmstadt	25L	CF	Pl	pl
407 Technicians	25L	CF	Pl	pl
605	12L	CF	Pl	pl

Janus

T50 Duplex	25L	Dup	Pl	pl
T51 Duplex	25L	Dup	Pl	pl

In addition there were:

S143-10	25L	CF	Pl	pl
S143-20	12L	CF	Pl	pl
S150-20	12L	CF	Pl	pl
904	25L	CF	Pl	pl
905	25L	CF	Pl	pl

From a BRL catalogue c 1950

Students No 1	10"L	CF	Pl	m/p	K,A/B,R,C/D,M
Students No 2 Log Log	10"L	CF	Pl	m/p	LL2,A/B,R,C/D,LL3
A.G.1 General Purpose	10"L	CF	Pl	m/p	ins\K,A/B,R,C/D,M
A.G.2 General Purpose	10"L	CF	Pl	m/p	ins\K,A/B,R,C/D,M; extensions to 5 scales
G.22 General Purpose	10"L	CF	Pl	m/p	ins\A/B,C/D; extensions to scales
G.23 General Purpose	10"L	CF	Pl	m/p	ins\K,A/B,R,C/D,M
G.27 Folded Scales	10"L	CF	Pl	m/p	ins\K,CF/DF,R,C/D,M
G.29 Simplified Reitz	10"L	CF	Pl	m/p	ins\A/B,R,C/D//S,L,T
A.G.5 Reitz	10"L	CF	Pl	m/p	ins\K,A/B,R,C/D,M//S,ST,T; extensions to 5 scales
A.L.4 Log-Log	10"L	CF	Pl	m/p	LL2,A/B,R,C/D,LL3; extensions to scales
A.L.6 Log-Log	10"L	CF	Pl	m/p	LL2,A/B,R,C/D,LL3//S,ST,T; extensions to scales
E.18 Electricians	10"L	CF	Pl	m/p	ins\V,D,A/B,C/D,°F,°C,cos//S,ST,T
E.25 Electricians	10"L	CF	Pl	m/p	ins\UL,A/B,R,C/D,LL,K//S,ST,T\\V,Dyn
D.26 Darmstadt	10"L	CF	Pl	m/p	ins\K,L,A/B,R,C/D,S,T,P//LL3,LL2,LL1
A.U.7 Wide Faced Technicians	10"L	CF	Pl	m/p	K,L,A/B,R,C/D,LL3,LL2,S,ST,T
C.24 Wide Face Cost & Profit Reckoner	10"L	CF	Pl	m/p	s,D/IR,IL,A/B,£
T.11 Technicians	10"L	CF	Pl	m/p	LL1,LL2,A/B,R,C/D,P,LL3//S,ST,D
T.12 Technicians	10"L	CF	Pl	m/p	K,LL1,LL2,CF/DF,CRF,CR,C/D,P,LL3,A//S,ST,M
E.13 Electricians	10"L	CF	Pl	m/p	K,LL1,LL2,DF/CF,CR,CD/A,LL3,A,CD//S,ST,T,M
E.13c Electricians	10"L	CF	Pl	m/p	K,LL1,LL2,DF/CF,CR,CD/A,LL3,A,CD//S,ST,T,M, E13c is as E13 with HP gauge point at 736
T.50 Log-Log Duplex	10"L	Dup	Pl	m/p	L,LL1,DF/CF,CIF,CI,C/D,LL2,LL3 K,A/B,DI,C/D,LL0,LL00,S,ST,T
T.51 Log-Log Duplex	10"L	Dup	Pl	m/p	L,LL1,DF/CF,CIF,CI,C/D,LL2,LL3 K,A/B,DI,C/D,LL0,LL00,S,ST,T; as T50 but lighter constr.
N.52 Navigators	10"L	Dup	Pl	m/p	K,M,A/B,C/D,S1,S2,ST,T1,T2 C/D,DI,true alt,dens alt,°F/°C,miles/km conversion
T.R.53 Traction Engineers	10"L	Dup	Pl	m/p	LL1,A/B,CI,C/D,LL2 special speed, time, effort, mean speed etc scales
T.20 Technicians	20"L	CF	Pl	m/p	LL1,LL2,A/B,R,C/D,L,LL3
P.14 General Purpose Reitz pattern	6"L	CF	Pl	m/p	K,A/B,R,C/D,L//S,ST,T; scales have extensions
P.15 General Purpose	6"L	CF	Pl	m/p	A/B,C/D//S,ST,T; scales with extensions
P.19 General Purpose	6"L	CF	Pl	m/p	A/B,R,C/D; scales with extensions
P.16 Electricians	6"L	CF	Pl	m/p	Volt,Dyn,A/B,C/D,LL1,LL2,Res,Cos//S,ST,T

a.

146

| P.17 Log-Log | 6"L | CF | Pl | m/p | LL1,A/B,R,C/D,LL2//S,ST,T |
| P.28 Darmstadt | 6"L | CF | Pl | m/p | K,L,A/B,R,C/D,S,T,P//LL1,LL2,LL3 |

An extremely simple BRL G.22/2, 5" slide rule has also been seen.

Well known manufacturers of specialist rules, also advertising covering specialist applications, for example:

The Ionospheric Slide Rule	25L	CF	Pl	pl	The Marconi Co Ltd, adjusters.	1962
Satellite & Star Coverage	20C	MD	Pl	n/a	The Marconi Co Ltd.	1966
BRL Radiac Calc No 1.	13C	MD	Pl	n/a	2x Reversible disc, oilcloth case. Ref 6665-99-91-0060	
BRL Omega Artillery slide rule MkIII	25L	CF	Pl	pl(2)	A/B,C/D. V6/1290-99-960-0934	
STD Calculator	13L	FP	Pl	nil		1992
STD Calculator	25L	FP	Pl	nil		1992
Teaching slide rules	6'L	CF	W	pl	6 foot, various.	
Humidity slide rule	25L	CF	Pl	pl	pat 1,114,917; specialist scales.	

It is not known where these fitted into the scheme.
Blundell Harling still make specialist slide rules to date.

<u>1964 Drawing Office Material Manufacturers and Dealers Association catalogue:</u>

Blundell "Omega" Series:

401 Reitz	10"L	CF	Pl	m/p	ins\K,A/B,CI,C/D,L//S,T,ST
402 Log-Log	10"L	CF	Pl	m/p	ins\LL2,A/B,CI,C/D,LL3//S,T,ST
403 Darmstadt	10"L	CF	Pl	m/p	ins\L,K,A/B,CI,C/D,P,S,T//LL1,LL2,LL3
404 Electro	10"L	CF	Pl	m/p	ins\LL2,A/B,CI,C/D,LL3//S,L,T
405 General Purpose	10"L	CF	Pl	m/p	ins\A/B,C/D
406 General Purpose	10"L	CF	Pl	m/p	ins\A/B,CI,C/D//S,T,ST
407 Technico	10"L	CF	Pl	m/p	ins\LL1,LL2,DF/CF,CIF,CI,C/D,P,LL3,A//S,ST,T,L

"Omega" Pocket slide rules:

601 Reitz	5"L	CF	Pl	m/p	ins\K,A/B,CI,C/D,L//S,T,ST
602 Log-Log	5"L	CF	Pl	m/p	ins\LL2,A/B,CI,C/D,LL3//S,T,ST
603 Darmstadt	5"L	CF	Pl	m/p	ins\L,K,A/B,CI,C/D,P,S,T//LL1,LL2,LL3
604 Electro	5"L	CF	Pl	m/p	ins\LL2,A/B,CI,C/D,LL3//S,L,T
605 General Purpose	5"L	CF	Pl	m/p	ins\A/B,C/D
606 General Purpose	5"L	CF	Pl	m/p	ins\A/B,CI,C/D//S,T,ST

Blundell "Academy" Series:

300 Simplified Reitz	10"L	OF	Pl	pl	K,A/B,CI,C/D,L
302 Log-Log Trig	10"L	OF	Pl	pl	L,LL2,LL3,A/B,CI,C/D,S,T,ST
303 Reitz	10"L	OF	Pl	pl	L,K,A/B,CI,C/D,S,ST,T
304 Darmstadt	10"L	OF	Pl	pl	L,K,A/B,CI,C/D,P,S,T//LL1,LL2,LL3
305 Electro	10"L	OF	Pl	pl	A,B,C/D,CI,K,LL2,LL3//S,L,T

Blundell "Janus" Series:

T50 Log-Log Duplex	10"L	Dup	Pl	m/p	LL01,LL02,LL03,CF/DF,CIF,CI,C/D,LL1,LL2,LL3
					L,P,A/B,T,ST,S/D,DI,K
T51 Log-Log Duplex (Junior)	10"L	Dup	Pl	m/p	LL01,LL02,LL03,CF/DF,CIF,CI,C/D,LL1,LL2,LL3
					L,P,A/B,T,ST,S/D,DI,K
N52 Navigation	10"L	Dup	Pl	m/p	S,A/B,T1,T2,ST,S,C/D,K,L
					Temp conv,Density & True Alt,Speed & Dist Conv, True Air speed, CI
JV56 Vector	10"L	Dup	Pl	m/p	LL01,LL02,LL03,DF,CF,CIF,CI,C/D,LL3,LL2,LL1
					Sh1,Sh2,Th,A/C,T,ST,S,D/DI,K,L

Variants:

| 409 Mercantile | 10"L | CF | Pl | pl | M1,months/days&weeks,profit,C/D,M2(10/- to £8),M3(£5 to £80); |

Comment:

a. Available with Perspex stands to enable it to be used as a desk calculator.
b. Range of magnifying cursors, half-round, and clip-on full view magnifier.
c. This range of slide rules with an additional suffix D, supplied with Diakon cursor as alternative to metal frame cursor.

BLUNDELL HARLING (GB)
Address: See **BLUNDELL**
History: See **BLUNDELL**.

BOOTS (GB)
Address: Nottingham, England.
History: Were not makers of rules, however they sold rules from **FABER-CASTELL, BLUNDELL-HARLING** (in the 1970's) and also various Japanese manufacturers under their "RINGPLAN" logo.

Slide Rules:

1	2	3	4	5	6	7
RingPlan Standard	25L	CF	Pl	pl	F-C?	
RingPlan Standard	12L	CF	PL	plm	F-C? Rev slide.	
RingPlan Standard	12L	CF	Pl	plm	Japan Rev slide.	
RingPlan Model B	12L	CF	Pl	pl	F-C?	
RingPlan Precision log-log	12L	CF	Pl	pl	Japan	
RingPlan	25L	CF	Pl	plm	Japan Rev slide.	

BOPP & REUTHER (Germany)
Address: Bopp & Reuther GmbH, Mannheim Waldhoff, Germany.
History: Not known, made a Venturi Slide Rule

Slide Rules:

1	2	3	4	5	6	7
Venturi Slide Rule						

BOUSFIELD (GB)
Address: George Bousfield, England.
History: Produced a Vade Mecum slide rule for the Timber trade.

BPM or BIPM (Holland)
Address: Betaafsche Petroleum Maatschappij (another name for Royal Dutch Shell).
History: Will have specified many different slide rules which were made by others, see **LINEX**.

BREITLING (Switzerland)
Address: Not known.
History: Amongst a number of watch manufacturers, they produced the Chronomat and Navitmar models which featured a circular slide rule on a bezel around the edge of the watch. These have the equivalent of a B and C scale on the watch and bezel.

BRITISH SLIDE RULE CO. (GB)
Address: British Slide Rule Company. 16, Barter Street, Holborn, London, WC1.
History: Proprietors: **MARINE & OVERSEAS SERVICES** (1939) Ltd.
Advertised in 1944.
May not have made rules, it is known that Unique supplied slide rules to **MARINE & OVERSEAS**. Sold a range which had paper/celluloid scales similar to Unique and Classic. Advertised 12" with 9 scales, and 5" scale rules, possibly:

Slide Rules:

1	2	3	4	5	6	7
Technical Standard Slide Rule	25L	CF	W/C	pl	M,A/B,C/D,N//S,ST,T	
Technical Surveyors Slide Rule	25L	CF	W/C	pl		
The Supreme Students Slide Rule	25L	CF	W/C	pl	M,S,A/B,CR,C/D,T,N	1949

BROWN (Australia)
Address: Collis O. Brown. Australia.
History: The inventer Collis O. Brown (1917 - 1988) who produced a Fullers type calculator with spiral scales without the handles c1960. He also invented a number of microscope lenses and microscope items. Sold by **DOBIE**.

BROWN (USA)
Address: Arthur Brown & Bros.
History: c1942 sold drafting materials including slide rules, in all probability made for them by others, as well as retailing other manufacturers instruments.

Slide Rules:

1	2	3	4	5	6	7
7993-B	10"L	?	?	?	??	

BROWN (GB?)
Address: William Brown, not known, UK?
History: Known from a Wood/Celluloid Textile Rule with special scales.

BRL (GB)
Address: see Blundell Rules Limited.
History: See **BLUNDELL**.

BRUNNING (USA)
Address: Charles Brunning Co.
History: Not known.

Slide Rules:

1	2	3	4	5	6	7
2420	10"L	?	?	?	ins/A/B,CI,C/D,K//S,L,T	
68-540	5"L	?	Pl	?	?	

BTU (USA)
Address: Not known.
History: c1911 there was a BTU slide rule in existence.

C-THRU (USA)
Address: C-Thru Ruler Company, Bloomfield, Connecticut, USA., also, Hartford, Connecticut, USA.
History: Made and sold slide rules. Some made in Japan.

Slide Rules:

1	2	3	4	5	6	7
396	C	MD	Pl	n/a	K,D/C,B/A/CI/L	
S888	L	??	Pl	??	S,K,A/B,CI,C/D,L,T; Japan	
889 Duplex	L	Dup	Pl	??	??	

CAL (Australia)
Address: CAL Slide Rule Co P/L. Sydney, Australia.
History: c1942 a CAL No 2 slide rule, similar to a **UNIQUE** slide rule from England with different fonts, apparently made by CAL.

CALCULATOR (USA)
Address: The Calculator Company,
Teaneck, New Jersey, USA.
History: c1945 a Mechanical and Hydraulic slide rule

CALCULIGRAPHE (France)
Address: Henri **CHATELAIN**, Paris, France.
History: the name of a Boucher style pocket-watch calculator made by **CHATELAIN**.

CAL TAPE (USA)
Address: Not known.
History: c1960 sold a measuring tape with a slide rule built into the tape, details not known.

CAMERON (USA)
Address: Cameron Iron Works, P.O. Box 1212, Houston, Texas, USA.
History: c1957 made a number of circular slide rules for oil production equipment.

CARBIC (GB)
Address: Carbic Limited (dept S.R.), 1921: 44 & 66 Bray's Lane, Coventry, England. (Otis C.F. King)
1922: "Ashdene", Emscote Road, Stoke, Coventry.
1923-38: 51 Holborn Viaduct, EC1
1942-45: 137 Conway Road, N14
1950: 171 Seymour Place, W1
1962-1971+: 54 Dundonald Road, SW19
>1962: Whitton Precision Limited, Bridge Works, Durnsford Road, Wimbledon London, SW19.
History: Sold the **OTIS-KING** Calculator, basic patent 1922, other patents 1923 including one with multi-scales and several cursors as well as one with multi scales and apertures in the cursor, also a desk models with replaceable scales.
Two main models: Model K and Model L.

Slide Rules:

1	2	3	4	5	6	7
Model A,	T	n/a	M	n/a	Retail Cash, Scales 1d. to £100, & 1 to 2,500 fractions	
Model B,	T	n/a	M	n/a	Retail cash, Scales 1d. to £100, & 1 to 2,500 decimals	
Model C,	T	n/a	M	n/a	Wholesale cash, Scales 1s. to £1,000 & 1 to 2,500 fractions	
Model D,	T	n/a	M	n/a	Wholesale cash, Scales 1s. to £1,000 & 1 to 2,500 decimals	
Model E,	T	n/a	M	n/a	Retail Cash, Scales 1d. to £100, & 1oz to 156 lbs.	
Model F,	T	n/a	M	n/a	Retail Cash, Scales 1d. to £100, & 1qr to 28 tons.	
Model G,	T	n/a	M	n/a	Wholesale cash, Scales 1s. to £1,000 & 1oz to 156 lbs.	
Model H,	T	n/a	M	n/a	Wholesale cash, Scales 1s. to £1,000 & 1qr to 28 tons.	
Model J,	T	n/a	M	n/a	Wages cash, Scales 1d to £8 and 1 hr. to 80 hrs.	
Model K,	T	n/a	M	n/a	Slide rule scales to 4 or 5 significant figures.	
Model L,	T	n/a	M	n/a	Slide rule scales plus a log scale.	
Model M,	T	n/a	M	n/a	Metrical Scales.	
Model N,	T	n/a	M	n/a	Sterling Cash numbers, English and French weights and measures ¼ to 2500 (later version)	
Fisher calculator	?	?	?	?	??	b.

Comment:

a. Came in 4 main styles, plus the ones from the patents, and one on a walking cane. Some all chrome/nickel, different features etc. White on Black and Black on White scales carry the same scale numbers.

b. Fisher calculator is mentioned in **DARGUE** catalogue under part No F.3676, otherwise nothing further known.

c. Scale numbers: K: 414, 423; L: 429, 430; Monetary (some) 17, 24.

CARBONNEL & LEGENDRE (France)
Address: Carbonnel & Legendre, Paris, France.
History: Mentioned as the maker on a **MARC, UNIS-FRANCE** slide rule, otherwise nothing known.

CARROLL (USA)
Address: J.B. Carroll Co. Chicago, IL, USA
History: c1950 onwards made a number of patented Flight Calculators.

CARROW (USA)
Address: A.M. Carrow, Houston 4, Texas, USA.
History: Simple folded plastic 5" slide rule. Probably manufactured by another company.

Slide Rules:

1	2	3	4	5	6	7
The "Midget"	12L	FP	Pl	plf	Folded Plastic rule.	

CASARTELLI (GB)

Address: 1917: Joseph Casartelli & Sons, 43 Market St. Manchester. 1927: 18 Brown St, Manchester. 1&3 Duke Street, Liverpool.
History: Established 1790. Sold the full range of **FOWLERS** watch type calculators for Fowlers.

Slide Rules:

1	2	3	4	5	6	7
Boucher's Calculating Circle.					(sold?)	
Fowler's circular calculators					sold by Casartelli	
Farmar's Profit Calculating rule					sold by Casartelli	c1938
R.F. Agnew's textile rule					made by Casartelli	c1927

CASELLA (GB)

Address: L.P. Casella, c1903, 147, Holborn Bars, London. Also, C.F. Casella & Co. Ltd., 1951, 49-50, Parliament Street, London SW1.
Factory: Wallsey St. Works, Walworth, London SE17.
History: One of the earliest manufacturers in England.

Slide Rules:

1	2	3	4	5	6	7
Anderson	10"L	CF	W/C	m/g	special extended multi-scales for additional accuracy, pat.	1903
Fuller calculator					made by Stanley, sold by Casella in 1951.	

CASSANI (Italy)

Address: Cassani Marcantoni, Milano, Italy.
History: See **MARCANTONI** otherwise not known.

CASIO (Japan)

Address: Casio Calculators. Japan.
History: A watch manufacturer who produced at least one "sports" model with a circular slide rule on a bezel round the face.

CASU (Spain)

Address: Barcelona, Spain.
History: Known from a wood/celluloid 12.5 Log-Log slide rule. Not known what their history is.

CENTRIC (Germany)

Address: Not known.
History: Took over the **KLAWUN** slide rule manufacturing activities on the death of Jos. Klawun in 1978.

CHADBURN (GB)

Address: Chadburn & Sons Ltd, London.
History: Not known, made Farmar's Spirit Rule.

CHADWICK (GB)

Address: J. Chadwick, London.

History: c1905, Chadwick's improved slide rule, otherwise history not known.

CHAPIN-STEPHENS (USA)

Address: Chapin-Stephens Co. Address not known
History: Made Routledge's Rule pre World War 1.

CHARVOZ (USA)

Address: Not known.
History: c1957 sold **ARISTO** slide rules in the USA. Also known to have sold other mathematical and drawing instruments. There was some link with **ROOS**. See **CHARVOZ-ROOS**.

CHARVOZ-ROOS (USA)
Address: Not known.
History: Known from a number of slide rules.

Slide Rules:

1	2	3	4	5	6	7
PSR-102	12.5L	?	?	?	?	
SR 105M	10"L	?	W	?	A/B,CI,C/D,K//S,L,T	
Trig-O-Log SR111	25L	Dup	Pl	?	LL0,LL00,A/B,K,CI,C/S,ST,T	
					L,LL1,DF/CF,CIF,C/D,LL3,LL2,LL1	

CHATELAIN (France)
Address: H. Chatelain, Paris, France.
History: According to Cajori, in 1878 they were making an Improved Boucher Calculator, the **CALCULIGRAPHE**.

Slide Rules:

1	2	3	4	5	6	7
Calculigraphe	2"W	n/a	M	n/a	2 sided watch calc	1878, a.

Comment:
a.   There were a number of variants with different cases, winders, hands, markings on the faces, but the same scales.

CHEMIC (USA)
Address: Not known.
History: A slide rule sold by **WEBER** c1923.

CIRCULATOR (GB)
Address: Circulator Co. Ltd., Grove Park, London N.W.9, England.
History: Also CIRCOLATOR. Made circular plastic slide rules. History not known.

CITIZEN (Japan)
Address: UK Office: Citizen Watch (U.K.) Ltd., P.O. Box 161, Wokingham, Berkshire, RG41 2FS.
Tel: 0118 989 0333; Fax: 0118 977 5110
History: Model WR100 and Model 8945-087836 slide rule watches with B and C scales on watch and bezel.

CLASSIC (GB)
Address: See **HARDTMUTH**
History: Made by Hardtmuth, specific Classic "Torch" logo (Advert December 1944). Slide rules very similar to Unique.

Slide Rules:

1	2	3	4	5	6	7
Series I	12L	CF	W/C	m/p	A/B,C/D	
Series I	25L	CF	W/C	m/p	A/B,C/D	
Series II	12L	CF	W/C	m/p	LU,A/B,C/D,LL	1.
Series II	25L	CF	W/C	m/p	LU,A/B,C/D,LL	1.
Series III	6"L	CF	W/C	m/p	Constants and gauge points for the timber trade on the back.	
Series III	25L	CF	W/C	m/p	Constants for the timber trade.	
Series IV	10"L	CF	W/C	m/p	S,A/B,CI,C/D,T	
Series V	10"L	CF	W/C	m/p	L,A/B,S,T,C/D,L	
(No Series Number):	25L	CF	W/C	m/p	LU,A/B,S,T,C/D,LL.	1.
-	12L	CF	Pl	pl	A/B/C/D	2.

Comment:
1.   The LU/LL scales are equivalent to LL2/LL3 scales on other slide rules.
2.   Identical format to the W/C, but made in a white plastic.

CLEMENTSON (USA)
Address: Not known
History: Not known

Slide Rules:

1	2	3	4	5	6	7
Pipe sizing rule.						

COLGATE (USA)
Address: A.E. Colgate. 36 Pine Street, New York, USA.
History: c1901 produced a slide rule for electrical calculations.

COLLARD (Australia)
Address: G.L. Collard & Sons. Australia.
History: Produced various slide charts for the motor trade.

Slide Rules:

1	2	3	4	5	6	7
Price Selector - BII					Miller Dial	
Price Selector - BIII-2						

COMPASS (USA)
Address: Compass Instrument and Optical Co., 104 East 25th. Street, New York 10, USA.
History: c1940. Sold **GILSON** circular rules,

Slide Rules:

1	2	3	4	5	6	7
"Binary" 8"	8"C	MD	Al	n/a	Double sided	1944
Gilson Midget	4"C	MD	Al	n/a		1931
SR-909	10"L	Dup	W/C	?	LL0,LL00,A/B,K,CI,C/S,S,T L,LL1,DF/CF,CIF,CI,C/D,LL3,LL2	

CONCENTRIC (GB)
Address: Concentric Indicator Laboratories,
36 Bourdon Street, Mayfair, London W1.

History: c1945. Sold a range of plastic multi-disc circular slide rules.

Slide Rules:

1	2	3	4	5	6	7
Fotogram, The Master photo exp. calc.	4"C	MD	Pl	pl	3D, 2C; photography scales, H&D, Scheiner; registered design applied for, 843,711	1944

CONCISE (Japan)
Address: Concise Corporation, Japan. also Concise Company Limited, Tokyo, Japan. Agent: Takeda Drawing Instruments Mfg Co Ltd.
History: Made a large range of Circular slide rules. These were of very high quality, and often included Company advertising material and/or Logos. c1960 were available in the USA with local offices. A number of slide rules designed by **SAMA & ETANI** were made by Concise.

Slide Rules:

1	2	3	4	5	6	7
28	8C	MD	Pl	n/a	Single sided, conversion tables on rear.	
320	9.6C	MD	Pl	n/a	as 28.	
260	9.6C	MD	Pl	n/a	Double sided.	
270	10C	MD	Pl	n/a	- " -	
271	10C	MD	Pl	n/a	- " -	
280	10C	MD	Pl	n/a	- " -	
300	11C	MD	Pl	n/a	Double sided, including log-log scales.	
Stadia Computer	10C	MD	Pl	n/a		
Type E,	2"C	MD	Pl	n/a	Japanese patent 381,371	1970
Type 28N	2.5"C	MD	Pl	n/a		
Type 57	2.5"C	MD	Pl	n/a		

EE-112	C	MD	Pl	n/a	Designed by Sama & Etani
CTCS 552	C	MD	Pl	n/a	Designed by Sama & Etani
600	C	MD	Pl	n/a	Designed by Sama & Etani

CONTROLLER (Germany)
Address: Controller, Munchen 15, Germany.
History: Made a number of circular slide rules.

Slide Rules:

1	2	3	4	5	6	7
-	C	MD	Pl	n/a	C/D,A,K	
Mod 75 R	C	MD	Pl	n/a	C/D,CI,B	
110 R	11C	MD	M	pl	3D,1C; C/D/CI,x.	
200 R	22C	MD	M	pl	C/D,CI,B	

COSMO (GB)
Address: Cosmo Rule, England.
History: Range of special converters under patent 2,563/69.

Slide Rules:

1	2	3	4	5	6	7
Series 1A	5"U	-	Pl	None	2 slides duplex/8 scales; U=5"x1.75" + 0.5" for slides	1969
Series 1B	5"U	-	Pl	None	8 slides duplex/8 scales; U=5"x1.75" + 0.5" for slides	1969

COX & STEVENS (USA)
Address: Cox & Stevens Aircraft Corp, Roosevelt Field, Mineola, New York, USA.
History: c1939, the Electronic Scales Division of the **REVERE** Corp of America, Wallingford, Connecticut, E1129, USA. Made a series of Aircraft Load Adjusters for a large range of American military aircraft. Known example of a metal flight direction and air speed calculator with Dutch markings which could show that they made rules for sale in the Benelux countries.

Slide Rules:

1	2	3	4	5	6	7
B-36D&F Load Adjuster	x"L	-	Pl	?	?	1969

CROWE (USA)
Address: Crowe Name Plate & Manufacturing Company, Chicago, Illinois, USA.
History: c1942/45 made a range of Airspeed Computers.

CRUVER (USA)
Address: Cruver Manufacturing Company, Chicago, Illinois, USA.
History: c1942/45 made a range of Airspeed Computers.

Slide Rules:

1	2	3	4	5	6	7
Time Distance D-4	4"C	MD	Pl	?	?	1969
Dead Reckoning E-6B	4"C	MD	Pl	?	special scales	
Altitude Correction AN 5837-1	8"C	MD	Pl	?	?	

DAEMEN-SCHMID (Switzerland)
Address: Johann Heinrich Daemen-Schmid, Zurich, Switzerland.
History: Formed c1900 and traded as Daemen-Schmid to c1915 when they became **LOGA**.

Slide Rules:

1	2	3	4	5	6	7
Rechenwalze, various sizes.					see **LOGA**	

DALLMEYER (GB)
Address: J.H. Dallmeyer, Church End Works, Willesden, London NW10.
History: Probably better known as scientific and optical instrument makers.

Slide Rules:

1	2	3	4	5	6	7
Wilson's double slide rule - shipping dimensions.						

DAMIEN (France)
Address: Ingénieur, A.M & E.S.E; 39 Rue des Mathurins, Paris, Usines, Saint Quentin (Seine).
History: Maurice Damien - Universal Rechenschieber Damien, Fabriziert in Frankreich. Unusual to have a French slide rule described in German, not known what the history or relationships are. The logo is the linked letters "d a m" in a circle. Thought to have only ever made one model of slide rule with an unusual metal back and a very fragile cursor. Possibly also used the brand name DAM.

Slide Rules:

1	2	3	4	5	6	7
-	27L	CF	W/C	pl	K,A/B,CI,C/D,L//S,ST,T; metal framework on rear.	

DANFOSS (Denmark)
Address: Norberg, Denmark.
History: They are pump and other central heating equipment manufacturers, in business to date.

Slide Rules:

1	2	3	4	5	6	7
Gas Pressure Conversion rule.						

DARGATZ (Germany)
Address: Albert Dargatz, Hamburg, Germany.
History: May have made the Visco calculator?

DARGUE (GB)
Address: Dargue Brothers Ltd. Simplon Works, South Street, Halifax. England.
also, New Simplon Works, South Parade, Halifax, England. (c1952?)
History: Were selling slide rules in 1931 including **SUN-HEMMI** rules, started making their own slide rules in about 1932 or 33. **SIMPLON** was the trade name used by Dargue.

Slide Rules:

1	2	3	4	5	6	7
In 1931 the Hemmi rules sold were:						
192/b/5/1	20"L	CF	B/C	m/g	Engineers (Standard), 55/-	
193/80/1	10"L	CF	B/C	m/g	Log Log Electrical, 30/-	
194/11	10"L	CF	B/C	m/g	Engineers, 21/-	
194a/8	5"L	CF	B/C	m/g	Divided as 10" rule with magnifying cursor, 21/-	
195/6	5"L	CF	B/C	m/g	Engineers, 12/6d	
195a	10"L	CF	P	??	Students cardboard slide rule, 2/6d	

Various spare cursors, magnifying, digit registering, cases etc available. The cursor was the semi-circular metal edged with glass.

Also A.W Faber's as follows:					
No 339 School pattern	10"L	CF	W/C	c	6/6d
No 360 Standard Engineers	10"L	CF	W/C	m/g	21/-
No 361 Students	10"L	CF	W/C	m/g	15/9d
No 369 Students	5"L	CF	W/C	m/g	14/6d
No 392 Rietz	12"L	CF	W/C	m/g	with L-L and CI scale, 28/-
No 398 Improved Electrical Engrs	12"L	CF	W/C	m/g	LL2,A/B,CI,C/D,LL1//S,T,L\\V,D, 31/6d
No 380 Standard Engineers	20"L	CF	W/C	m/g	55/-

The 1935 Dargue range was:					
SP10 "SIMPLON" Primary (Log-log)	10"L	CF	W/C	m/g	LU/A/B,CI,C/D,LL; 5/-
SP5 "SIMPLON" Primary (Log-log)	5"L	CF	W/C	m/g	LU/A/B,CI,C/D,LL; 3/9d

SB10 "SIMPLON" Bilateral (Log-log)	10"L	CFD	W/C	m/g	LU/A/B,CI,C/D,LL A,S,L,T,D; 7/6d	a.
SB5 "SIMPLON" Bilateral (Log-log)	5"L	CFD	W/C	m/g	LU/A/B,CI,C/D,LL A,S,L,T,D; 5/-	a.
WR1A "SIMPLON" Selecta	10"L	CF	W/C	pl	A/B,C/D; 8/6d	
WR2A "SIMPLON" Service	10"L	CF	W/C	pl	A/B,C/D; Constants table; 10/6d	
SE10 "SIMPLON" Electro	10"L	CF	W/C	pl	LL2,A/B,CI,C/D,LL;\\V,D/S,T; 12/6d; pat34598	
SD10 "SIMPLON" Duplex	10"L	CFD	W/C	pld	A/B,C/D S,L,T; 15/-	
WR5 Uniface Pattern	10"L	CF	W/C	pl	A/B,C/D,S,L,T; 20/-	
WR6 "SIMPLON" Standard	10"L	CF	W/C	m/g	A/B,C/D,L; 20/-	
WR7 Reitz	10"L	CF	W/C	m/g	K,A/B,C/D,L; 24/-	
SR1 "SIMPLON" Sub-Ivory	5"L	CF	W/C	m/g	K,A/B,C/D,L; 8/6d	
SR2 "SIMPLON" Sub-Ivory	5"L	CF	W/C	m/g	K,A/B,C/D,L; 10/6d; stouter material	
SR3 "SIMPLON" Sub-Ivory Uniface	5"L	CF	W/C	m/g	A/B,C/D,S,L,T; 12/6d	
SR4 "SIMPLON" Sub-Ivory Student's	5"L	CF	W/C	m/g	A/B,C/D, 17/6d	

Later catalogue shows only:

SP10 "SIMPLON" Primary Log-Log	10"L	CF	W/C	m/g	LU,A/B,CI,C/D,LL	
SB10 "SIMPLON" Bi-Lateral Log-Log	10"L	CFD	W/C	m/g	LU,A/B,CI,C/D,LL C,L,S,T	
Hemmi No 32	4"L	CF	Pl	plm		
Hemmi No 34RK	5"L	CF	Pl	pl		
Hemmi No 74 Rietz	5"L	CF	Pl	pl		
Hemmi No 136 Darmstadt	6"L	CF	PL	pl		
Hemmi No 40RK Students pattern	10"L	CF	Pl	pl		
Hemmi No 64 Rietz	10"L	CF	Pl	pl		
Hemmi No 80K Electrical	10"L	CF	Pl	pl		
Hemmi No 130 Darmstadt	10"L	CF	Pl	pl		
SIMPLON XL-10 Log-Log	10"L	CFD	W/C	m/g	LU,A/B,CI,C/D,LL C,L,S,T; identical to BILATERAL	b.
SIMPLON Major Log-Log	10"L	CF	W/C	pl	LU,A/B,CI,C/D,LL;//A,S,T,D/L,D	

Range of **FULLERTON** ACUDIAL Circular slide rules, made in Japan.

F3212 (No 280J) Trig 2	3.75"C	MD	Pl	3D2C	A,D/C,CI,B,K // L,D,S,T_1,T_2,ST; 10" Scale length, Fullerton 1454
F3211 (No 285) Basic	3.25"C	MD	Pl	3D2C	D/C,CI,A,K // Conv tables; 8" Scale length, Fullerton No 1450
F3213 (No 30J) Log Log	4.38"C	MD	Pl	3D2C	K,A,D,C,CI,B,L // LL_3,LL_2,D,C,S,T_1,T_2,ST; Fullerton 1458
F3214 Stadia	3.75"C	MD	Pl	3D2C	D/S,C_2, // D/C,S,T,ST,L; Fullerton No 1499

Also a range of **OTIS-KING** calculators (see **CARBIC**) as follows:

F3673 Model K
F3674 Model L
F3675 Leather sheath for above
F3676 Fisher Calculator — Nothing known about this calculator.

Comment:

a. The Bilateral is a strange Duplex rule. All rules are covered by patent 413,308 of 1934. They were still making slide rules to the late 1950's or early 1960's.

b. It is not known when the XL-10 appeared.

DATALIZER (USA)

Address: Datalizer Slide Charts Inc, 501 Westgate St, Addison, Illinois 60601-4524, USA. also Box 494, Addison, Illinois 60101, USA.
Phone: 708-543-6000; Fax: 708-543-1616

History: Producers of specialist calculators the same as **PERRYGRAF, SLIDE CHARTS LTD.** and others.

Slide Rules:

1	2	3	4	5	6	7
VSWR Calculator (Global Microwave)		P	none	specialist		?
Amplifier Performance Calculator (Amplifonix)		P	none	specialist		1986
Cablemate Helper (Mohawk/CDT)		P	none	specialist		1992

DAVIS (GB)

Address: John Davis & Son (Derby) Ltd., All Saints Works, Derby. also London Office at 39 Victoria Street., Westminster, SW1.

History: According to various adverts produced a wide range of slide rules. Also known to have sold other manufacturers rules including **DENNERT AND PAPE** and **THORNTON P.I.C.** Ceased manufacture c1941, may have sold other manufacturers slide rules afterwards.

Slide Rules:

1	2	3	4	5	6	7
Mannheim	10"L	CF	CF	?		
Simplex	L	?	?	?		1902
Electrical (Trician)	10"L	CF	C	?	Dynamo,LL,A/B,K,C/D,LL,Volt;	c1915
Stelfox	5"L	?	W/C	?	5" rule with 10" slide (jointed) A/B, C/D in 2 sections	c1915
Jakin	11"L	?	W/C	?	for surveyors, uses short subsidiary scales	c1915
Davis-Lee-Bottomley					circle spacing	c1915
Davis-Scurfield Split Unit slide rule						c1925
Cuntz					2.25" wide with 11 scales above A,	
					slide near bottom of rule	c1915
Davis log-log						c1911-38
Jackson Davis Double slide						c1915-38
Dunlop Jackson celluloid slide rule	10"L					1901
Dunlop-Jackson	10"L				two slides, additional Log-log;	c1910
Dunlop-Jackson	20"L				two slides, additional Log-log;	c1910
Yokota	10"L	CF	C	m/g	first real log-log slide rule	c1915-38
Autocrat	10"L	CF	C	m/g	A/B,C/D//S,T,L	c1925
Aristocrat	5"L	CF	C	m/gm	A/B,C/D//S,T,L	c1925
Bureaucrat	5"L	CF	W/C	m/g	A/B,C/D	c1925
Democrat	15"L	CF	C	m/g	A/B,C/D//S,T,L	c1925
Plutocrat	10"L	CF	C	m/g	(effectively Reitz); E,A/B,C/D,F//S,T	c1925
Durocrat	20"L	CF	C	m/g	A,/B,C/D//S,T,L	c1925
Monocrat	10"L	CF	C	m/g	S,A/B,T,C/D,L	c1925
Daffield	10"L	CF	C	m/g	as Monocrat with a 20th scale alongside L for £,s,d calcs	c1938
Commerce					A/B,L,£sd,C/D. Slide is wider than usual.	c1938
Smith-Davis Piecework Calculator	11'U				circular/cylindrical, scales for commercial applications	c1915-38
Smith-Davis Premium Calculator	12.5"U				12.5"dia, gives 4'6" scales similar to previous	c1915-38
Glider	10"L	CF	C	m/g	F,A/B,C/D,E,//S,T,L;	
					Patent Slotted Slide, No 232,037. (1925)	c1930
Davis Stokes Field Gunnery	8"L	CF	W/C	?	2 slides; 19 scales "encounter" & "entrenched"	c1938
Davis Martin Wireless	10"L	CF	W/C	?	wireless scales	c1938-45
Wingham's slide rule	L				for furnace charges;	c1910
Student	10"L	CF	W/C	m/g	A/B,C/D	c1910
Celluloid	10"L	CF	C	?	3 or 5 adjusting screws;	c1910
Celluloid	15"L	CF	C	?	3 or 5 adjusting screws;	c1910
Celluloid	20"L	CF	C	?	3 or 5 adjusting screws;	c1910
Tacheometric	20"L	CF	W	?	360 or 400 degree, rule designed by Werner, Vienna.	c1910
1C (Onesee)	10"L	CF	W?	?	no adjusting screws;	c1910
1C (Twensee)	20"L	CF	W?	?	no adjusting screws;	c1910
Grinstead	15"L	CF	W/C	?	complex calculations	c1925
Pickworth Power Computer	L	CF	P	n/a	card version;	c1910
Pickworth Power Computer	L	CF	W/C	n/a	Ivorine scales version;	c1910
Machine Gun slide rule	U	CF	W	n/a	U=6"x3", two slides, for M.G. Mk.III.	c1941
PIC3644	10"L	CF	W/C	m/g	has John Davis & Son stamped into the well of the rule	

DECHESNE (Belgium)

Address: A. Dechesne, Belgium.

History: Not known.

Slide Rules:

1	2	3	4	5	6	7
Modèle Déposé	25L	CF?	P	?	K,A/B,C/D,L	

DEMPSTER (USA)

Address: J.R. Dempster, Manufacturer, Berkeley, California, USA.

History: c1929, 50" pocket rule, the **ROTARULE** range of slide rules with patents 1,849,058 and 1,899,616(?) approx 1930. Two models, A and AA, details not known.

Slide Rules:

1	2	3	4	5	6	7
Rotarule Pocket Slide Rule	4.5"C	MD	M	C	multi-scale with and without magnifiers	

DENNERT & PAPE or D&P (Germany)

Address: Dennert & Pape, Altona, Hamburg.

History: Often seen as a simple **D&P** or **DUPA** on their early slide rules. In 1925 became better known as Aristo, that being the name given to their new range of all-plastic slide rules. - see **ARISTO**.

1872 - first boxwood slide rules produced to designs of A. Goering; under the name Dennert & Pape Mechanical-Mathematical Institute

1879 - brass slide rules offered in addition to wooden designs.

1886 - DRP 34583 for wooden scales veneered with celluloid.

1888 - first wood/celluloid slide rules produced.

1890 - metal/glass cursors replace all-metal wing cursors.

1902 - first system Rietz slide rules.

1905 - First catalogue with 12 slide rules.

1924 - DUPA introduced as the D&P trademark.

1926 - Rietz standard enhanced with CI & ST scales, and with extensions to A,B,C,D scales.

1934 - Darmstadt standard added.

1936 - ARISTO introduced as the trademark for a new range of plastic slide rules.

1872 - 1910, no catalogue or identity numbers used.

1910 - 1920, catalogue numbers used, 32 rules in the catalogue, some use same numbers for different sizes.

Slide Rules:

1	2	3	4	5	6	7
Rechenstab von Dennert und Pape	L	?	?	?		1874
Jordan Rechenschieber	L	?	?	?		1897
-	5"L	CF	W/C	Pl	A/B,C/D//S,L,T,LL DRP126,499	1901
Simplex Taschenrechenschieber	5"L	CF	W/C	?		1903
Neuer Prazisions-Rechenstab	L	CF	W/C	?		1908
Mannheim Slide Rule	25L	CF	W/C	m	A/B,C/D//S,L,T	1872-1914
No 1, Mannheim slide rule	25L	CF	W	m	A/B,C/D//S,L,T	
2.28	25L	CF	W	?	A/B,C/D//??; DUPA tm,	1920-36
2/28, Mannheim?	25L	CF	W	?	A/B,C/D//??; DUPA tm,	1920-36
No 8, Cube slide rule System Rietz	12.5L	CF	W/C	?	K,A/B,C/D,L//S,T; self adjusting body,	1902-20, a.
No 8, Cube slide rule System Rietz	25L	CF	W/C	?	K,A/B,C/D,L//S,T; self adjusting body,	1902-20
No 8, Cube slide rule System Rietz	50L	CF	W/C	?	K,A/B,C/D,L//S,T; self adjusting body,	1902-20
No 9, Cube slide rule System Rietz	12.5L	CF	W/C	?	K,A/B,C/D,L//S,T; screw adjustment,	1902-20, a.
No 9, Cube slide rule System Rietz	25L	CF	W/C	?	K,A/B,C/D,L//S,T; screw adjustment,	1902-20
No 9, Cube slide rule System Rietz	50L	CF	W/C	?	K,A/B,C/D,L//S,T; screw adjustment,	1902-20
Yokota	25L	CF	W/C	m/g	6 log-log scales; exported to GB,	1907-18
Improved exponential	25L	CF	W/C	m/g	as Yokata,	1918-38
-	50L	CF	W	m	surveying scales, 360^0 & 400^g,	1882
No 8/28, System Rietz	25L	CF	W/C	m/g	K,A/B,C/D//S,T; self adjusting body;	1920-34
No 9/28, System Rietz	25L	CF	W/C	m/g	K,A/B,C/D//S,T; screw adjusters;	1920-34
No 9/28, System Rietz	25L	CF	W/C	m/g	A/B,CI,C/D//S,ST,T; extns to A,B,C,D; screw adjusters;	1925
B8c Rietz	15L	CF	W/C	m/g		1920-26
K&E1744 (made by D&P)	25L	Dup	W/C	m	A/B,C/D; double 2 wing cursor A/BI,CI/D	1895
Azimuth Slide Rule	L	??	W/C	?	?? by R. Nelting; Patent?	
Rietz slide rule	25L	CF	W/C	m/g	K,A/B,C/D,L//S,T	1902
No 15 Log-Log	25L	CF	W/C	m/g	first log-log design	1905
Rietz slide rule (improved)	25L	CF	W/C	m/g	K,A/B,CI,C/D,L//S,ST,T	1926
Dufix 955	25L	Dup	?	?	?	1925-38
Dufix 966	25L	Dup	?	?	?	1925-38

Comment:

a.   K is called F, L is called E. An example is labelled R.Reiss, Liebenwerda.

DEVCO (USA)

Address: Device Development Co. New York, USA.

History: Not known.

Slide Rules:

1	2	3	4	5	6	7
Pencil slide rule	P	n/a	Pl	pl	ins,A/B,CI,C/D,L,K, slides in both directions. $5.00	1956

DIETRICH-POST (USA)
Address: Not known
History: c1959, associated with Post, also known as **POST** separately with a range of slide rules. See also **TELEDYNE POST**.

DIETZGEN (USA)
Address: Eugene Dietzgen Co. New York, USA., also, Chicago, USA.
History: Probably the second slide rule supplier in the USA, set up in 1887, sold **GILSON** circular slide rules as well as many other manufacturers rules as well as their own large range of rules. In existence to at least late 1972 selling slide rules, and continue to date.

Slide Rules:

1	2	3	4	5	6	7
Mack's slide rule	10"L					1898
Thacher-Schofield	10"L					1901
Union (Mannheim)	10"L					1905
Rosenthal's slide rule	10"L					1906
1731, Maniphase Log-Log Trig	10"L			fv		1941-52
1732, Maniphase Log-Log decimal trig	10"L			fv		1941-52
1732, Log-Log Duplex Decitrig	10"L					
1733, Decimal Trig Log-Log microglide	10"L					1970
1735, Maniphase Log-Log Vector	10"L			fv		1941-52
1736, Multilog Multiplex Dec Trig L-Log	10"L			ffv		1955
1737	10"L					1972
1738, Decimal Trig type Log-Log (metal)	10"L			fv		1955-70
1739L, Clear scale Log-Log (plastic)	10"L		Pl			1959-72
1741	10"L					1972
1742, Maniphase Multiplex Decimal Trig	10"L					
1743, Maniphase Multiplex Trig	10"L			fv		1941-59
1744, Maniphase Multiplex Decimal Trig	10"L			fv		1941-55
1744, Polymath	10"L					1955-70
1745, Maniphase (Duplex)	10"L					
1746, Maniphase Multiplex	10"L			fv		1941-52
1748P, Maniphase	8"L			f		1928-41
1749,	10"L?					
1750P, Maniphase	10"L			f		1928-41
1751, Industrial	10"L			ffv		1941-59
1752P, Maniphase	20"L			f		1928-41
1755P, Mannheim	5"L			f		1928-39
1756, Commerce	10"L					1928, 49-52
1756-20, Commerce desk model	20"L					1928
1757, Prima Trig (plastic)	10"L		Pl			1955-59
1758P, Mannheim	8"L			f		1928-41
1759A, 1738P, Philips	8"L			f		1919-27, 28-41
1759B, 1772P, 1740P, Philips	10"L			f		1919-28, 28-41
1759C, 1742P, Philips	20"L			f		1919-27, 28-41
1760P, 1766P, Mannheim	10"L			f		1928-41
1760A, Multiplex, Dietzgen adjuster	5"L			g		1910-21
1760AL, Multiplex, Dietzgen adjuster	8"L			g		1910-12
1760B, Multiplex, Dietzgen adjuster	10"L			g		1910-21
1760BL, Multiplex, Dietzgen adjuster	16"L			g		1910-12
1760C, Multiplex, Dietzgen adjuster	20"L			g		1910-12
1761, Maniphase Mannheim	10"L			f		1941-52
1761-20, Maniphase Mannheim	20"L			f		1941-52
1761A, Multiplex	5"L			g	No cube scale	1910-12
1761AL, Multiplex	8"L			g	No cube scale	1910-12
1761B, Multiplex	10"L			g	No cube scale	1910-21
1761BL, Multiplex	16"L			g	No cube scale	1910-12
1761C, Multiplex	20"L			g	No cube scale	1910-12
1762A, Multiplex	5"L			g	Mack adjustment	1905-12
1762B, Multiplex	10"L			g	Mack adjustment	1905-12

Name	Length		Material	Type	Description	Years
1762C, Multiplex	20"L			g	Mack adjustment	1905-12
1762P, Mannheim	20"L			f		1928-41
1762AL, Multiplex	8"L			g	Mack adjustment	1907-12
1762BL, Multiplex	16"L			g	Mack adjustment	1907-12
1763A, Multiplex, reciprocal scale	5"L			g		1905-12
1763B, Multiplex, reciprocal scale	10"L			g		1905-12
1763C, Multiplex, reciprocal scale	20"L			g		1905-12
1763P, Monarch	10"L			f		1938/9
1763AL, Multiplex, reciprocal scale	8"L			g		1907-12
1763BL, Multiplex, reciprocal scale	16"L			g		1907-12
1764, Mack Improved	5"L					1902-12
1764, Maniplex Mannheim	10"L					1949-52
1764A, 1764M, Pocket rule, metal	5"L		Al	nb		1939, 1938/9
1765, Mack Improved	10"L					1902-12
1765F, Pocket rule, celluloid	5"L		C	nb		1928-41
1765L, Mack Improved	16"L			g		1907-12
1765P, Pocket rule, boxwood	5"L		W	nb		1938-41
1765P, Basik (plastic)	10"L		Pl			1955-72
1766, Scholar	10"L					1949
1766P, Scholar, white wood	10"L		W	f		1939, 55-59
1766PN, Beginners	10"L			f		1938/9
1767, National, plastic	10"L		Pl	n/a		1955-59
1767, Mack Improved	20"L					1902-12
1767-P, National, plastic	10"L		Pl	nb		1941-52
1767-H, Apprentice	10"L			n/a		1941
1768, Dietzgen Improved	5"L			g		1910-12
1768L, Dietzgen Improved	8"L			g		1910-12
1768, -L, Mannheim	5"L			f		1919-28
1768P, Reliance	10"L			nb		1941-52
1769, Dietzgen Improved	10"L			g		1910-12
1769L, Dietzgen Improved	16"L			g		1910-12
1769P, Federal	10"L		W/C	nb	Celluloid on cherrywood	1941-46
1769, Mannheim	10"L			f		1919-28
1770, Mannheim	20"L			f		1919-28
1770, Standard adjustable	5"L			g		1904/5
1770, Dietzgen Improved	20"L			g		1910-12
1770P, Union	10"L			f		1928-36
1770T, Langsner Industrial	10"L	CF	W/C	f	Feed,Diam/R.P.M,Time(BI),Cutting Speed/D\K	1939, 41, 47-55
1771, Standard adjustable	10"L			g		1904/5
1771, 1776, Redirule, plastic	5"L		Pl	pl		1941-72
1771A, Economy (Ivorene)	10"L		Pl	g		1910-12
1771B, Economy (hardwood)	10"L		W	g		1910-12
1772, Standard adjustable	20"L			g		1904/5
1772A, Union	5"L			g		1910-12
1772A, Union Pocket	5"L			f		1919-28
1772B, Union	10"L			g		1910-12
1772B, Union	10"L			f		1919-28
1772A	10"L					1972
1773, Log-Log, plastic	5"L		Pl	nb		1955-72
1773A, Scholar, cardboard	5"L		P	m		1926-28
1773B, Scholar, cardboard	10"L		P	m		1926-28
1774F, Celluloid pocket rule	6"L		C	n/a		1952
1774M, Light metal pocket rule	5"L		Al	n/a		1952
1775P, Scholar, cardboard	5"L		P	m		1928
1776, Union	5"L					1902-8
1777, Union	10"L					1902-8
1778, College, Paper scales	10"L					1910-12
1778, Redilog, plastic	5"L		Pl	n/a		1970-72
1778A, Hazen-Williams	10"L			g		1910-12
1778B, Universal	10"L			g		1910-12
1779, Faber	5"L			g		1910-12
1780, Faber	10"L			g		1907-12
1780P, Scholar, cardboard	10"L		P	m		1928
1781, Faber self adjusting	10"L			g		1907-12
1782, Faber without lateral lines	10"L			g		1907-12
1783, Faber with digit reg cursor	10"L			g/dr		1907-12

1783A, Faber electrical & mechanical	10"L		g			1910-12
1784, Faber with digit reg cursor	20"L		g/dr			1907-12
1785/95, College	10"L		g			1902-08
1787, Engineers slide rule	24"L	CF	W	n/a	A/B,C/B//A/B,C/B; Thatcher-Schofield	1901-30
1792, Engineers slide rule					as 1787	1928
1793, Mack Improved	10"L					1897/8
1793K&L, Otis-King	T				see **CARBIC**	1938 & 41
1794, Mack Improved	20"L					1897/8
1794, Fuller	T					1902-41
1795A, Halden Calculex	2.3"C				see **HALDEN**	1910-55,b.
1795B, Rotorule	3"C				see **DEMPSTER**	1912-19 & 21
1796, Thacher	T					1897/8
1796, Charpentier Calculator	6C	MD	M	m	C/D,√,//L,T,S	1904-31
1797, Boucher, card dial	2"W	n/a	n/a	n/a		1902-36
1797 1/2, Boucher, silvered metal dials	2"W					1904-31
1797A, Atlas	8.3"C				See **GILSON**	1948-72
1797B, Binary	8"C				See **GILSON**	1941-72
1797M, Midget	4"C				See **GILSON**	1938-72
1798, Rota-Rule	5"C				see **DEMPSTER**	1936-39 & 41-49
Multiplex	10"L	CF?	??	??	A/B(rev),C/D\K	c1915-45
N1725, Vector type Log-Log	10"L			ffv		1955-59
N1725, Vector type Log-Log microglide	10"L					1970
N1731, Maniphase Trig type Log-Log	10"L			ffv		1955-59
N1732, Maniphase decimal trig Log-Log	10"L			ffv		1955-59
N1733, Decimal Trig type Log-Log	10"L			ffv		1955-70
N1749, Maniphase decimal trig (metal)	10"L		Al	fv		1955-70
N1767	10"L					1972
B1734	10"L					
B1734L	10"L					1972

Comments:

a.   Introduced 1896, shown in Dietzgen catalogues 1907-12.
b.   Sold with some breaks for this period.
c.   fv is ful-vue, ffv is frameless ful-vue for cursors.

DIMIER (France)

Address: Dimier, 18 Rue E. Bloch, Epinay Sur Seine, France.
History: Not known, other than that they made an Electrical slide rule.

Slide Rules:

1	2	3	4	5	6	7
Electrical Slide Rule	15L					

DING FEN (China)

Address: Shanghai? China.
History: Not known, other than that they were a major Chinese manufacturer. Logo is a stylised flower in a triangle.

Slide Rules:

1	2	3	4	5	6	7
-x-	12.5L	Dup	B/C	m/p		
5471 Duplex Vector	25L	Dup	B/C	m/p	Th,L,K,A/B,S,H,T,C/D,DI,Sh2,Sh1 LL01,LL02,LL03,DF/CF,CIF,H,CI,C/D,LL3,LL2,LL1	
-	25L	Dup	B/C	m/p	LL01,LL02,LL03,DF/CF,CIF,H,CI,C/D,LL3,LL2,LL1 2D,L,K,A/T,S,H,CI,C/D,Th,Sh1,Sh2	

DINSDALE (GB)

Address: James Dinsdale Limited, 1960, New Station St, Leeds 1, England, also, 94, Woodhouse Lane, Leeds 2.
History: Wholesalers only, established 1883. In Sept 1960 sold **PIC, FABER-CASTELL, BRL, ARISTO, FOWLER, UNIQUE** and **O-K** calculators (at 62/6d) as follows:

Slide Rules:

1	2	3	4	5	6	7
P.I.C. Slide rules as follows:						
F3654 Standard	10"L	CF	W/C	m/g	A/B,CI,C/D//S,L,T; 50/-	
VP3656 Standard	5"L	CF	Pl	m/g	A/B,CI,C/D//S,L,T; 33/-	
F3756 Standard	20"L	CF	W/C	m/g	A/B,CI,C/D//S,L,T; £7/17/6d	
F4458 Rietz	10"L	CF	W/C	m/g	K,A/B,CI,C/D,L//S,ST,T; 60/-	
VP114 Rietz	5"L	CF	Pl	m/g	K,A/B,CI,C/D,L//S,ST,T; 30/-	
F4459 Rietz	20"L	CF	W/C	m/g	K,A/B,CI,C/D,L//S,ST,T; £8/15/0d	
No131 Comprehensive Log-Log	10"L	CF	W/C	m/g	LL1,LL2,A/B,CI,Is,Vs,C/D,P//S,ST,T\\Volt,Dyn; 72/6d	
VP134 Comprehensive Log-Log	5"L	CF	Pl	m/g	LL1,LL2,A/B,CI,Is,Vs,C/D,P//S,ST,T\\Volt,Dyn; 39/-	
No132 Comprehensive Log-Log	20"L	CF	W/C	m/g	LL1,LL2,A/B,CI,Is,Vs,C/D,P//S,ST,T\\Volt,Dyn; £9/5/0d	
No121 Comprehensive Log-Log	10"L	CF	W/C	m/g	LL1,LL2,A/B,CI,Is,Vs,C/D,P//S,ST,T\\Vector; 60/-	
VP124 Comprehensive Log-Log	5"L	CF	Pl	m/g	LL1,LL2,A/B,CI,Is,Vs,C/D,P//S,ST,T\\Volt,Dyn; 32/6d	
No122 Comprehensive Log-Log	20"L	CF	W/C	m/g	LL1,LL2,A/B,CI,Is,Vs,C/D,P//S,ST,T\\Volt,Dyn; £8/10/0d	
P170 Standard Student	10"L	CF	Pl	m/g	A/B,CI,C/D,L; 19/3d	
P171 Rietz Student	10"L	CF	Pl	m/g	K,A/B,CI,C/D,L//S,ST,T; 23/9d	
P172 Log-Log Student (Different'l)	10"L	CF	Pl	m/g	K,A/B,CI,C/D,L//S,ST,T; 36/-	
P172E Log-Log Student (Electrical)	10"L	CF	Pl	m/g	K,A/B,CI,C/D,L//S,ST,T; 36/-	
Faber-Castell rules as follows:						
1/60 Standard	10"L	CF	W/C	m/g	A/B,C/D; 50/-	a.
1/87 Rietz	10"L	CF	W/C	pl	K,A/D,CI,C/D,L//S,ST,T; 58/6d	a.
4/87 Rietz	20"L	CF	W/C	pl	K,A/D,CI,C/D,L//S,ST,T; 160/-	
1/98 Electrical	10"L	CF	W/C	m/g	LL1,A/B,CI,C/D,LL2//S,T,L\\Volt,Dynamo; 65/-	a.
4/98 Electrical	20"L	CF	W/C	m/g	LL1,A/B,CI,C/D,LL2//S,T,L\\V,D; 175/-	
1/54 Darmstadt	10"L	CF	W/C	m/g	L\K,A/B,CI,C/D,P//S,ST,T; 72/-	a.
4/54 Darmstadt	20"L	CF	W/C	m/g	L\K,A/B,CI,C/D,P//S,ST,T; 186/-	
1/92 Precision Log-Log	10"L	CF	W/C	?	A/B,CI,C/D//S,ST,T; 65/-	
1/22 Business	10"L	CF	W/C	?	60/-	
4/22 Business	20"L	CF	W/C	?	120/-	
57/74 Textile	10"L	CF	Pl	pl	65/-	
111/66 Demegraph	10"L	CF	Pl	pl	Printing & Paper trade; 82/-	
57/87 Students Rietz	10"L	CF	Pl	pl	27/6d	
57/92 Log-Log	10"L	CF	Pl	pl	33/-	
57/88 Advanced Rietz	10"L	CF	Pl	pl	33/-	
57/89 Super Log-Log	10"L	CF	Pl	pl	38/-	
52/82 Duplo Log-Log	10"L	Dup	Pl	pl	L,K,A/B,CI,C/D,LL2,LL3 DF/CF,-----; 45/-	
67/39 Standard	5"L	CF	Pl	pl	A/B,C/D; 26/-	
67/98B Electrical	5"L	CF	Pl	pl	LL1,A/B,CI,C/D,LL2//S,L,T\\Volt,Dynamo; 36/-	
67/54B Darmstadt	5"L	CF	Pl	pl	L\K,A/B,CI,C/D,P//S,ST,T; 40/-	
67/91 Standard	5"L	CF	Pl	pl	A/B,C/D//S,L,T; 28/-	
67/87 Rietz	5"L	CF	Pl	pl	K,A/D,CI,C/D,L//S,ST,T; 30/-	
67/22 Business	5"L	CF	Pl	pl	Commercial slide rule; 25/-	

Availability of calculators with the Addiator mentioned.

1	2	3	4	5	6	7
B.R.L. Rules as follows:						
401 Rietz	10"L	CF	Pl	pl	K,A/B,CI,C/D/L//S,ST,T; 58/-	b.
601 Rietz	5"L	CF	Pl	pl	K,A/B,CI,C/D/L//S,ST,T; 28/-	b.
405 General Purpose	10"L	CF	Pl	pl	A/B,C/D; 38/-	b.
605 General Purpose	5"L	CF	Pl	pl	A/B,C/D; 21/-	b.
406 General Purpose	10"L	CF	Pl	pl	A/B,CI,C/D//S,ST,T; 49/6d	b.
606 General Purpose	5"L	CF	Pl	pl	A/B,CI,C/D//S,ST,T; 25/-	b.
402 Log-Log	10"L	CF	Pl	pl	LL2,A/B,CI,C/D,LL3//S,ST,T; 58/-	b.
602 Log-Log	5"L	CF	Pl	pl	LL2,A/B,CI,C/D,LL3//S,ST,T; 28/-	b.
407 Technico	10"L	CF	Pl	pl	K,LL1,LL2,DF/CF,CIF,CI,C/D,P,LL3,A//S,ST,T,L; 75/-	b.
403 Darmstadt	10"L	CF	Pl	pl	L,K,A/B,CI,C/D,P,S,T//LL1,LL2,LL3; 67/6d	b.
603 Darmstadt	5"L	CF	Pl	pl	L,K,A/B,CI,C/D,P,S,T//LL1,LL2,LL3; 35/-	b.
404 Electro	10"L	CF	Pl	pl	LL2,A/B,CI,C/D,LL3//S,L,T\\Volt,Dyn; 65/-	b.
604 Electro	5"L	CF	Pl	pl	LL2,A/B,CI,C/D,LL3//S,L,T\\Volt,Dyn; 31/6d	b.
Academy 300 Standard	10"L	CF	Pl	pl	22/6d	c.
Academy 302 Log-Log	10"L	CF	Pl	pl	LL2,A/B,CI,C/D,LL3//S,ST,T; 30/-	c.
Academy 303 Rietz	10"L	CF	Pl	pl	K,A/B,CI,C/D/L//S,ST,T; 25/-	c.
Academy 304 Darmstadt	10"L	CF	Pl	pl	L,K,A/B,CI,C/D,P,S,T//LL1,LL2,LL3; 37/6d	c.

Academy 305 Electrical	10"L	CF	Pl	pl	42/6d	c.
Academy 307 Navigational	10"L	CF	Pl	pl	35/-	c.

Aristo rules as follows:

810 Lilliput	5"L	CF	Pl	pl	27/6d
812 Simplex	5"L	CF	PL	pl	A/B,C/D; 27/6d
89 Rietz	5"L	CF	Pl	pl	K,A/B,CI,C/D/L//S,ST,T; 32/6d
814 Electrical	5/6L	CF	Pl	pl	43/-
867 Darmstadt	5"L	CF	Pl	pl	K,A/B,BI,CI,C/D,P,S,T//L,LL1,LL2,LL3; 47/6d
868 Studio	5"L	Dup	Pl	pl	T,ST,DF/CF,CIF,CI,C/D,L LL01,LL02,LL03,A/B,L,K,C/D,LL3,LL2.LL1; 51/6d
829 Multirietz	5"L	Dup	Pl	pl	K,DF/CF,CIF,CI,C/D,L P,A/B,T,ST,S/D,DI; 45/-
903 Scholar	10"L	CF	Pl	pl	27/6d
903 Scholar Log-Log	10"L	CF	Pl	pl	37/6d
0911 Simplex	10"L	CF	Pl	pl	19/6d
929 Multirietz	10"L	Dup	Pl	pl	K,DF/CF,CIF,CI,C/D,L P,A/B,T,ST,S/D,DI; 70/-
945 Commercial	10"L	CF	Pl	pl	63/-
965 Commercial	10"L	CF	Pl	pl	72/-
99 Rietz	10"L	CF	Pl	pl	K,A/B,CI,C/D/L//S,ST,T; 59/-
915E Electrical	10"L	CF	Pl	pl	75/-
967 Darmstadt	10"L	CF	Pl	pl	K,A/B,BI,CI,C/D,P,S,T//L,LL1,LL2,LL3; 77/6d
868 Studio Duplex	10"L	Dup	Pl	pl	T,ST,DF/CF,CIF,CI,C/D,L LL01,LL02,LL03,A/B,L,K,C/D,LL3,LL2.LL1; 84/-
970 Multilog	10"L	Dup	Pl	pl	LL01,LL02,LL03,DF/CF,CIF,L,CI,C/D,LL3,LL2,LL1 LL00,K,A/B,T,ST,S,C/D,DI,LL0; 95/-
971 Hyperbolog vector	10"L	CF	Pl	pl	105/-
958 Surveyor	10"L	CF	Pl	pl	110/-
939 Reinforced concrete	10"L	CF	Pl	pl	117/6d
1045 Commercial	20"L	CF	Pl	pl	195/-
109 Rietz	20"L	CF	Pl	pl	K,A/B,CI,C/D/L//S,ST,T; 195/-
1067 Darmstadt	20"L	CF	Pl	pl	K,A/B,BI,CI,C/D,P,S,T//L,LL1,LL2,LL3; 245/-
1068 Studio Duplex	20"L	Dup	Pl	pl	T,ST,DF/CF,CIF,CI,C/D,L LL01,LL02,LL03,A/B,L,K,C/D,LL3,LL2.LL1; 275/-
3/150 Scholar	72"L	CF	Pl	pl	239/-
9/150 Rietz	72"L	CF	Pl	pl	K,A/B,CI,C/D/L//S,ST,T; 295/-
5/150 Commercial	72"L	CF	Pl	pl	239/-
67/150 Darmstadt	72"L	CF	Pl	pl	K,A/B,BI,CI,C/D,P,S,T//L,LL1,LL2,LL3; 317/-
68/150 Studio Duplex	72"L	Dup	Pl	pl	T,ST,DF/CF,CIF,CI,C/D,L LL01,LL02,LL03,A/B,L,K,C/D,LL3,LL2.LL1; 425/-

Also **Fowlers** calculators

Vest pocket model	2.5"W	2 dials; 55/-
Universal Model	3.25"W	55/-
"12-10"	3.25"W	for architects and builders, 55/-
Magnum Long scale	4.5"W	72/-
Jubilee Magnum	4.5"W	79" scale length, 72/-

Unique slide rules

Universal 1	10"L	CF	W/C	pl	13/-
Universal 1/1	5"L	CF	W/C	pl	14/-
Universal II	10"L	CF	W/C	pl	13/-
Universal 1/2	5"L	CF	W/C	pl	9/6d
Log-Log	10"L	CF	W/C	pl	9/6d
Log-Log	5"L	CF	W/C	pl	6/9d
Legible	10"L	CF	W/C	pl	9/6d
Legible	5"L	CF	W/C	pl	6/9d
10/20 Precision	10"L	CF	W/C	pl	13/-
5/10 Precision	5"L	CF	W/C	pl	6/9d
Commercial	10"L	CF	W/C	pl	13/-
Electrical	10"L	CF	W/C	pl	13/-
Navigational	10"L	CF	W/C	pl	13/6d
D1 Dualistic	5"L	CF	W/C	pl	14/-

D2 Dualistic	5"L	CF	W/C	pl	9/6d
D3 Dualistic	10"L	CF	W/C	pl	23/-
Brighton	10"L	CF	W/C	pl	27/6d
Thin pocket (5 models)	5"L	CF	W/C	pl	11/6d, scaled as L-L, 5-10, 5G, U1/2 and D2

Otis-King

Standard Model K		62/6d
Logarithm Model L		62/6d

Comments:

a.   All available with magnifying cursor at extra price

b.   B.R.L. rules were all supplied with "Personal Tension Adjustment" (Patent 644,944)

c.   Academy series introduced about 1956, all available with various types of magnifier.

DIWA (Denmark)

Address: Diwa Manufacturing Company, Copenhagen, Gentofte, Denmark.

History: Not known.

Slide Rules:

1	2	3	4	5	6	7
From the Lawes Rabjohns Technical Catalogue - 1952						
DIWA Slide Rules						
Commercial Pattern	10"L	CF	W/C	pl	A/B,CI,C/D//S,T,L	
Rietz Pattern	10"L	CF	W/C	pl	K,A/B,CI,C/D,L//S,T,L	
Rietz Pattern	10"L	CF	Pl	pl	K,A/B,CI,C/D,L//S,T,L	
Log-Log pattern	10"L	CF	W/C	pl	LL1,A/B,CI,C/D,LL2//S,T,L	
Darmstadt pattern	10"L	CF	W/C	pl	LL0,LL1,A/B,CI,C/D,LL2//S,T,L	
Electro Pattern	10"L	CF	W/C	pl	LL0,LL1,A/B,CI,C/D,LL2//S,T,L\\V,D.	
Reinforced Concrete	10"L	CF	W/C	pl	as Rietz + concrete scales	
Duplex - Mod 901	10"L	Dup	Pl	pl	L,LL1,DF/CF,CIF,CI,C/D,LL3,LL2 LL0,LL00,A/B,T,ST,S/D,DI,K	
Duplex - Mod 911	10"L	Dup	Pl	pl	L,LL1,DF/CF,CIF,CI,C/D,LL3,LL2 LL0,LL00,A/B,T,ST,S/D,DI,K Trig in decimal parts of deg. 10" plastic	
Commercial	5"L					
Technical	5"L					
Electro	5"L					
No 101 Log-Log	10"L	CF	W/C			
No 141 Darmstadt	10"L	CF	W/C	m/g		
No 351 Darmstadt						
No 401 System Rohrburg	10"L				Uses Faber patents.	
No 501 Rietz	10"L					
No 601	6"L					
No 621	5"L					
No 741 Businessman	10"L					
DIWA Sales Leaflet						
No 101 Log-Log	25L	CF	W/C	m/g		
No 111 Electro	25L	CF	W/C	m/g		
No 141 Darmstadt	25L	CF	W/C	m/g		
No 201 Technical	25L	CF	W/C	m/g		
No 221 Reinforced Concrete	25L	CF	W/C	m/g	System Rietz	
No 231 Topographical	25L	CF	W/C	m/g	300°	
No 241 Topographical	25L	CF	W/C	m/g	400°	
No 301 Commercial	25L	CF	W/C	m/g		
No 611 Electro	12.5L	CF	Pl	m/g?		
No 621 Technical	12.5L	CF	Pl	m/g?		
No 631 Commercial	12.5L	CF	Pl	m/g		

DOBIE (Australia)
Address: Dobie Instrument, 18-20 George Street, Sandringham, Victoria, Australia.
History: Formed in 1921 and still in business in 1963. Sold the **COLLIS BROWN** tubular calculator c1960. These are the Fuller Calculator made under licence. One version is produced without the handle, a second version is very similar to the Fuller.

DUALFACE (Australia)
Address: See W&G.
History: Dualface is the Trade name for **W&G** slide rules.

DUJARDIN (France)
Address: Dujardin (Successor to Salleron, Paris.), 24 Rue Parée, Paris, France.
History: Other than taking over from Salleron nothing is known. see **SALLERON**.

Slide Rules:

1	2	3	4	5	6	7
Regle de Corrections, Alconometriques						

DU MONT (USA)
Address: Du Mont, Zanesville, Ohio, USA.
History: Appear to have sold **LAWRENCE**, or similar, under their name. History not known.

DUPA (Germany)
Address: Dennert & Pape, Altona, Hamburg, Germany.
History: See **DENNERT & PAPE**, DUPA was the trade mark used from 1924 to 1936 prior to formation of **ARISTO**.

DUVAL (France)
Address: Monsieur Roger Duval, Maitre-Artisan, 30 Rue Des Sports, Drancy, Seine.
History: Régle Electro Duval. An unconventional 6cm wide Electro slide rule in heavy plastic. Well constructed. Duval describes himself as "Constructor" and Master-Craftsman". History not known.

DWARF (USA)
Address: Not known.
History: c1923 a slide rule sold by **WEBER**.

DYSON (GB?)
Address: Not known
History: Not known

Slide Rules:

1	2	3	4	5	6	7
Dyson's Textile Calculator	W	n/a	M	n/a	Single sided watch type calculator	

E (Switzerland)
Address: Not known, Switzerland.
History: Produced cheap pocket-sized plastic slide rules. Logo is lower case 'e' in a triangle.

E&S (GB)
Address: E&S Publishers Ltd, London.
History: Not known

Slide Rules:

1	2	3	4	5	6	7
The Learner slide rule.						

ECOBRA (Germany)
Address: Ecobra Instruments, Nurnberg, Germany.
History: May have been a slide rule maker, known to make other mathematical instruments.
Present day agents in the UK are **TECHNICAL SALES (TS)**. The logo is ECO(flower)BRA.

Slide Rules:

1	2	3	4	5	6	7
Rietz R141	5"L	CF	Pl	pl	A/B,CI,C/D,K//S,L,T; reversible slide	
Darmstadt R143	15L	?	Pl	pl	LL3,LL2,A/B,CI,C/D,K,LL1//P,T,L,S	
Rietz R151	30L	CF	Pl	pl	25cm\K,A/B,CI,C/D,L//S,ST,T;	
Perfect R152	30L	?	Pl	?	L,K,A/B,CI,C/D,S,ST,T//25cm/S,LL2,LL3/10inch	
160	8C	MD	Pl	?	A,D/C,CI	
-	15L	?	Al	?		
1261	14L	?	Al	?	14cm/A/B,C/D	
1461 System Rietz	12L	?	Al	?	K,A/B,CI,C/D,L//12cm/S,T/	
1511 System Rietz	25L	OF	Al	pl	K,A/B,CI,C/D,L/25cm/S,T/10inch	
1611 System Darmstadt	15L	Dup	Al	pl	K,A/B,L,C/LL3,LL2,LL1 S,T,DF/CF,CI,C/D,P	

EDGE (USA)
Address: Edge Computer Sales, New York, USA.
History: c1908 Edge's Weight Computer for Structural Shapes. Not known what the later history was.

EICHHORN (USA)
Address: M.J. Eichhorn, Chicago, USA.
History: c1908 Eichhorn's Trigonometric slide rule. Not known what the later history was.

ELCOMA (Holland)
Address: Part of Philips, Holland.
History: Not known, ELCOMA stands for the ELectrical COmponents and MAterials division of Philips.

ELLIOTT (USA)
Address: B.K. Elliott Co.
History: c1925, in existence to at least 1941 and probably later. Not known what they sold.

ELLIOTT BROS. (GB)
Address: Elliott Bros., London, various addresses.
History: A well known firm of mathematical instrument makers with a long pedigree of making specialist slide rules.

Slide Rules:

1	2	3	4	5	6	7
Lord's Calculator		C	MD	W	??	mounted on a wooden box, probably a textile calculator. early?

EMBLEM GB)
Address: See **TECHNICAL SALES**.
History: EMBLEM is the brand name of **TECHNICAL SALES (T.S.)** a range of slide rules made in Japan.

EMPIRE (USA)
Address: Empire Pencil Co (?), USA.
History: Not known, made a range of inexpensive plastic slide rules in 5" and 10" lengths.

Slide Rules:

1	2	3	4	5	6	7
Pedigree slide rule	5"L	OF	Pl	pl	K,A/B,CI,C/D,L//S,ST,T; reversible slide, slide scales in red.	

ENDACOTT (GB)

Address: Endacott Scientific Instrument Co. Verulam St, Greys Inn Rd, London WC1.
History: Identified as a manufacturer, but no slide rule types identified in 1921 *Directory of Instrument Makers*.

ENGINEERING INSTRUMENTS (USA)

Address: Engineering Instruments Inc., 750 North Street, Peru, Indiana, USA.
History: Was the successor to Lawrence Engineering Services, set up in about 1947 and continued to mid 1967 when the factory burnt down. As **LAWRENCE**, they made cheap slide rules which were sold under their own name as well as by others. Most rules were painted wood with printed scales, some plastic rules.

Slide Rules:

1	2	3	4	5	6	7
5-B Students 'Learner' Rule	5"L	CF	W/P	m/p	A/B,CI,C/D,K	a.
8-B	8"L	CF	W/P	m/p	A/B,CI,C/D,K	
Students	10"L	CF	W/P	m/p	A/B,CI,C/D,K; equivalents on reverse.	
10-B	10"L	CF	W/P	m/p	A/B,CI,C/D,K; Equivalents, weights and strengths on back	
250-V	10"L	CF	W/C	m/p	as 10-B	
250-BT	10"L	CF	W/P	m/p	A/B,CI,C/D,K//S,L,T; Equivs etc on back.	
-	10"L	CF	W/P	m/p	as 250-BT with advertising on back	
77A American Log Log	10"L	Dup	Pl	pl		b.
Poker Meter	10"L	CF	W/P	??	novelty rule for Poker players	
Air & Hydraulic Cylinder rule	10"L	CF	W/P	m/p	Copyright Miller Motor Company, pipe size chart on back	1950

Comments:
a.   One model called "DRAFT-A-PLAN".
b.   May have been manufactured for them.

ENGINEERS SLIDE RULE CO. (USA)

Address: Engineers Slide Rule Co,, Philadelphia, USA.
History: Not known, see also **DIETZGEN**

Slide Rules:

1	2	3	4	5	6	7
Schofield's Rule						1890

EQUATIONER (USA)

Address: Equationer Company, 114 Liberty Street, New York.
History: in existence from approx 1888, the date of Hart's original patent.

Slide Rules:

1	2	3	4	5	6	7
Hart Equationer or Universal calculator	8"C	MD	M	m	E,B',A'/A,B,C,D	1888
Hart Proportior					is this the same as Equationer?	

FABER (Germany)

Address: A.W. Faber, Stein bei Nurnberg, Germany; *Also:* c1904, 41-47 Dickerson Street, Newark, New Jersey, USA.
History: "Manufactory" Established 1761, started making slide rules 1882. Became **FABER-CASTELL** after 1906.
Sold in England by the main agents **BERRICK BROS** Limited, London EC3, also others e.g.**DINSDALE** and **LAWRENCE RABJOHNS**.

Slide Rules:

1	2	3	4	5	6	7
Pickworth's Slide Rule	10"L	CF	W/C	m/g	made to Pickworth's design, includes a K scale in the rear of the stock, read by a mark on the back of the slide.	c1915
Faber log-log	10"L	CF	W/C	m/g	LL1,A/B,C,D,LL2//V,D	1915-55

<u>From Instructions for A.W. Faber's Improved "Castell" Precision Calculating Rule 6th edn, 1905?</u>
Numbers (part on bottom LH of stock. Wooden pegs hold celluloid to wood.

1/60/360 Standard	4"L	CF	W/C	m/g	Identified as Castell 1/60
1/60/360 Standard	5"L	CF	W/C	m/g	Identified as Castell 1/60
1/60/360 Standard	6"L	CF	W/C	m/g	Identified as Castell 1/60
1/60/360 Standard	10"L	CF	W/C	m/g	Identified as Castell 1/60
1/60/360 Standard	20"L	CF	W/C	m/g	Identified as Castell 1/60
1/61/361 Students					as 1/63 but no trig or log scales. Identified as Castell 1/61
1/63/363 Standard with Decimals					as 1/60 with extra markings. Identified as 363 A.W. Faber-Castell
1/67/367 with Digit Registering Cursor					Identified as 367 A.W. Faber-Castell
1/78/378 For Elect. & Mech. Engineers					with LL, Efficiency & Volt drop scales & prong. Free-sight cursor. Identified as Castell 1/78
1/98/398 For Elect. & Mech. Engineers	10"L				as 1/78/378 with CI & K on lower edge.
61/98/319 For Elect. & Mech. Engineers	5"L				as 1/98/398
1/75/375 Standard with .... scales					K,A/B,C/D,L. Identified as No 375
1/76/376 Polyphase					As 1/60/360 with CI & K on lower edge. Identified as 1/76
1/87/387					as 1/75/375 plus K,A/B,CI,C/D,L. Number in well?
1/92/392					with LL and Reciprocal, K on lower edge. Identified as 392 on LH of slide

Different Cursors:
Aluminium Frame
Glass to Edge Guides "Free-sight" (later type)
Digit Registering

Electrical	10"L	CF	W/C	m/g	DRP206428, DRGM271169,247514	1914
63/39 Castell	5"L	CF	W/C	m/g	A/B,C/D.	1937
319 Castell	5"L	CF	W/C	m/g	DRP206428.	1930
343 Castell System BAUR	10"L	CF	W/C	m/g	A/B,C/D//S,L,T; DRP206428.	1908
360 Castell	10"L	CF	W/C	m/g	A/B,C/D//S,L,T; DRP206428.	1945
362	5"L	CF	Pl/I	m/g	A/B,C/D//S,L,T; Pl/ivory, DRGM881735.	1933
364	10"L	CF	W/C	m/g	A/B,C/D//S,L,T	
369 Castell	5"L	CF	W/C	m/g	A/B,C/D//SLT, DRP206428.	1931
370	20"L	CF	W	?	A/B,C/D//S,L,T (possibly early Pickworth design)	c1906
378 Castell (elec)	10"L	CF	W/C	m/g	DRP206428.	1909
380	20"L	CF	W/C	m/g	A/B,C/D//S,L,T; DRP206428, DRGM371190	1910 1.
380N	20"L	CF	W/C	m/g	A/B,C/D//S,L,T; DRP206428, DRGM371190	1930-35
4/60/380	20"L	CF	W/C	m/g	A/B,C/D//S,L,T;	1935-40
4/60	20"L	CF	W/C	m/g	A/B,C/D//S,L,T;	>1940
382 Log-Log Engineers	20"L	CF	W/C	?	LL2,A/B,C,CI,C/D,LL3//S,L,T	1930-35 2.
4/92/382 Log-Log Engineers	20"L	CF	W/C	?	LL2,A/B,C,CI,C/D,LL3//S,L,T	1935-40 2.
4/92 Log-Log Engineers	20"L	CF	W/C	?	LL2,A/B,C,CI,C/D,LL3//S,L,T	>1940 2.
384 Pickworth	20"L	CF	W/C	?	A/B,C/D//S,L,T; also special K	c1912
385N Rietz	20"L	CF	W/C	?	K,A/B,CI,C/D	1930-35 2.
4/87/385 Rietz	20"L	CF	W/C	?	K,A/B,CI,C/D	1935-40 2.
4/87 Rietz	20"L	CF	W/C	?	K,A/B,CI,C/D	1940-57 2.
388N Electro	20"L	CF	W/C	?	LL2,A/B,C/D,LL3	1930-35 2.
4/98/388 Electro	20"L	CF	W/C	?	LL2,A/B,C/D,LL3	1935-40 2.
4/98 Electro	20"L	CF	W/C	?	LL2,A/B,C/D,LL3	1940-57 2.
398 Castell (elec)	10"L	CF	W/C	m/g	LL2,A/B,CI,C/D,LL3\D,V//S,L,T	
System Rietz 1/87	10"L	CF	B/C	m/g	Sun-Hemmi pat 107,562.	1917
51/87 System Rietz	10"L	CF	W/C	m/g	K,A/B,CI,C/D,L//S,ST,T.	

Comments:
1.   From c1902 to 1912 these were boxwood/celluloid, after were mahogany/celluloid
2.   Pearwood/celluloid.

FABER, Johan (Germany)
Address: Johan Faber, Bleistiftfabric, Nurnberg, Germany.
History: 1878 Founded, 1885 became Johan Faber AG; 1931 became part of the A.W. Faber organisation. Used crossed hammers as a logo.

Slide Rules:

1	2	3	4	5	6	7
No 9202	5"L	CF	W/C	?	A/B,C/D//S,L,T	

FABER-CASTELL (Germany)

Address: A.W. Faber-Castell GmbH & Co,, D-8504, Stein near Nuremberg, Germany.

History: After 1952, F-C made 10" versions of the major types in plastic as well as the traditional pearwood with celluloid facings (wood/celluloid).

Factories: Apart from the main factory in Stein, they also had a factory in Switzerland (Grabs TG) which produced slide rules for the EFTA market. Also a factory in Engelhartzell.

Slide Rules:

1	2	3	4	5	6	7
Special rules from 1945						
<u>wood/celluloid</u>						
1/38 Tachymeter	10"L	CF	W/C	m/g		
4/38 Tachymeter	20"L	CF	W/C	m/g		
1/44 Ekagnost	10"L	CF	W/C	m/g		
1/48 Machine time	10"L	CF	W/C	m/g		
3/11 Steel-Concrete	10"L	CF	W/C	m/g		
3/31 Steel-Concrete	10"L	CF	W/C	m/g		
<u>plastic</u>						
2/62 DYWIDIG	10"L	CF	Pl	m/g	control of concrete mix	
2/66 DEMEGRAPH	10"L	CF	Pl	m/g	for graphical use	
111/38 Tachymeter	10"L	CF	Pl	m/g		
111/48 Machine time	10"L	CF	Pl	m/g		
111/66 DEMEGRAPH	10"L	CF	Pl	m/g		
67/32 System Kramer	5"L	CF	Pl?	m/g	paper/cardboard	
67/34 System Dr Vogel Hämognost	5"L	CF	Pl	m/g		
67/38 Tachymeter	5"L	CF	Pl	m/g		
67/56 Welding Engineering	5"L	CF	Pl	m/g		
<u>From the Lawes Rabjohns Technical Catalogue - 1952</u>						
<u>Wood/celluloid</u>						
1/60 Standard	10"L	CF	W/C	pl	A/B,CI,C/D//S,T,L Freesight cursor	
1/87 Rietz	10"L	CF	W/C	pl	K,A/B,CI,C/D,L//S,T,L; Freesight cursor	
4/87 Rietz	20"L	CF	W/C	pl	K,A/B,CI,C/D,L//S,T,L; Freesight cursor	
1/98 Electrical	10"L	CF	W/C	pl	K\LL1,A/B,CI,C/D,LL2\V,HP//S,T,L; Freesight cursor	
4/98 Electrical	20"L	CF	W/C	pl	K\LL1,A/B,CI,C/D,LL2\V,HP//S,T,L; Freesight cursor	
1/92 Log-Log	10"L	CF	W/C	pl	K\LL1,A/B,CI,C/D,LL2//S,T,L; Freesight cursor	
1/54 Darmstadt	10"L	CF	W/C	pl	L\K,A/B,CI,C/D,P\S,T//LL; Freesight cursor	
4/54 Darmstadt	20"L	CF	W/C	pl	L\K,A/B,CI,C/D,P\S,T//LL; Freesight cursor	
3/11 Reinforced Concrete	10"L	CF	W/C	?	A/B,C/D\K//S,T,SC,L; sl cursor	
1/22 Business	10"L	CF	W/C	?	DF/CF,CI,C/D (Folded at 3.6); sl cursor	
4/22 Business (Disponent)	20"L	CF	W/C	?	DF/CF,CI,C/D (Folded at 3.6); sl cursor	
<u>Astralon 10"</u>						
111/87 Rietz	10"	CF	Pl	pl	K,A/B,CI,C/D,L//S,T,L	
111/54 Darmstadt	10"	CF	Pl	pl	L\K,A/B,CI,C/D,P\S,T//LL	
<u>Astralon 5"</u>						
67/91 Standard	5"L	CF	Pl	pl	A/B,C/D//S,T,L; sl cursor	
67/39 Standard	5"L	CF	Pl	pl	A/B,C/D; sl cursor	
67/87 Rietz	5"L	CF	Pl	pl	K,A/B,CI,C/D,L//S,T,L	
67/98 Electrical	5"L	CF	Pl	pl	K\LL1,A/B,CI,C/D,LL2\V,HP//S,T,L	
67/54 Darmstadt	5"L	CF	Pl	pl	L\K,A/B,CI,C/D,P\S,T//LL	
67/22 Business	5"L	CF	Pl	pl	DF/CF,CI,C/D (Folded at 3.6)	
<u>With Addiator</u>						
1/87A Rietz	10"L	CF	W/C	pl	K,A/B,CI,C/D,L//S,T,L; Addiator	
67/87R Rietz	5"L	CF	Pl	pl	K,A/B,CI,C/D,L//S,T,L; Addiator	
1/54A Darmstadt	10"L	CF	W/C	pl	L\K,A/B,CI,C/D,P\S,T//LL; Addiator	
67/54R Darmstadt	5"L	CF	Pl	pl	L\K,A/B,CI,C/D,P\S,T//LL; Addiator	
1/22A Business	10"L	CF	W/C	pl	DF/CF,CI,C/D (Folded at 3.6); Addiator	
67/22R Business	5"L	CF	Pl	pl	DF/CF,CI,C/D (Folded at 3.6); Addiator	
67/98R Electrical	5"L	CF	Pl	pl	K\LL1,A/B,CI,C/D,LL2\V,HP//S,T,L; Addiator	

From the 1958 Faber Castell catalogue:

1/22 Castell Business	10"L	CF	W/C	pl		
111/22 Castell Business	10"L	CF	Pl	pl		
4/22 Castell Business	20"L	CF	W/C	pl		
67/22 Castell Business	5"L	CF	Pl	pl		
1/22A Castell Business	10"L	CF	W/C	pl	/Addiator.	
67/22R Castell Business	5"L	CF	Pl	pl	/Addiator.	
1/28 Castell Super Business	10"L	CF	W/C	pl		
1/87 Castell System Rietz	10"L	CF	W/C	pl		
111/87 Castell System Rietz	10"L	CF	Pl	pl		
4/87 Castell System Rietz	20"L	CF	W/C	pl		
67/87 Castell System Rietz	5"L	CF	Pl	pl		
1/87A Castell System Rietz	10"L	CF	W/C	pl	/Addiator.	
67/87R Castell System Rietz	5"L	CF	Pl	pl	/Addiator.	1.
1/54 Castell System Darmstadt	10"L	CF	W/C	pl		
111/54 Castell System Darmstadt	10"L	CF	Pl	pl		
4/54 Castell System Darmstadt	20"L	CF	W/C	pl		
67/54 Castell System Darmstadt	5"L	CF	Pl	pl		
1/54A Castell System Darmstadt	10"L	CF	W/C	pl	/Addiator.	
67/54R Castell System Darmstadt	5"L	CF	Pl	pl	/Addiator.	
1/98 Castell Electro	10"L	CF	W/C	pl		
111/98 Castell Electro	10"L	CF	Pl	pl		
67/98 Castell Electro	5"L	CF	Pl	pl		
1/38 Castell Stadia	10"L	CF	W/C	pl		2.
4/38 Castell Stadia	20"L	CF	W/C	pl		c.
1/48 Castell Machine Time	10"L	CF	W/C	pl		
3/31 Castell Static	10"L	CF	W/C	pl		
2/82 Castell Duplex	10"L	Dup	W/C	pl		
62/82 Castell Duplex	10"L	Dup	Pl	pl		

From the Stanley "A" catalogue - 1958.
Wood based

Business 1/22	10"L	CF	W/C	pl	basic CD, Comp CF, DF, R, M £conv, L-L, conv
Darmstadt 1/54	10"L	CF	W/C	pl	M, Cub, Sq, R, CD, Pythag, Trig, L-L.
Basic 1/60	10"L	CF	W/C	pl	A/B, C/D.
Rietz 1/87	10"L	CF	W/C	pl	Cub, Sq, Recip, CD, Mant, Trig
Engineers 1/92	10"L	CF	W/C	pl	L-L, Sq, Recip, CD, Cub, Mant, Trig.
Electro 1/98	10"L	CF	W/C	pl	L-L, Sq, Recip, CD, Cub, Mant, Trig, VD, %M&D
Darmstadt 4/54	20"L	CF	W/C	pl	M, Cub, Sq, R, CD, Pythag, Trig, L-L.
Rietz 4/87	20"L	CF	W/C	pl	Cub, Sq, Recip, CD, Mant, Trig
Electro 4/98	20"L	CF	W/C	pl	L-L, Sq, Recip, CD, Cub, Mant, Trig, VD,

Geroplast.

Darmstadt 111/54	10"L	CF	Pl	pl	M, Cub, Sq, R, CD, Pythag, Trig, L-L.	
Rietz 111/87	10"L	CF	Pl	pl	Cub, Sq, Recip, CD, Mant, Trig	
Electro 111/98	10"L	CF	Pl	pl	L-L, Sq, Recip, CD, Cub, Mant, Trig, VD,	
Students Rietz 57/87	10"L	CF	Pl	pl	Cub, Sq, Recip, CD, Mant, Trig.	
Students Log-Log 57/92	10"L	CF	Pl	pl	L-L, Sq, Recip, CD, Cub, Mant, Trig	
Primary 67/39	5"L	CF	Pl	pl	A/B, C/D.	
Darmstadt 67/54B	5"L	CF	Pl	pl	M, Cub, Sq, R, CD, Pythag, Trig, L-L.	
Rietz 67/87	5"L	CF	Pl	pl	Cub, Sq, Recip, CD, Mant, Trig	
Basic 67/91	5"L	CF	Pl	pl	A/B, C/D.	
Electro 67/98B	5"L	CF	Pl	pl	L-L, Sq, Recip, CD, Cub, Mant, Trig, VD,	
Student Rietz 167/87	5"L	CF	Pl	pl	F, Cub, Recip, Mant, Trig.	
1/60/360	10"L	CF	W/C	pl		1939?
67/87 Rb Rietz	5"L	CF	Pl	pl	Addiator	1970
57/87 Rietz Schul Rechenstab	10"L	CF	Pl	pl		1955
52/82 Student Duplo	10"L	Dup	Pl	pl	Swiss	
57/87 Rietz	10"L	CF	Pl	pl		1962
57/89 (Super log-log)	10"L	CF	Pl	pl	K,A/B,CI,C/D,S,ST,T1,T2	1965
167/98b Elektro	5"L	CF	Pl	pl		1971
52/82 Duplex	10"L	Dup	Pl	pl	Swiss	
57/87 Rietz Schul Rechenstab	10"L	CF	Pl	pl		1968
57/92	10"L	CF	Pl	pl		1969
57/89	10"L	CF	Pl	pl	K,A/B,L,CI,C/D,S,ST,T1,T2,Pl, Austria	
52/82 Duplo	10"L	Dup	Pl	pl	18 scales	1961

170

57/22 Business	10"L	CF	Pl	pl		1959
1/98 Elektro	10"L	CF	W/C	pl	Black prong on slide	1963
57/88 Students Advanced Rietz	10"L	CF	Pl	pl	K,A/B,L,CI,C/D,S,ST,T1,T2	1967
57/189	10"L	CF	Pl	pl	K,A/B,CI,C/D,S,ST,T1,T2	1961
57/87 Rietz	10"L	CF	Pl	pl	K.A/B,R,C/D,L//S,ST,T	1955
57/87 Rietz	10"L	CF	Pl	pl	K.A/B,R,C/D,L//S,ST,T Swiss	1971
57/92	10"L	CF	Pl	pl		1966
57/87 Rietz 1973	10"L	CF	Pl	pl	K,A/B,R,C/D,L//S,ST,T, (green/white) Swiss	
57/87 Rietz	10"L	CF	Pl	pl	different logo.	1972
67/98b Elektro	5"L	CF	Pl	pl		1959
62/83	5"L	Dup	Pl	pl	Duplex	1965?

<u>1966 Faber-Castell Catalogue</u>

2/83 Novo-Duplex	25L	Dup	Pl	pl	K,T$_1$,T$_2$,DF/CF,CIF,CI,C/D,ST,S,P LL03,LL02,LL01,W$_2$/W$_2$',L,C,W$_1$/W$_1$',LL1,LL2,LL3	
62/83 Novo-Duplex	12.5L	Dup	Pl	pl	K,T$_1$,T$_2$,DF/CF,CIF,CI,C/D,ST,S,P LL03,LL02,LL01,W$_2$/W$_2$',L,C,W$_1$/W$_1$',LL1,LL2,LL3	
334/83 Novo-Duplex Demonstration	100L	Dup	W/C	pl	K,T$_1$,T$_2$,DF/CF,CIF,CI,C/D,ST,S,P LL03,LL02,LL01,W$_2$/W$_2$',L,C,W$_1$/W$_1$',LL1,LL2,LL3	
310/83 Novo-Duplex Projection	xxL	Dup	Pl	pl	K,T$_1$,T$_2$,DF/CF,CIF,CI,C/D,ST,S,P; xx = 37 x 22 LL03,LL02,LL01,W$_2$/W$_2$',L,C,W$_1$/W$_1$',LL1,LL2,LL3	
2/82 Duplex	25L	Dup	Pl	pl	L,T,T1,T2,A/B,BI,CI,C/D,ST,S,P LL03,LL02,LL01,K/K',C,CIF,CF/DF,LL1,LL2,LL3	
62/82 Duplex	12.5L	Dup	Pl	pl	L,T,T1,T2,A/B,BI,CI,C/D,ST,S,P LL03,LL02,LL01,K/K',C,CIF,CF/DF,LL1,LL2,LL3	
334/82 Duplex Demonstration	100L	Dup	W/C	pl	L,T,T1,T2,A/B,BI,CI,C/D,ST,S,P LL03,LL02,LL01,K/K',C,CIF,CF/DF,LL1,LL2,LL3	
310/82 Duplex Projection	xxL	Dup	Pl	pl	L,T,T1,T2,A/B,BI,CI,C/D,ST,S,P; xx = 37 x 22 LL03,LL02,LL01,K/K',C,CIF,CF/DF,LL1,LL2,LL3	
111/54 Darmstadt	25L	CF	Pl	pl	K,A/B,CI,C/D,P,S,T//LL1,LL2,LL3	c.
111/54A Darmstadt, Addiator	25L	CF	Pl	pl	K,A/B,CI,C/D,P,S,T//LL1,LL2,LL3; with Addiator	
1/54 Darmstadt	25L	CF	W/C	pl	K,A/B,CI,C/D,P,S,T//LL1,LL2,LL3	c.
4/54 Darmstadt	50L	CF	W/C	pl	K,A/B,CI,C/D,P,S,T//LL1,LL2,LL3	c.
315/54 Darmstadt Demonstration	150L	CF	W/C	pl	K,A/B,CI,C/D,P,S,T//LL1,LL2,LL3	
334/54 Darmstadt Demonstration	100L	CF	W/C	pl	K,A/B,CI,C/D,P,S,T//LL1,LL2,LL3	
310/54 Darmstadt Projection	xxL	CF	W/C	pl	K,A/B,CI,C/D,P,S,T//LL1,LL2,LL3; xx = 37 x 22	
67/54b Darmstadt	12.5L	CF	Pl	pl	L,K,A/B,CI,C/D,P,S,T//LL1,LL2,LL3	
67/54R Darmstadt, Addiator	12.5L	CF	Pl	pl	L,K,A/B,CI,C/D,P,S,T//LL1,LL2,LL3; with Addiator	
111/87 Rietz	25L	CF	Pl	pl	K,A/B,CI,C/D,L//S,ST,T	c.
111/87A Rietz, Addiator	25L	CF	Pl	pl	K,A/B,CI,C/D,L//S,ST,T; with Addiator	
1/87 Rietz	25L	CF	W/C	pl	K,A/B,CI,C/D,L//S,ST,T	c.
4/87 Rietz	50L	CF	W/C	pl	K,A/B,CI,C/D,L//S,ST,T	
315/87 Rietz Demonstration	150L	CF	W/C	pl	K,A/B,CI,C/D,L//S,ST,T	
334/87 Rietz Demonstration	100L	CF	W/C	pl	K,A/B,CI,C/D,L//S,ST,T	
310/87 Rietz Projection	xxL	CF	W/C	pl	K,A/B,CI,C/D,L//S,ST,T; xx = 37 x 22	
67/87 Rietz	12.5L	CF	Pl	pl	K,A/B,CI,C/D,L//S,ST,T	
67/87R Rietz, Addiator	12.5L	CF	Pl	pl	K,A/B,CI,C/D,L,S,ST,T; with Addiator	
1/60 Basic Trig	25L	CF	W/C	pl	A/B,C/D//S,L,T	
67/91 Basic Trig	12.5L	CF	Pl	pl	A/B,C/D//S,L,T	
334/60 Basic Trig Demonstration	100L	CF	Pl	pl	A/B,C/D//S,L,T	
67/39 Basic	12.5L	CF	Pl	pl	A/B,C/D	
1/92 Engineers Log Log	25L	CF	W/C	pl	LL2,A/B,CI,C/D,LL3//S,L,T	
111/98 Electro	25L	CF	Pl	pl	LL2,A/B.CI,C/D,LL3,V,D//S,L,T	
1/98 Electro	25L	CF	W/C	pl	LL2,A/B.CI,C/D,LL3\\V,D//S,L,T	
4/98 Electro	50L	CF	W/C	pl	LL2,A/B.CI,C/D,LL3\\V,D//S,L,T	
334/98 Electro Demonstration	100L	CF	W/C	pl	LL2,A/B.CI,C/D,LL3,V,D//S,L,T	
67/98b Electro	12.5L	CF	Pl	pl	LL2,A/B.CI,C/D,LL3,V,D//S,L,T	
67/98R Electro, Addiator	12.5L	CF	Pl	pl	LL2,A/B.CI,C/D,LL3,V,D//S,L,T; with Addiator	
111/22 Business	25L	CF	Pl	pl	conv,DF/CF,CI,C/D,s.d,£//LL1,LL2,C	
1/22 Business	25L	CF	W/C	pl	conv,DF/CF,CI,C/D,s.d,£//LL1,LL2,C	
4/22 Business	50L	CF	W/C	pl	conv,DF/CF,CI,C/D,s.d,£//LL1,LL2,C	
111/22A Business, Addiator	25L	CF	Pl	pl	conv,DF/CF,CI,C/D,s.d,£//LL1,LL2,C; with Addiator	
67/22 Business	12.5L	CF	Pl	pl	conv,DF/CF,CI,C/D//s.d,£	
67/22R Business, Addiator	12.5L	CF	Pl	pl	conv,DF/CF,CI,C/D,s.d,£; with Addiator	

1/28 Super-Business	25L	CF	W/C	pl	K,DF/CF,CI,C/D,A//LL1,LL2,C	
57/22 Students Business	25L	CF	Pl	pl	conv,DF/CF,CI,C/D,L//LL1,LL2,C	
334/22 Students Business Demonstration	100L	CF	Pl	pl	conv,DF/CF,CI,C/D,L//LL1,LL2,C	
315/22 Students Business Demonstration	150L	CF	Pl	pl	conv,DF/CF,CI,C/D,L//LL1,LL2,C	
52/82 Students Duplo	25L	Dup	Pl	pl	L,K,A/B,CI,C/D,LL1,LL2,LL3	
					T1,T2,DF/CF,CIF,CI,C/D,S,ST	
334/52 Students Duplo Demonstration	100L	Dup	W/C	pl	L,K,A/B,CI,C/D,LL1,LL2,LL3	
					T1,T2,DF/CF,CIF,CI,C/D,S,ST	
310/52 Students Duplo Projection	xx	Dup	W/C	pl	L,K,A/B,CI,C/D,LL1,LL2,LL3; xx = 37 x 22	
					T1,T2,DF/CF,CIF,CI,C/D,S,ST	
52/80 Mentor	25L	Dup	Pl	pl	A,DF/CF,CI,C/D,K	
					cm, conv, ins	
334/80 Mentor Demonstration	100L	Dup	Pl	pl	A,DF/CF,CI,C/D,K	
					cm, conv, ins	
310/80 Mentor Projection	xxL	Dup	Pl	pl	A,DF/CF,CI,C/D,K; xx = 37 x 22	
					cm, conv, ins	
57/87 Students Rietz	25L	CF	Pl	pl	K,A/B,CI,C/D,L//S,ST,T	
57/88 Students Advanced Rietz	25L	CF	Pl	pl	K,A/B,L,CI,C/D,S,ST,T1,T2	
334/88 Students Advanced Rietz Demo	100L	CF	Pl	pl	K,A/B,L,CI,C/D,S,ST,T1,T2	
310/88 Students Advanced Rietz Proj	xxL	CF	Pl	pl	K,A/B,L,CI,C/D,S,ST,T1,T2; xx = 37 x 22	
57/89 Students Super Log Log	25L	CF	Pl	pl	K,A/B,L,CI,C/D,S,ST,T1,T2//LL2,S,LL3	
334/89 Students Super Log Log Demo	100L	CF	Pl	pl	K,A/B,L,CI,C/D,S,ST,T1,T2//LL2,S,LL3	
315/89 Students Super Log Log Demo	150L	CF	Pl	pl	K,A/B,L,CI,C/D,S,ST,T1,T2//LL2,S,LL3	
310/89 Students Super Log Log Proj	xxL	CF	Pl	pl	K,A/B,L,CI,C/D,S,ST,T1,T2//LL2,S,LL3; xx = 37 x 22	
57/92 Students Log Log	25L	CF	Pl	pl	LL2,A/B,CI,C/D,LL3//S,L,T	
57/86 Students Columbus	25L	CF	Pl	pl	A/B,C/D	
334/86 Students Columbus Demonstration	100L	CF	Pl	pl	A/B,C/D	
2/84 Mathema	25L	Dup	Pl	pl	ln.tanX^8,(1+Y^2)^i,Y^2/y^2,y$^{2-1}$,y^{-1},y/Y,(1-Y^2)î,gsinY,lnY	
					e^{-10Y},e^{-Y},e$^{-0.1Y}$,^{8}tanhy,^{8}cosh10Y,^{8}sinhY,e$^{0.1Y}$,e^Y,e^{10Y}	
111/38 Stadia/Tachymeter	25L	CF	Pl	pl	L,A/B,CI,C/D,P,S,T//1-cos^2,sin.cos	d.
67/38b Stadia/Tachymeter	12.5L	CF	Pl	pl	L,A/B,CI,C/D,P,S,T//1-cos^2,sin.cos	
4/38 Stadia	50L	CF	W/C	pl	L,A/B,CI,C/D,P,S,T//1-cos^2,sin.cos	c.
57/74 Textile	25L	CF	Pl	pl	LL,A/B,BI,C/D,Z1//Z3,BI,Z2, System Schirdewan.	
111/48 Machine Time	25L	CF	Pl	pl	mc times/oper.dist,cut speed/basic scale,rpm//belt feed,belt width	
3/11 Steel Concrete	25L	CF	W/C	pl	A/B,CI,K_n,K_d,K_{At},C/D\K//S,ST,T,L	
111/66 Demegraph	25L	CF	Pl	pl	DF/CF,CI,C/D,wt of paper//LL1,LL2,C	
8/10 Disc Calculator	12.5C	MD	Pl	pl	K,A,L,D/C,CI,B,T1,T2,S,ST	
8/20 Disc Calculator	9.5C	MD	Pl	pl	C/D	

An earlier 1961 catalogue is changed from 1966 as follows:

62/83 - not included.

All projection rules excluded

No disc calculators included

67/39 now called "Primary"

Plus a number of new rules, particularly special function:

3/31 Steel-Concrete	25L	CF	W/C	pl	A/B,CI,a,C/D	3.
67/21b Steel Concrete	12.5L	CF	Pl	pl	L,K,A/B,a,b/D,P,S,T	4.
315/60 Basic Trig demonstration	150L	CF	W/C	pl	A/B,C/D//S,L,T	
57/62 Dywidag	25L	CF	Pl	pl	Calculations for checking the quality of concrete	
67/56b Welding Technics	12.5L	CF	Pl	pl	L,K,A/H,CI,C/D,P,S,T//B,CI,C; System Titscher	5.
20/66 Typometer	xxL	n/a	Pl	n/a	typographical rule	
20/66 SL Typometer	xxL	n/a	Pl	pl	typographical rule with cursor	
20/68 Parifix	xxL	n/a	Pl	pl	ready reckoner for various conversions	
991 Statifix Evaluator	xxL	n/a	Pl	n/a	Sliding calculator for statistics and quality control	
989 Complex calculator	xxL	n/a	Pl	n/a	Complex hyperbolic function calculator as well as normal calcs	
1080 Function Slide Board	xxL	n/a	Pl	n/a	System Dr Marzani, demo of complex functions and area calcs	

A 1968 catalogue of Students Slide Rules differs from 1966 as follows:

52/80 becomes Students Mentor	25L	Dup	Pl	pl	A,%,DF/CF,CIF,CI,C/D,K//setting,conversions	

some other terminology changes, also Projection slide rules and demonstration slide rules now known as "Students" ....

Another 1968 catalogue differs from 1966, and 1968 above as follows:

52/80 becomes Mentor	25L	Dup	Pl	pl	A,%,DF/CF,CIF,CI,C/D,K//setting,conversions	
67/21b becomes 67/21 Stahlbeton	12.5L	CF	Pl	pl	L,K,A/B,a,b/D,P,S,T (cf 1961)	
3/31 Steel-Concrete to 2/31 Stahlbeton	25L	CF	W/C	pl	A/B,CI,a,C/D (cf 1961)	3.
67/56b Schweitstechnik	12.5L	CF	Pl	pl	L,K,A/H,CI,C/D,P,S,T//B,CI,C; System Titscher (cf 1961)	5.

172

20/66 Typometer	xxL	n/a	Pl	n/a	typographical rule (cf 1961)
20/66 SL Typometer	xxL	n/a	Pl	pl	typographical rule with cursor (cf 1961)

New rules:

52/81 Novo-Mentor	25L	Dup	Pl	pl	A,DF/CF,CIF,CI,C/D,K,S,S',ST,T1,T2
57/86 Columbus	25L	CF	Pl	pl	A/B,C/D
157/80 Mentor Fix	25L	CF	Pl	pl	K,DF/CF,CIF,CI,C/D,A
1/28 Bivius	25L	CF	W/C	pl	K,DF/CF,CI,C/D,A,L,LL1,LL2,£

Comments:

c.   Also available in 400g decimal scales.
d.   Also available in 360⁰ scales.
e.   All available with the NC type open-view cursor, HL semi cylindrical lens cursor, or periscopic folding magnifier.
1.   This was also available as the 67/87Rb, the difference is not known.
2.   This was provided with 360⁰ scale, a version was also available with 400⁰ scales, type number unknown.
3.   a are special scales for steel and concrete tensions and round bar calculations.
4.   a and b are special scales for steel and concrete tensions and round bar-iron tables.
5.   H is a special scale for welding, half scale, right half is as B.

FAé-LuM (France)

Address: FAé-LuM, France.
History: Not known, may have only made one slide rule model. Logo is "HSB".

Slide Rules:

1	2	3	4	5	6	7
Hono-Rietz 300	L	?	?	?	L,K,A/B,CI,C/D,S,ST,T.	

FARMAR (GB)

Address: F.C. Farmar, 100 Station Approach, Ramsgate, Kent; also later (?) 10 Gloucester Terrace, Thorpe Bay, Essex.
History: Range of many specialist slide rules, including:

Slide Rules:

1	2	3	4	5	6	7
Farmar's Publicans Rule, Size 1, 2328.						
Farmar's Profit Calculating rule						c1938
Farmar's Entire Wine & Spirit	50L	OF	W/C	pl	Spirit scales; desk size.	

FEARNS (GB)

Address: Fearns Calculators, 7, Oakfield Drive, Whickham, Newcastle upon Tyne, England.
History: Range of specialist Circular calculators supplied in a plastic wallet, range includes:

Slide Rules:

1	2	3	4	5	6	7
A6 Screw Thread Metric Converter	5.5"C	MC	Pl	pl	1 Disc 2 Cursors	
A20 Trigonometry Calculator	5"C	MD	Pl	pl	2 Disc 1 Cursor	
A25 Date Calculator	5"C	MD	Pl	none	2 Disc	
A62 Economic Order Qty Calc	7.5"C	MD	Pl	none	5 Disc	
D.18M Metric Conveyor Calculator	7.5"C	MD	Pl	pl	3 Disc 2 Cursors	
Production Engineers Calculator	7.5"C	MD	Pl	none	5 Disc	
Machining Time Calculator	7.5"C	MD	Pl	none	4 Disc	
Circular Slide Rule	5"C	MD	Pl	pl	2 Disc 1 Cursor; A/B,C/D	a.
International Conversions Calculator						
Hardness Conversions Calculator						
Inches/Millimetres Conversions Calculator						
Machining Time Calculators - Lathe work, Milling, Drilling						
Ratefixers Calculator						
Economic Batch Size Calculator						
Gear Horsepower Calculators - Spurs & Helicals, Bevals, Worm Gears						
Costing and Estimating Calculator						
Conveyor Calculator						
Horsepower Calculator						

Shaft Size Calculator
Buyers Calculator
Surveyors Calculator
Weight Calculator for Castings & Forgings
Weight Calculator for Plates & Sheets
Inertia Calculator for Rotating Masses
Tank Capacity Calculator
Transmission Belt Drive Calculator
Time Clerks Calculator
Fearns Odontograph for Gear Tooth Geometrical Construction

Comments:
a.   Circular slide rules were sold with advertising names and logos.

FEDRA (Germany)
Address: Rudolf Bauer & Co, Norderstedt, Germany.
History: The trade name for **BAUER** slide rules.

FELSENTHAL (USA)
Address: Felsenthal Instrument Co, Chicago, Illinois, USA
also Felsenthal & Sons Inc, same address, unknown date.

History: from c1951 to 1976 made Air Navigation Computers and other logarithmic calculators.

FISCHINGER (Germany)
Address: Stuttgart, Germany.
History: c1950 in existence, not known what they made or sold.

FISHER (GB)
Address: Fisher Governor Company Limited, Airport Works, Rochester, Kent.
History: Known from 1967 (Copyright) Hydraulic Control Valve sizing rule, probably made no others. Fishers Control Valve Sizing rule was made by many manufacturers, including **PICKETT & ECKEL** and others.

Slide Rules:

1	2	3	4	5	6	7
Control valve sizing rule MK V.	8"L	Dup	Pl	pl	specialist	

FLYING FISH (China)
Address: Nanian Road 578, Shanghai, China.
History: Various slide rules with the logo of a flying fish in various sizes and in various materials.

Slide Rules:

1	2	3	4	5	6	7
1001	25L	Dup	Pl	m/p	LL01,LL02,LL03,A/B,L,K,C/D,LL3,LL2,LL1 T,ST,DF/CF,CIF,CI,C/D,P,S	
1002	25L	Dup	Pl	m/p	SH2,SH3,K,A/B,S,P²,T1,T2,C/D,DI,L,TH LL01,LL02,LL03,DF/CF,CIF,P2,P1,CI,C/D,LL3,LL2,LL1	
1003	25L	Dup	Pl	m/p	SH0,SH1,CU1,CU2,CU3,LT0,T1,T2,S,Cos,C/D,SQ2,SQ1,TH0,CH1 LL03,LL02,LL01,LL00,DF/CF,CIF,H0,H1,CI,C/D,LL0,LL1,LL2,LL3	
1018 Electricians slide rule	25L	Dup	Pl	m/p	LL01,LL0,LL1,K,A/Q,QI,L,I/D,DI,LL02,LL2 Φ,RΦ,P',P/B,T,S,C/J,J',T,GΦ	
1203	5"L	OF	Pl	pl/m	K,A/B,CI,L,C/D,ST//S,T	

FOUNTAIN & SEENEY (GB)
Address: Address not known, England.
History: Not known, made an Applied Observation of shot slide rule as **NICHOLSON** and others.

Slide Rules:

1	2	3	4	5	6	7
Applied Observation of Fall (Artillery)	20"L	CF	Al	m	special	1917

FOWLER (GB)

Address: Fowler and Company

1900: 53 New Bailey St, Salford, Manchester, Lancashire. (as the **SCIENTIFIC PUBLISHING CO.**)

1920: "Station Works", Sale, Cheshire.

1938: Hampson St, Sale, Cheshire.

(Fowler's (Calculators) Limited), Hampson Street Works, Sale, Cheshire.

History: Founded in 1898 by William Henry Fowler, when an article on circular calculators appeared in the "The Mechanical Engineer", published by Scientific Publishing Co owned by Fowler. His son Harold, born in 1879, joined Scientific Publishing in 1905, also designing circular calculators. Patents were issued in 1910, 1914 and 1923. The Company went into liquidation in 1988.

Slide Rules:

1	2	3	4	5	6	7
Fowlers Circular Calculator					Scientific Publishing, *Cajori*.	1898
Circular watch type calc with centre knob and side knob						1904. 1
Circular calculator with single centre knob.						1910
Circular calculator with two knobs on rim.						1914
Circular calc. with single knob on rear, and rotating bezel, single sided.						1914
Long Scale calculator.						1927
The Magnum calculator, 50" scale length.						1927
12-10 Circular calculator, scales in decimal and duo-decimal (1/12).						1936
Jubilee Magnum with 76" scale.						1948
Universal single dial.						1948
Textile Calculator.						1948
Textile calculator Type E1	2.5"W	n/a	n/a	n/a	C/D//table of weft,looms & reeds.	
Comprehensive List of Fowlers Calculators.						
-	2.7"W	d/s	cast	n/a	short, 4 turn; one winder, one centre knob	
Type B	3.4"W	s/s	pres	n/a	short, 8's scale	
Type B Textile	3.2"W	s/s	pres	n/a	short, 8's scale	
Type E1	2.6"W	d/s	cast	n/a	short, 8's scale//	
Type E1 (Textile)	2.7"W	s/s	cast	n/a	short, 8's scale//consts.	
Type H (early)	2¼"W	d/s	cast	n/a	log,short,√,sin//cubes,cube root	<1915 1.
Type H	2.6"W	d/s	cast	n/a	short,CI,L,√,S//cubes,cube root	1915
Type O	2.7"W	s/s	cast	n/a	short, 4 turn	
Type T (Textile)	2¼"W	s/s	?	n/a	short//consts	
Short Scale Textile	2.7"W	d/s	pres	n/a	short, 8's scale//	
"Magnum" Textile	4.7"W	s/s	cast	n/a	short, 8's scale, 16's scale	1927?
Vest Pocket Calculator, Type MD	2½"W	s/s	?	n/a	short, 6 turn	
Vest Pocket Calculator, Type H	2½"W	s/s	?	n/a	short,CI,L,√,LS,LT,LS	
Vest Pocket Calculator, Type CSR	2½"W	s/s	?	n/a	short/short (C/D)	
Universal Calculator	3.4"W	s/s	pres	n/a	short,CI,L,3 turn,S,T	
Universal Calculator	3.5"W	s/s	cast	n/a	short,CI,L,3 turn,S,T	1950
"Twelve-Ten" Calculator	3.4"W	s/s	pres	n/a	short,CI,duo-dec,S,S,T	1936
"Magnum" Long Scale	4.6"W	s/s	pres	n/a	short,CI,√,√,6 turn,L,S,S,T	
"Magnum" Long Scale	4.6"W	s/s	pres	n/a	short,CI,√,√,L,6 turn,S,S,T	1927
Jubilee "Magnum" Extra Long scale	4.7"W	s/s	cast	n/a	short,CI,L,11 turn	1948
Nautical Calculator	4¼"W	s/s	?	n/a	6 x Sine/Long	1957
Artillery Calculator	?W	s/s	?	n/a	"single type scales"//consts	2.
Artillery Calculator	?W	s/s	?	n/a	short/short (C/D)//conv	
Long Scale Type RX	2.4"W	d/s	?	n/a	short,6 turn//short,CI,L,√,LS,LT,LS	
Long Scale	2.7"W	d/s	pres	n/a	short,6 turn//short,CI,L,√,LS,LT,LS	
Circular Slide Rule	?W	d/s	?	n/a	short/short (C/D)//CI,cubes,3xcube root	
Circular Slide Rule	?W	s/s	?	n/a	short/short (C/D)//consts	
"The Mechanical Engineer" Pocket Calc	2"W?	s/s	cast	n/a	L,short,√,√,S,S	c1900
Fowler's Calculator	2.6"W	d/s	pres	n/a	short,CI,L,√LS,LT,LS//cubes,3 x cube root	c1935
Junior Calculator	4 ½"C	MD	pl	pl	short,short (F/D),L,√,√,S,T//short,CI,L,3 turn,S,S,T	3.

Notes to Table above:

General: s/s is single sided, d/s is double sided; cast is a cast case and pres is a pressed metal case.

1. The rear scale is worked from a centre knob.

2. Advertised with "single type" dials, not known whether these were as Vest Pocket Calculator types or other.

3. This is not a true watch style calculator, but a circular slide rule which used watch-type scales

Slide Rules:

1	2	3	4	5	6	7
From a Fowlers catalogue of 1957.						
Vest Pocket Calculator	2.5"W	MC	M	n/a	two sided calculator, also known as "Long Scale Calculator"	
Universal Calculator	3.25"W		MC	M	n/a	6 scales, D,CI,L,K,S,T; single sided	
"12-10" calculator	3.25"W		MC	M	n/a	7 scales, D,CI,a,b,S1,S2,T; single sided.	a.
"Magnum" Long Scale	4.25"W		MC	M	n/a	7 scales, B,BI,C,L,c,S,T; single sided	b.
"Jubilee Magnum" Calculator	4.25"W		MC	M	n/a	4 scales, C,CI,L,d; single sided	c.
Nautical Calculator	4.25"W		MC	M	n/a	2 effective scales, e,C,	d.

From the Lawes Rabjohns Technical Catalogue - 1952

Vest Pocket	
Textile Vest Pocket	2 5/8"W
Textile Type B	3½"W
Textile Magnum	4½"W
Universal	3"W
12-10	3"W
Jubilee Magnum	4 5/8"W
Junior calc and circular slide rule	3"C

Comments:

a.	a is "Twelfths" from 10 to 100; b is "Twelfths" from 1 to 10.	
b.	c is the "Long" scale equivalent to 50" in 6 turns, the normal B scale is 13.5" equivalent in one turn.	
c.	d is the "Extra Long" scale equivalent to 79" in 10 turns, C is the normal scale, 13.5" equivalent.	
d.	e is the "Comprehensive sine scale" over 6 concentric circles, C shares the same circles working from smallest to largest.	
1.	The 1904 Fowlers Mechanical Engineers Pocket Book has an advert for a watch type calc supplied by Scientific Publishing Co.	

FUJIKEIKO - FUJI (Japan)

Address: Fujikeiko Co, Japan
History: also **FUJI Slide Rule Manufacturing Co Ltd**, Tokyo.
Established in 1949. Sold in Holland as **WOLTERS-NOORDHOFF**. Logo is a stylised Mount Fuji with Fuji written underneath.

Slide Rules:

1	2	3	4	5	6	7
Darmstadt Special No 1200	25L	Dup	Pl	pl	LL1,LL2,LL3,DF/CF,CIF,CI,C/D,K,L,LL0 T1,T2,A/B,BI,CI,C/D,P,S,ST	

FULLER

Address: No address

History: No slide rule manufacturers are known with this name, however the name is well known for slide rules as follows:
- The Fuller tubular calculator invented by Prof. George Fuller of Queens College Belfast and patented in 1878, was made by **STANLEY** and sold by many companies including **THORNTON** and **KEUFFEL & ESSER**.
- John Fullers Circular Slide Rule & Telegraph Computer were engraved by various American companies.

FULLERTON (Japan)

Address: K.K. Konsaiso, Tokyo, Japan.
History: Made a range of **ACUDIAL** circular slide rules sold by **DARGUE** in the later part of their existence.

Slide Rules:

1	2	3	4	5	6	7
Acudial 1454 Trig 2	3.8"C	MD	Pl	3D2C	A,D/C,CI,B,K // L,D,S,T₁,T₂,ST; 10" Scale length,	
Acudial 1450 Basic	3¼"C	MD	Pl	3D2C	D/C,CI,A,K // Conv tables; 8" Scale length,	a.
Acudial 1458 Log Log	4.38"C	MD	Pl	3D2C	K,A,D,C,CI,B,L // LL₃,LL₂,D,C,S,T₁,T₂,ST;	
Acudial 1499 Stadia	3.75"C	MD	Pl	3D2C	D/S,C₂, // D/C,S,T,ST,L;	

Comment: a.   All, except 1450, had CI scale in red.

FYFE (GB)

Address: Fyfe Carnegie of Chichester, Chichester, England.
History: Specialist Converters and Calculators.

176

Slide Rules:

1	2	3	4	5	6	7
Readers Digest Fletcher Metric Converter						c1980

FYSMA (Czechoslovakia)

Address: Prague, Czechoslovakia.
History: Not known. Made a Circular Aluminium/Brass calculator.

GAMMA (Hungary)

Address: Budapest, Hungary.
History: Not known, made a Model 1252 Rietz slide rule.

GARFIELD (USA)

Address: Oliver Garfield Co Inc., 126 Lexington Avenue, New York 16, N.Y.
History: Sold the Otis King calculator under the name **GENIAC** Pocket Calculator pre 1963 in America.

GASIOROWSKI (Germany)

Address: K. Gasiorowski, Hamburg, Germany.
History: Gasiorowski was an engineer of Polish extraction. Produced a set of two slide rules with scales based on Chemical Formulae for British Petroleum (BP) in Germany.

GENIAC (USA)

Address: GENIAC Pocket Calculator, Oliver Garfield Co Inc, New York, USA
History: c1950, sold a version of the **OTIS-KING** calculator under the name Geniac Pocket Calculator.

GILSON (USA)

Address: Gilson Slide Rule Co. 1915: Niles, Michigan; 1927: Stuart, Florida; 1960: Sold. Became P.O. Box 134, Newark, New Jersey.
History: Established 1915 in the Gilson family home, initially using paper scales glued to steel discs, later making cardboard slide rules then printing or silk screening on white enamelled aluminium. May have bought the **RICHARDSON** Rule Works.
Used the "Owl" logo - as Richardson - see also **COMPASS** and **DIETZGEN**.
Made circular slide rules.

Slide Rules:

1	2	3	4	5	6	7
The "Midget"	4"C	MC	P	pl	Early rules paper, later silk screened. Initially their only product. C and CI scales only.	1915-27
Midget	3.5"C	MC	W/P	pl	C/L//S,S; crows foot cursors.	
Atlas	10"C	MC	M	pl	square format base; D, 30 turn spiral scale.	1927
Apex	8"C	MC	M	pl	as Midget, C,CI,A//S,T scales.	1932
Binary	8"C	MC	M	pl	C,CI,A//S,T; as Apex.	
Atlas	8.5"C	MC	M	pl	circular format, spiral scale	1935
Apex renamed Binary	8.5"C	MC	M	pl	C,LL scales.	1936
Commercial	C	MC	M	pl	C, days between two dates	>1960
The Richardson	11"L	?	?	?	Made for ICSS, had different slides for use on different courses.	
Horse handicapping						
Planting some plant						
Speed of sound in sea water					made for Woods Hole Oceanographic Institution.	

GLOBA (USA?)

Address: Globa, address not known.
History: Made a slide rule watch, similar in design to a number of other designs with a rotating bezel carrying one scale, working against a second fixed scale.

GKN (GB)
Address: Guest Keen & Nettlefolds Ltd,, England.
History: Probably had slide rules made for them, they are better known for the supply of woodscrews and other fixings and fasteners.

GOBLE (USA)
Address: Goble Aircraft Specialities Inc. Montauk, New York, USA.
History: c1946 made a Flight Computer to a 1936 patent.

GOLDSCHMIDT (USA)
Address: D. Goldschmidt, New York, USA.
History: c1904, Goldschmidt's slide rule with two runners, otherwise history not known.

GRAFIA (Germany)
Address: see NORMA
History: The tradename of a range of NORMA circular plastic and metal slide rules.

GRANT (USA)
Address: Houston, Texas, USA.
History: made a range of oil production slide rules.

GRAPHOPLEX (France)
Address: Graphoplex, c1962/3: Paris, France.
History: Made extremely high quality slide rules, some have a clever tensioner built into the stock.

Slide Rules:

1	2	3	4	5	6	7
Bonna	25L	CF	Pl	n/a	Q/D/J; no cursor, Q is flow, D is diameter, J is pressure	
Gaz de France	xxL	CF	Pl	pl	PA-PB,P,PA^2-PB^2/L,D/Q,LogP; xx = 297 x 39.5 x 5.5	
Instruction	100L	CF	Pl	pl	LL2,LL3,L,A/B,K,CI,C/D,S,ST	
Instruction	100L	CF	Pl	pl	P,B^3,B^2/b^2,L,a,b/B,LL3,LL2,LL1//S,T,ST,b	
Calculateur Roplex	xxC	MD	Pl	pl	n/n,n^2/n^2/nI; xx = 12.8 x 12.8 x 0.6; 3 disc, 2 cursor n^3,S,L,T,S&T	
Show Model Large	100L	CF	Pl	pl	P,B^3,B^2/b^2,L,a,b/B,LL3,LL2,LL1	
Show Model Small	58.8L	CF	Pl	pl	P,B^3,B^2/b^2,L,a,b/B,LL3,LL2,LL1	
Regle a Calcul de Vannes	20L	Dup	Pl	pl	special scales for gas and vapour pressures	
Graphoplex 01	25L	CF	Pl	m/p	special metric and imperial conversion scales	
Graphoplex 02	25L	CF	Pl	pl	L,B^3,B^2/b^2,L,a,b/B,S,ST,T	
Graphoplex 03	25L	CF	Pl	??	K,A/B,CI,C/D,L//S,ST,T; metal adjusting screws	
Graphoplex 04	15L	CF	Pl	plm	B^3,B^2/b^2,L,a,b/B,L\\$^0$F,^{0}C	
Graphoplex 05	12.5L	CF	Pl	plm	P,K,A,DF/CF,L,CI,C/D,S,ST,T	
Graphoplex 06	25L	CF	Pl	??	B^3,B^2/b^2,L,a,b/B,L//S,ST,T	
Fedra Rietz 63	12.5L	CF	Pl	plm	B^3,B^2/b^2,L,a,b/B,L//S,ST,T, identical to 612	
Lignomètre - Typomètre 604	xxL	n/a	Pl	pl	no slide, converter for font sizes; xx = 33.6 X 4.15 X .55	
Rietz 610	25L	CF	Pl	plm	K,A/B,CI,C/D,L	
612 (Rietz)	12.5L	CF	Pl	pl	B^3,B^2/b^2,a,b/B,L//S,ST,T	1.
No 614 Systeme Log-Log C.A.P	L	CF	Pl	pl	??	1962/3
615	15L	CF	Pl	plm	K,A/B,CI,C/D,L//S,ST,T	2.
No 600 Systeme Rietz C.A.P	L	CF	Pl	pl	K,A/B,CI,C/D,L//S,ST,T	1962/3
Rietz 620	25L	CF	Pl	??	B^3,B^2/b^2,a,b/B,L//S,ST,T	3.
620 d	25L	CF	Pl	m/p	B^3,B^2/b^2,a,b/B,L//S,ST,T	4.
Rietz S 621	25L	CF	Pl	plm	P, ST,A/B,T,S,CI,C/D,K,L	
630 Géomètre - Topographe	25L	CF	Pl	plm	L,B^3,B/sc,c^2,ST,b/B,B^2,B^{2}I//S,ST,T,b^2	5.
Electric Log-Log 640	25L	CF	Pl	m/p	L,B^3,B^2/b^2,a,b/B,LL3,LL2,LL1//C,S,T,b	6.
Log-Log S641	25L	CF	Pl	plm	L,K,P,ST,A/B,T,S,CI,C/D,LL3,LL2,LL1	7.
Electro Log-Log 643	12.5L	CF	Pl	plm	L,c,B^3,B^2/b^2,a,b/B,LL3,LL2,LL1//ST,S,T,b	
645	25L	CF	Pl	plm	%,DF/CF,CI,C/D,£,sd,units	
647	25L	CF	Pl	plm	K,P,A/B,AI,CI,C/D,S,ST,T//L,LL1,LL2,LL3	
Electro 650	12.5L	CF	Pl	plm	L,B^3,c,B^2/b^2,a,b/B,c,Dyn,Mot,Volt//ST,S,T,b	
No 650 Regle de L'electricien, "Electro"	12.5L	CF	Pl	plm	L,B^3,c,B^2/b^2,a,b/B,c,Dyn,Mot,Volt//ST,S,T,b; Rietz sys.	1962/3

178

Neperlog 690	25L	Dup	Pl	plm	L,P,AI,A/B,T,ST,S,C/D,T,DI,K LL01,LL02,LL03,DF/CF,CIF,K,CI,C/D,LL3,LL2,LL1	
Neperlog 690 a	25L	Dup	Pl	plm	LL00,L,T,A/B,T,ST,S,C/D,P,DI,LL0 LL01,LL02,LL03,DF/CF,CIF,K,CI,C/D,LL3,LL2,LL1	
Neperlog Hyperbolic 691 a	25L	Dup	Pl	plm	L,P,Ch,A/B,T,ST,S,C/D,Th,Sh2,Sh1 LL01,LL02,LL03,DF/CF,CIF,K,CI,C/D,LL3,LL2,LL1	
692 a	25L	Dup	Pl	plm	LL00,L,T,A/B,T,ST,S,C/D,P,DI,LL0 LL01,LL02,LL03,DF/CF,CIF,K,CI,C/D,LL3,LL2,LL1	
692 b	12.5L	Dup	Pl	plm	LL00,L,T,A/B,T,ST,S,C/D,P,DI,LL0 LL01,LL02,LL03,DF/CF,CIF,K,CI,C/D,LL3,LL2,LL1	
Tecnilog 694 a	25L	Dup	Pl	plm	P,T,A/B,T,ST,S,C/D,DI,K L,LL3,DF/CF,CIF,CI,C/D,LL2,LL1	
695	25L	Dup	Pl	plm	P1,P2,T,A/B,T,ST,S,C/D,DI,L(Dd-Rd),degrees LL01,LL02,LL03,DF/CF,CIF,K,CI,C/D,LL3,LL2,LL1	
697	25L	Dup	Pl	plm	L,dB,Dyn,Mot,Volt,A/B,Cap,Ind,CI,C/D,f,λ,DI ST,T,S,DF/CF,CIF,K,CI,C/D,LL3,LL2,LL1	
Electronicien 698	25L	Dup	Pl	plm	f,XC,f,XL,A/B,L,Cap,CI,C/D,f',f,Log x dB ST,T,S,DF/CF,CIF,K,CI,C/D,LL3,LL2,LL1	
Décilog 699	25L	Dup	Pl	plm	LL01,L,A/B,T,SRT,S,C/D,DI,LL1 LL02,LL03,DF/CF,CIF,K,CI,C/D,LL3,LL2,	
1600	25L	CF	Pl	plm	L,T,A/B,K,CI,C/D,S,ST,T	8.
1612	12.5L	CF	Pl	plm	K,A/B,CI,C/D,L//S,ST,T	9.
Log-Log 1614	25L	CF	Pl	plm	LL2,LL3,L,A/B,K,CI,C/D,S,ST,T	
Techni-math 1694	25L	CF	Pl	plm	P,K,A,DF/CF,L,CI,C/D,S,ST,T	
Rietz 6250	50L	CF	Pl	m/p	$B^3,B^2/b^2$,a,b/B,L//S,ST,T	
1600 Vers 4 (PMH)	25L	CF	Pl	plm	L,K,A/B,CI,C/D,S,ST,T	
1146 Instruction slide rule	?L	CF	?	?	made in Germany?	
No 6250 Pour bureau d'etudes	50L	CF	Pl	pl	??	1962/3
No 620 Pour bureau d'etudes	25L	CF	Pl	pl	??	1962/3
No 612 Modele pour la poche (Rietz)	12.5L	CF	Pl	pl	$B^3,B^2/b^2$,a,b/B,L//S,ST,T	1962/3
No 615 Mixte poche et bureau	15L	CF	Pl	plm	K,A/B,CI,C/D,L//S,ST,T	1962/3
STATOS - Reinforced concrete	12.5L	CF	Pl	plm		
Règle à calculs polycombustibles (ELF)	15L	CF	Pl	pl*	B^3,FOD/E.A,CH,-/B,CO_2/b^2,a,b; parallelogram cursor	1976

Comments:

1.	5 variants known with minor text differences, minor size variants and with/without metal inserts.
2.	1 known variant with different windows in stock and number on case, not ruler.
3.	1 known variant called 620 with variants in windows, scales annotated K,A/B,CI,C/D,L.
4.	2 known variants, one called Rietz 620d, rear of slide has Scos,ST,Tctg.
5.	1 known variant called 630 Geometre 400G.
6.	4 known variants, two called 640, some with m/p cursor, other minor text changes.
7.	1 minor variant known with matt finish instead of the normal high gloss finish.
8.	5 variants, minor text and format changes, minor font differences, some all one colour scales.
9.	1 variant with scales annotated $B^3,B^2/b^2$,a,b/B,L.

GRAVET (France)

Address: Gravet et Lenoir, 14R Cassette, Paris France.

History: France's premier slide rule maker, see also **TAVERNIER-GRAVET**, made a huge range of slide rules.

Slide Rules:

1	2	3	4	5	6	7
-	26L	?	W	?	A/B,B/D//L(0-90),L(0-45),A	c1860

GRIFFIN (GB)

Address: Messrs John J. Griffin & Sons, London; c1908: J.J. Griffin & Sons, Kingsway, London. also Charles Griffin & Co Ltd, Exeter Street, Strand, London WC.

History: Range of Scientific supplies, slide rules were only a side line, however they are also mentioned in Cajori.

Slide Rules:

1	2	3	4	5	6	7
Griffin and Son Card calculator	U	n/a	P	n/a		1908
Proell's Pocket Calculator	U	Oth	P	n/a	Card with 15 logarithmic scales for accuracy, transparent overlay with same scales for calculation.	c1915, a.

"121"	3.25"C MD Pl	pl	Circular; 2 disc, 1 cursor.			1968
Horse Power Computer for Steam Engines C	MD P	n/a	3D; Golding's Patent			

Comment:
a.   R. Proell of Dresden is shown as designing Proell's Rechentafel in 1902 (*Cajori*).

GUEDON (USA)
Address: Guedon, Camden, New Jersey, New York, USA.; also possibly France.
History: Not known.

Slide Rules:

1	2	3	4	5	6	7
-	12.5L	CF	M	?	A/B,CI,C/D	

GUERRA (Switzerland)
Address: D. Guerra-Moser & Co,
Neuhausen am Rheinfall, Switzerland.
History: Not known.

GURLEY (USA)
Address: W & L.E. Gurley, Troy, New York, USA.
History: Founded 1852, better known as a general instrument manufacturer. They advertised slide rules from 1895, though these are generally made by others including **K&E**.

Slide Rules:

1	2	3	4	5	6	7
Circular slide rule for Coast Guard.	7"C	MD	M	m	ST1,D,SI/S,CI; Metal (German Silver)	1920
Cox's Stadia Computer	20C	MD	Pl	?	special scales	c1900
Circular Stadia computer						
K&E/Gurley Mannheim slide rule	?L	CF	W/C	m?		

HADFIELDS (GB)
Address: Hadfields Ltd. England.
History: Manufacturers of armour, slide rules do not appear to be a normal product.

Slide Rules:

1	2	3	4	5	6	7
Hadfields slide rule			pl		for Armour penetration calculations.	1947

HALL (GB)
Address: Originally B.J. Hall,
Then Hall-Harding,
Stourton House, Dacre Street, London SW5.
History: Became Hall-Harding and then GAF.
Mr Harding of Hall-Harding was part of the UK presence in Germany post World War II to bring back slide rule technology for use by UK Industry. It is doubtful if they actually made slide rules, sold **A.G. THORNTON PIC** slide rules under their name.

HALDEN (GB)
Address: J. Halden & Co Ltd.,
c1877, Albert Square, Manchester.
1889, workshop at East Street, Manchester.
1900, new workshop at Lund Street, Cornbrook, Manchester.
1908, new works at Reddish Stockport, near Manchester.
By 1920 branches also at:
8&9 Chapel Street, Victoria Street, London SW,
8 Grainger St, Newcastle-on-Tyne,
65 Moor St, Birmingham,
31 Cadogan Street, Glasgow,
14 Park Row, Leeds, and Bristol.

Also agencies in Berlin (Gebr **WICHMANN**) and Paris.

History: Established in 1877 in partnership with **THORNTON**. Slide rules 1906? Taken over by Ozalid in 1969 when all production was stopped. Range of rectilinear Gravet/Tavernier-Gravet/Mannheim slide rules using seasoned cane and metal knife edges and metal/celluloid cursors. 1951, Rectilinear rules and Calculex available. See also **GRAVET, TAVERNIER-GRAVET**. Probably also sold **DAVIS** rules.

Slide Rules:

1	2	3	4	5	6	7
Halden's Calculating Slide Rule	10"L	CF	C	m/c	A/B,C/D//S,L,T; 10/6d	c1910?
Halden's Calculating Slide Rule	20"L	CF	C	m/c	A/B,C/D//S,L,T; 50/-d	c1910?
Gravet's Calculating Rule, No 3.	26L	CF	W	m/c	A/B,C/D; 10/6d boxwood	c1910?
Gravet's Calculating Rule, No3A.	26L	CF	C	m/c	A/B,C/D; 12/6d celluloid	c1910?
Gravet's Calculating Rule, No7	36L	CF	C	m/c	A/B,C/D, ins. plain; 28/-d celluloid	c1910?
Gravet's Calculating Rule, No8	36L	CF	C	m/c	A/B,C/D, mm; 28/-d celluloid	c1910?
Gravet's Calculating Rule, No12	50L	CF	C	m/c	A/B,C/D, plain; 48/6d celluloid	c1910?
Gravet's Calculating Rule, No13	50L	CF	C	m/c	A/B,C/D, mm; 58/-d celluloid	c1910?
Steel weight slide rule	?L	CF	W	?	Steel weight, length, thickness, breadth etc.	
Rule Range No 1; No 1051	?L	?	?	?	Artillery rule	1916
Halden Autocrat	10"L	CF	W/C	?	10inch/A/B,C/D\28cm//S,L,T	

Best known for their CALCULEX watch type calculator available from about 1910:

Calculex 2120; Glass cursors	1.5"W	n/a	n/a	n/a	L,A/B,x//S,A/B,x; double sided;	
Calculex 2121; Celluloid Cursors						
Calculex 2122; Solid silver, Glass cursors						
Calculex 2123; Solid silver, Celluloid Cursors						
Desk Model Calculex, 5.25" dia						c1915

HANSA (Germany)

Address: Hansa Rechenschiebe, Germany?
History: made metal circular slide rules, history not known.

HARDTMUTH (GB)

Address: L. & C. Hardtmuth (Great Britain) Limited, (1944) Temporary address: 44 Alexandra Road, Epsom, Surrey.
History: made **CLASSIC** slide rules, also others without any Logo.
There were also Koh-i-Noor-Hardtmuth companies in Italy and Czechoslovakia, relationship not known, and Hardtmuth in Germany and Austria, and possibly the USA.

Slide Rules:

1	2	3	4	5	6	7
Classic Series 1	5"L	CF	W/C	m/p	A/B,C/D wood/celluloid with Brass/celluloid cursor	>1944

HARLING (GB)

Address: W.H. Harling, 1914: 47 Finsbury Pavement, London.
History: Established in 1851, taken over by **BLUNDELL** in 1966.

Slide Rules:

1	2	3	4	5	6	7
Improved calculating slide rule	10"L	CF	W/C	?	specially selected cane faced with celluloid, aluminium back, normal scales.	1915
Duplex adjustable	12"L	Dup	W/C	?		1915
Favourite Slide rule	10"L	CF	W/C	?	celluloid faced boxwood	1915-17
Mannheim adjustable	10"L	CF	W/C	m	celluloid face mahogany, adjusting screw for slide tension	1917
Polyphase adjustable	10"L	CF	W/C	?	celluloid/mahogany, adjusting screw for slide tension	1915-17
Polyphase duplex	12"L	CF	W/C	?	celluloid face mahogany	1915-17
Log Log duplex	10"L	CF	W/C	?	celluloid face mahogany	1917
Roylance Electrical	8"L	CF	W/C	?	celluloid face mahogany	1917
All wood 10"L		CF	W	?		1917
These all appear to be part of the K&E range of slide rules at that time. Also:						
Masthead height & range distance off calc.	?L	??	Al	?	No 1526?	
Students slide rule	10"L	CF	W/C	m/g	26cm/A/B,C/D,11inch	
-	20"L	CF	W/C	m/g	20inch/A/B,C/D\20inch//S,T	
No 11	20"L	CF	W/C	m/g	20inch/A/B,C/D,L\50cm//S,T	

HEATH (GB)

Address: Heath & Co Ltd, New Eltham Scientific Instrument Works, New Eltham, London SE9.
History: Established 1850. Absorbed into W.F. Stanley in 1926.
1951, Heath, Hicks and Perken Ltd, formed from:
Perken, Son & Co, est 1852 - specialists in Thermometers.
James J. Hicks and Heath & Co were also part of **W.F. STANLEY**
In 1922 *The Dictionary Of British Scientific Instruments* showed the range included:

Slide Rules:

1	2	3	4	5	6	7
Anderson Slide Rule	10"L	CF	W/C	pl	Scale in 4 sections.	
Atmospheric					Gunnery, corrections for temp and pressure.	
Boucher's Calculating Circle.	W	n/a	M			
Chemists.						
Fuller's Calculating Slide Scale.						
Jakin slide rule.						
Multiplex					Cubes / cube roots	
Whitley slide rule					Calculating altitudes.	
Wilson's double slide rule					Shipping dimensions.	
Hezzanith Carmody Navigation slide rule	11"L	?	Pl	?	LL2,A/B,CI,C/D,LL3//S,ST,T; Blundel pat 644,944	

HELIX (GB)

Address: Helix International Ltd., P.O. Box 15, Lye, Stourbridge, DY8 7AJ, England.
History: May also have been the Helix (Universal) Co Ltd of Stourbridge, Worcestershire at some stage.
All slide rules are marked Japan, not known whether they made slide rules, or only sold them.

Slide Rules:

1	2	3	4	5	6	7
The range in the 1970's was:						
International Slide Rule Series:						
U01	5"L	CF	Pl	pl	Log-log slide rule with 2 log-log scales in addition to the normal 4 scales.	
U02	5"L	CF	Pl	pl	Basic Reitz rule.	
U11	10"L	CF	Pl	pl	"Top Precision" slide rule.	
U12	10"L	CF	Pl	pl	"Standard", basic Reitz rule.	
U13	10"L	CF	Pl	pl	As U12, but with scales differently arranged and metric and imperial ruler graduations.	
U15	10"L	CF	Pl	pl	Comprehensive log-log rule with 16 scales, 4 log-log.	
U16	10"L	CF	Pl	pl	Beginners rule as U13, no rulers.	
Top Precision Slide Rule Series:						
A101	10"L	CF	Pl	pl	LL3,LL2,LL1,A/B,BI,CI,C/D,DI,K,LL0//T2,T1,L,S	
A102S)						
A102D) Students	10"L	CF	Pl	pl	S,K,A/B,CI,C/D,L,T	
A103W	10"L	CF	Pl	pl	A/B,CI,C/D,K//S,L,T, cms, ins.	
F106	10"L	CF	Pl	pl	K,DF/CF,CIF,CI,C/D,A//L,T,ST,S	
A13 Log-Log	10"L	CF	Pl	pl	LL2,A/B,C/D,LL3, cms.	
A50S	5"L	CF	Pl	pl	A/B,CI,C/D,K//S,L,T	
A50LL	5"L	CF	Pl	pl	LL3,LL2,A/B,C/D, cms.	
also:						
U06 Modern Maths	10"L	CF	Pl	pl	K,A/B,CI,C/D,L;	
U07 Log-Log	10"L	CF	Pl	pl	L,LL2,LL3,A/B,CI,C/D,A//L,T,ST,S; made in England.	
U12 Standard	10"L	CF	Pl	pl	S,K,A/B,CI,C/D,L,T;	

HELLERMAN (GB)

Address: Esmond Hellerman Limited, Hellerman House, Sunbury Trading Estate, Windmill Road, Sunbury-on-Thames, Middlesex, TW16 7EW.
History: Importer of slide rules, including **LOGAREX**. Also known as sellers of electrical insulation equipment and sleeving.

HEMLY HUGHES (GB)

Address: Hemly Hughes & Sons, 59 Fenchurch St, London EC3.
History: In 1922 *The Dictionary Of British Scientific Instruments* showed the range included:

Slide Rules:

1	2	3	4	5	6	7
Anderson Slide Rule	10"L				Scale in 4 sections.	
Atmospheric	10"L				Gunnery, corrections for temp and pressure.	
Bain's and Beghin Slide Rule.						
Boucher's Calculating Circle.	W					
Electro.						
Profit Calculating Slide Rule.						
Fuller's Calculating Slide Scale.	T					
Hall's nautical slide rule.						
Jakin slide rule.						
Log - duplex.						
Shepherd's cubing slide rule					Quantity surveyors.	
Whitley slide rule					Calculating altitudes.	
Wilson's double slide rule					Shipping dimensions.	
Artillery Map Registering Protractor						
Addison Luard Course & Wind calc.	8"C	MD	M	n/a	Patent 253,758	1926

HEMMI (Japan)

Address: Hemmi Slide Rule Co. Ltd. Tokyo, Japan.
Hemmi Bamboo Slide Rule Manufacturing Co. Ltd.; 20 Konno, Shibuya, Tokyo, Japan.
to date: Hemmi Slide Rule Co Ltd, 4-4 Kanda-Surugadai, Chiyoda-Ku, Tokyo 101, Japan.
     Telephone: 03-3253-2631; Fax: 03-3253-2633
History: See also **Sun.**
Slide rules from old stock can still be bought from the latest address.
1895: Established as Hemmi Jirou
1917: Registration of Sun trademark
1928: Renamed Hemmi Seisakusho & Co
1937: Hemmi registered as trademark
1946: renamed Hemmi Keisanjaku Company

SUN-HEMMI SLIDE RULE CHRONOLOGY.

1895, Hemmi Jirou & Co established as first Japanese manufacturer
1910, Patent No 22129. May 11, 1912 applied for.
1917, Registration of "SUN" trademark
1923, Original dividing engine devised
1925, Patent for $\sqrt{(a^2 + b^2)}$ scales as used on various slide rules e.g. Mod 152 etc.
1928, Hemmi Jirou & Co. renamed Hemmi Seisakusho & Co.
1933, Hemmi Seisakusho incorporated, manufacturing 40,000 slide rules p.a.
1935, Hideaki Hirano joins as R&D director
1937, HEMMI registered as trademark
1940, New factory
1941, Manufacturing 200,000 slide rules p.a.
1944, Manufacturing 360,000 slide rule p.a.
1946, Renamed HEMMI KEISANJAKU COMPANY
1949, Fujikeiki Co - another Japanese slide rule manufacturer established
1951, Relay Keisanjaku Co - another slide rule manufacturer established. Later merged into Ricoh, renamed Ricoh Keisanjaku Co.
1951, Hemmi manufacturing 300,000 slide rules p.a.
1951, Slide rule curriculum in Japanese schools
1953, Jirou Hemmi dies
1960, Hemmi manufacturing 600,000 slide rules p.a.
1963, Hemmi peaks at 1,000,000 slide rules p.a.
1963, Sharp announce electronic calculator. Slide rule production starts to decline at 10% per annum
1967, Hemmi starts to manufacture electronic products
1971, Mass produced pocket calculators. Slide rule production decreases rapidly
1972, H-P 35 "Electronic Slide rule" at $395. K&E cease slide rule manufacture.
Total slide rules manufactured 1895 - 1973, approx 15 Million.

Slide Rules:

1	2	3	4	5	6	7
Hemi No P280	10"L		Pl	pl		
Hemi No P281	10"L		Pl	pl		
Hemi No P283	10"L		Pl	pl	reverse slide.	

Hemi	5"L	B/C	pl	wood/cell.	
Hemi No P281	10"L	Pl	pl	pale blue bridge	
Hemi No 40RK	10"L	B/C	pl	bamboo/cell.	
-	10"L	B/C	pl	A/B,C/D//S,L,T; bamboo/cel Pat107562, Brevette SGDG	1917
Hemi No 40RK	10"L	B/C	pl	100,1000 on K; bamboo/cel	
- *	10"L	B/C	pl	A/B,C/D//S,L,T; no Japanese chars on slide.	1917

<u>From Chronology:</u>

Model 152	10"L	B/C		1925
Model 154	20"L	B/C		1925
Duplex bamboo "UNIVERSAL"	10"L	Dup B/C	Exported to USA, distributed by Post of Chicago.	1929
Model 153 for electrical engineers.	10"L	B/C	As 152 with additional G_0 hyperbolic scale.	1931
(HH) ULTRA ACCURACY	10"L	B/C		1936
Model 200	10"L	B/C		1936
ACCURATE STADIA	10"L	B/C		1938
ASTRONAVIGATION	10"L	B/C		1938
Model 120 MIGHT	10"L	B/C		1939
PERCENTAGE	10"L	B/C		1939
CALORIE	10"L	B/C		1939
NAVIGATION	10"L	B/C		1939
ARTILLERY	10"L	B/C		1942
TRANSMISSION	10"L	B/C		1943
COMMUNICATION ENGINEERING	10"L	B/C		1943
HEIGHT ADJUSTMENT	10"L	B/C		1943
Model 250	10"L	Dup B/C	bamboo duplex for general and engineering	1950
Model 255	10"L	Dup B/C	bamboo duplex for expert electrical engineering	1950
Model 275	20"L	Dup B/C	version of 255	1950
Model 256	10"L	Dup B/C	bamboo duplex for communication engineers	1950
Model 259	10"L	Dup B/C	bamboo duplex for expert mechanical engineer	1950
Model 279	20"L	Dup B/C	version of 259	1950
Model 2664	10"L	B/C	bamboo for general business use	1950
Model 2634	5"L	B/C	version of 2664	1950
Model 165	10"L	B/C	for statistical sampling	1953
Model 257	10"L	Dup B/C	bamboo duplex for chemical engineering	1954
Model 2664S	10"L	B/C		1958
Model 253	10"L	Dup Pl	PVC/plastic duplex for students	1961
Model 266	10"L	Dup B/C	bamboo duplex for electrical engineers	1961
Model 260	10"L	B/C	for advanced mechanical engineers	1964
Model P261	10"L	Dup Pl	PVC/plastic duplex for advanced engineer	1965
Model 269	10"L	Dup B/C	bamboo duplex for civil engineering	1965
Model P263	10"L	B/C	for business applications	1966

<u>Slide rule types sold in Japan:</u>

143	5"L	Dup B/C	?	P1,P2,DF/CF,CIF,CI,C/D,LL3,LL2 A,N/M,£.s.d,C/D,G	
149A	5"L	Dup B/C	?	LL01,LL02,LL03,DF/CF,CIF,CI,C/D,LL3,LL2,LL1 L,K,A/B,TI,SI,C/D,LL0,LL00	
153	10"L	Dup B/C	?	L,K,A/B,CI,C/D,T,GΦ Φ,RΦ,P/Q/Q',C/D,LL3,LL2.LL1	
154	20"L	Dup B/C pl		DF,P'/P,Q,CF,CI,SΦ/A,D,K Sx,Tx/TΦ⁰,Th,Sh,C/D,L,X	
201	20"L	Dup B/C	?	L2.4,L1.3,A2.4,A1.3/C4,C3,C2,C1/D1,D2,D3,D4 K4,K3,K2,K1/CF4,CF3,CF2,CF1/DF1,DF2,DF3,DF4	
22	8"L	CF B/C	?	CI/C/D,A	
250	10"L	Dup B/C	?	DF/CF,CIF,CI,C/D,L K,A/B,TI2,TI1,SI/D,DI	a.
251	10"L	Dup B/C	?	L,DF/CF,CIF,C/D,LL3,LL2,LL1 K,A/B,T,S,ST,C/D,DI	
254W	10"L	Dup B/C	?	LL01,LL02,LL03,DF/CF,CIF,CI,C/D,LL3,LL2,LL1 K,A/B,TI,SI,C/D,DI,L	
255	10"L	Dup B/C pl		L,K,DF/CF,CIF,CI,C/D,X,Φ Sh1,Sh2,Th,A/B,T1,T2,S,C/D,LL3,LL2,LL1	a.
255D	10"L	Dup B/C pl		L,K,DF/CF,CIF,CI,C/D,LL3,LL2,LL1 Sh1,Sh2,Th,A/B,TI2,TI1,SI,C/D,LL3,LL2,LL1	
257L	10"L	Dup B/C pl		L,K,A,DF/CF,CIF,CI,C/D,LL3,LL2,LL1 Ch,M_1/M_2,Xw,X,T(⁰F,⁰C,⁰K),F(t)/P,tw	
259D	10"L	Dup B/C pl		LL01,LL02,LL03,DF/CF,CIF,CI,C/D,LL3,LL2,LL1 L,K,A/B,T,S,ST,C/D,DI,LL0,LL00	a.
260	10"L	Dup B/C pl		LL01,LL02,LL03,DF/CF,CIF,CI,C/D,LL3,LL2,LL1 LL00,LL0,K,A/B,T,S,ST,C/D,DI,P,L	
2634	5"L	CF B/C pl		cm\K,DF/CF,CI,C/D,A//T2,T1,L,S	

266	10"L	Dup	B/C	pl	LL03,LL01,LL02,LL2,A/B,BI,CI,C/D,dBL,S,T
					XL,XC,Γ1/Γ2,L,C,CZ/L,Z,fo
2662	10"L	CF	B/C	pl	cm\K,A,DF/CF,CIF,CI,C/D,DI,L//B,T,S,C\13-0-13cm
2664S (1)	10"L	CF	B/C	pl	cm\K,DF/CF,CIF,C/D,A//T1,T2,L,S1\13-0-13cm
2664S (2)	10"L	CF	B/C	pl	cm\K,DF/CF,CIF,C/D,A//T1,T2,L,S1
2690	10"L	CF	B/C	pl	cm\L,D/SC,CI,C/D,A//S,ST,T\ins
279	20"L	Dup	B/C	pl	LL01,LL02,LL03,DF/CF,CIF,CI,C/D,LL3,LL2,LL1
					L,K,A/B,T,S,ST,C/D,LL0,LL00
30	4"L	CF	B/C	pl	cm\A/B,CI,C/D//S,L,T
32	4"L	CF	B/C	pl	cm\A/B,CI,C/D//S,L,T
34RK	5"L	CF	B/C	pl	cm\A/B,CI,C/D,K//S,L,T
400	10"L	Dup	B/C	pl	Y,A/B,CI,C/D,LL3,LL2,LL1
					chemical scales
45	8"L	CF	B/C	pl	DF/CF,CI,C/D,A//T,L,S
45K	8"L	CF	B/C	pl	K,DF/CF,CI,C/D,A//T,L,S
50W	10"L	CF	B/C	pl	in\A/B,CI,C/D,K//S,L,T\cm
640S	10"L	CF	B/C	pl	K,DF/CF,CIF,CI,C/D,A,L//ST1,TI2,TI1,SI,CI
66	6"L	CF	B/C	pl	cm\K,A/B,CI,C/D,L//S,ST,T\8-0-8cm
70	20"L	CF	B/C	pl	in\K,A/B,CI,C/D,L//S,ST,T\cm
74	5"L	CF	B/C	pl	K,A/B,CI,C/D,L//S,ST,T\cm
80K	10"L	CF	B/C	pl	dyn-mot,KW/PS,CI,C/D,LL//S,L,T\K
86K	6"L	CF	B/C	pl	dyn-mot,KW/PS,CI,C/D,LL//S,L,T\K
Mannheim	5"L	CF	B/C	pl	ins\A/B,CI,C/D\\S,L,T\cm
Mannheim	4"L	CF	B/C	pl	cm\A/B,C/D\\S,L,T\in
Mannheim	10"L	CF	B/C	n/a	cm\A/B,C/D,K\\S,L,T\in; no cursor.
P24	7"L	CF	B/C	pl	DF/CF,CI,C/D,A
P253	10"L	Dup	B/C	pl	LL3,LL2,DF/CF,CIF,CI,C/D,L,LL1
					K,A/B,T2,T1,SI/D,DI
P261	10"L	Dup	B/C	pl	LL01,LL02,LL03,DF/CF,CIF,CI,C/D,LL3,LL2,LL1
					T,ST,K,A/B,BI,CI,C/D,L,P,S
P263	10"L	Dup	B/C	pl	P1,P2,DF/CD,CIF,CI,C/D,LL
					A,N/M,£.s.d,C/D,G
P271 Moist Air Slide Rule	10"L	Dup	B/C	pl	hygrometer scales
P35	4.6"L	CF	B/C	pl	K,DF/CF,CI,C/D,A
P36	4.6"L	CF	B/C	pl	C/conv factors/C//conv factors
P37 Sales	5"L	CF	B/C	pl	profit1,profit2,D/C,CI,no of days/day-month,day-
month//days,profit					
P43D	8"L	CF	B/C	pl	A/B,CI,C/D,K//T,L,S
P45D	8"L	CF	B/C	pl	K,DF/CF,CI,C/D,A//T,L,S
Real Estate	8"L	CF	B/C	pl	Tsubo,m²/ken,m/m,ken
P23 Slide Rule for Students	7"L	CF	Pl	pl	? 1959

In 1931 the Hemmi rules sold by Dargue were:

192/b/5/1	20"L	CF	B/C	m/g	Engineers (Standard), 55/-
193/80/1	10"L	CF	B/C	m/g	Log Log Electrical, 30/-
194/11	10"L	CF	B/C	m/g	Engineers, 21/-
194a/8	5"L	CF	B/C	m/g	Divided as 10" rule with magnifying cursor, 21/-
195/6	5"L	CF	B/C	m/g	Engineers, 12/6d
195a	10"L	CF	P	??	Students cardboard slide rule, 2/6d

In 1935 Hemmi rules sold by Dargue were.

Hemmi No 32	4"L	CF	Pl	plm
Hemmi No 34RK	5"L	CF	Pl	pl
Hemmi No 74 Rietz	5"L	CF	Pl	pl
Hemmi No 136 Darmstadt	6"L	CF	Pl	pl
Hemmi No 40RK Students pattern	10"L	CF	Pl	pl
Hemmi No 64 Rietz	10"L	CF	Pl	pl
Hemmi No 80K Electrical	10"L	CF	Pl	pl
Hemmi No 130 Darmstadt	10"L	CF	Pl	pl

1962 Range of Hemmi rules sold via the Drawing Office Manufacturers and Dealers Association were:
as 1935, with different names and numbers as follows:

Hemmi No 30 Mannheim Basic	4"L	CF	B/C	m/p
Hemmi No 34RK Mannheim Basic	5"L	CF	Pl	pl
Hemmi No 40RK Mannheim Students	10"L	CF	Pl	pl

with the following additional rules:

| Hemmi 251 Mechanical Engineering | 10"L | Dup | B/C | m/p |
| Hemmi 255 Electrical Engineering | 10"L | Dup | B/C | m/p |

Hemmi 256 Communication Engineering   10"L   Dup B/C   m/p
Hemmi 257 Chemical Engineering   10"L   Dup B/C   m/p
Hemmi 259 Mechanical Engineering   10"L   Dup B/C   m/p
Hemmi 154 Electrical Engineering   20"L   Dup B/C   m/p
Hemmi 275 Advanced Engineering   20"L   Dup B/C   m/p
Hemmi 279 Advanced Mechanical Eng.   20"L   Dup B/C   m/p

December 1962 Belgian sales leaflet for Hemmi Bamboo rules:

Model	Length				Scales	
No 30 Hemmi Normal	4"L	CF	B/C	pl		
No 74 Hemmi Rietz	5"L	CF	B/C	pl	K,A/B,CI,C/D,L//S,ST,T	
No 66 Hemmi Rietz	6"L	CF	B/C	pl	K,A/B,CI,C/D,L//S,ST,T	
No 64 Hemmi Rietz	10"L	CF	B/C	pl	K,A/B,CI,C/D,L//S,ST,T	
No 70 Hemmi Rietz	20"L	CF	B/C	pl	K,A/B,CI,C/D,L//S,ST,T	
No 86 Hemmi Electro Log-Log	6"L	CF	B/C	pl	LL2,A/B,CI,C/D,LL3//???\Dynamo	
No 80 Hemmi Electro Log-Log	10"L	CF	B/C	pl	LL2,A/B,CI,C/D,LL3//???\Dynamo	
No 136 Hemmi Darmstadt	6"L	CF	B/C	pl		
No 130 Hemmi Darmstadt	10"L	CF	B/C	pl		
No 34RK Hemmi Normal	5"L	CF	B/C	pl		
No 2640 Hemmi Etudient	8"L	Dup	B/C	pl		
No 153 Hemmi Gudermanian	10"L	Dup	B/C	pl		
No 250 Hemmi Usage General	10"L	CF	B/C	pl		
No 400 Hemmi radioisotope	10"L	CF	B/C	pl		
No 410 Hemmi Calcul Engrenages	10"L	CF	B/C	pl		

Hemmi Bamboo Slide Rule Catalogue, undated but estimate late 1960's

Model	Length				Scales	
No 32	4"L	CF	B/C	plm	A/B,CI,C/D//???; magnifier is semi cylindrical lens.	
No 30	4"L	CF	B/C	pl	A/B,CI,C/D//???; plain cursor.	
No 34Rk	5"L	CF	B/C	pl	A/B,CI,C/D,K//S,L,T;	b.
No 50W Advanced Mannheim slide rule	10"L	CF	B/C	pl	A/B,CI,C/D,K//S,L,T;	
No 2640 Students Advanced Mannheim	8"L	CF	B/C	pl	A/B,CI,C/D,K//S,L,T;	
No 74 Rietz	5"L	CF	B/C	pl	K,A/B,CI,C/D,L//S,ST,T;	b.
No 66 Rietz	6"L	CF	B/C	pl	K,A/B,CI,C/D,L//S,ST,T;	b.
No 64 Rietz	10"L	CF	B/C	pl	K,A/B,CI,C/D,L//S,ST,T;	
No 70 Rietz	20"L	CF	B/C	pl	K,A/B,CI,C/D,L//S,ST,T;	
No 2634	5"L	CF	B/C	pl	K,DF/CF,CIF,CI,C/D,A//T1,T2,L,S1	b.c.
No 2664S	10"L	CF	B/C	pl	K,DF/CF,CIF,CI,C/D,A//T1,T2,L,S1	c.
No 135 Darmstadt	5"L	CF	B/C	pl	T,K,A/B,CI,C/D,P,S//LL1,LL2,LL3,T	
No 136 Darmstadt	6"L	CF	B/C	pl	L,K,A/B,CI,C/D,Sin,Cos//LL1,LL2,LL3,T	
No 130 Darmstadt	10"L	CF	B/C	pl	L,K,A/B,CI,C/D,Sin,Cos//LL1,LL2,LL3,T	
No 259 Mechanical Engineering	10"L	Dup	B/C	pl	LL01,LL02,LL03,DF/CF,CIF,CI,C/D,LL3,LL2,LL1 L,K,A/B,T,S,ST,C/D,LL0,LL01	
No 279 Mechanical Engineering	20"L	Dup	B/C	pl	LL01,LL02,LL03,DF/CF,CIF,CI,C/D,LL3,LL2,LL1 L,K,A/B,T,S,ST,C/D,LL0,LL01	
No 149 Mechanical Engineering	5"L	Dup	B/C	pl	LL01,LL02,LL03,DF/CF,CIF,CI,C/D,LL3,LL2,LL1 L,K,A/B,T,S,ST,C/D,LL0,LL01	
No 251 Mechanical Engineering	10"L	Dup	B/C	pl	L,DF/CF,CIF,CI,C/D,LL3,LL2,LL1 K,A/B,T,S,ST,C/D,DI	
No 256 Communication Engineering	10"L	Dup	B/C	pl	dB(L),Neper,DF/CF,CIF,CI,C/D,LL3,LL2,LL1 A,I/Cap,TI,SI/D,FY	
No 410 Involute of Gear Wheel	10"L	Dup	B/C	pl	Special scales for gear calculations	
No 255 Electrical Engineering	10"L	Dup	B/C	pl	L,K,DF/CF,CIF,CI,C/D,LL3,LL2,LL1 Sh1,Sh2,Th,A/B,T1,T2,S,C/D,X,Φ	
No 153 Electrical Engineering	10"L	Dup	B/C	pl	L,K,A/B,CI,C/D,T,GΦ Φ,RΦ,P/Q/Q',C/D,LL3,LL2.LL1	
No 154 Specialist Electrical Engineering	20"L	Dup	B/C	pl	DF,P'/P,Q,CF,CI,SΦ/A,D,K Sx,Tx/TΦ⁰,Th,Sh,C/D,L,X	
No 257 Chemical Engineering	10"L	Dup	B/C	pl	L,A,DF/CF,CIF,CI,C/D,LL3,LL2,LL1 Ch,M_1/M_2,Xw,X,T(^{0}F,^{0}C,^{0}K),F(t)/P,tw	
No 250 General Computation	10"L	Dup	B/C	pl	DF/CF,CIF,CI,C/D,L L,K,A/B,T1,T2,S,D/DI	c.
No 200 Precision Duplex	16"L	Dup	B/C	pl	C_1- C_6, D_1 - D_6, CF_1 - CF_6, DF_1 - DF_6; allows 4 figure accuracy	
No 400 Radioisotope Engineering	10"L	Dup	B/C	pl	special scales for isotope engineering.	
No 80K Electrical	10"L	CF	B/C	pl	Eff,Volt Drop,A/B,CI,C/D,LL2,LL3//S,L,T\K	
No 86K Electrical	6"L	CF	B/C	pl	Eff,Volt Drop,A/B,CI,C/D,LL2,LL3//S,L,T\K	
No 2690 Stadia	10"L	CF	B/C	pl	L,D/S,Cos,Cos^2,C/D,A//S,ST,T	
No 2664S Beginners	10"L	CF	B/C	pl	K,DF/CF,CIF,CI,C/D,A//T1,T2,L,S	
No 45 Beginners	8"L	CF	B/C	pl	K,DF/CF,CIF,CI,C/D,A//T1,T2,L,S	
No 40RK Students	10"L	CF	B/C	pl	A/B,CI,C/D,K//S,L,T	
No 301A Control engineering	10"L	CF	B/C	4pl	i,w/CI,f,r,C/D,A,dB	

Hemmi slide rule leaflet, c1970?

P253 Duplex	25L	Dup Pl	pl	LL3,LL2,DF/CF,CIF,CI,C/D,L,LL1
				K,A/B,T2,T1,S,D/DL
P261 Duplex	25L	Dup Pl	pl	LL01,LL02,LL03,DF/CF,CIF,CI,C/D,LL3,LL2,LL1
				T,ST,K,A/B,BI,CI,C/D,L,P,S
P262 Duplex	25L	Dup Pl	pl	LL01,LL02,LL03,DF/T2,T1,ST,S,C/D,LL3,LL2,LL1
				K,A,R1/R2,CF,CIF,CI,C/L,LL00,LL0
259D Duplex	25L	Dup B/C	pl	LL01,LL02,LL03,DF/CF,CIF,CI,C/D,LL3,LL2
				LL1,K,A/B,T,S,ST,C/D,DI,LL0,LL00
64T Super Rietz	25L	CF B/C	pl	ins\K,A/B,BI,CI,C/D,DI,L//T2,T1,ST,S
130W Electric Log-log	25L	CF B/C	pl	T,K,A/B,BI,CI,C/D,P,S//L,LL0,LL1,LL2,LL3
149A Duplex	17L	Dup B/C	pl	LL01,LL02,LL03,DF/CF,CIF,CI,C/D,LL3
				Ll2,LL1,L,K,A/B,T1,SI,C/D,LL0,LL00
30 Rietz	12L	CF B/C	pl	A/B,CI,C/D//S,L,T
32 Rietz	12L	CF B/C	plm	A/B,CI,C/D//S,L,T
P45S Rietz	20L	CF Pl	pl	K,DF/CF,CI,C/D,A//TI,L,SI
P281 Rietz	25L	CF Pl	pl	T,K,A/B,CI,C/D,L,S,ST
P280 Log-Log	25L	CF Pl	pl	T,K,A/B,CI,C/D,L,S,ST//LL1,LL2,LL3
P283	25L	CF Pl	pl	K,A,DF/CF,CI,C/D,L,T,S//LL1,LL2,LL3
P33S				
P33SS				
86K	15L	CF Pl	pl	15cm/Dyn,Mot,VA/B,CI,C/D,LL3,LL2\K//S,L,T
80/3 (Electro)	25L	CF B/Al	pl	27cm/LL2,A/B,CI,C/D,LL2\29cm\Dyn,Mot,V//S,L,T

Comments:

a.   Developed with **POST** as part of the VERSALOG series for the US market.

b.   Semi-cylindrical lens available as an option.

c.   CF/DF folded at √10.

HEL (Germany)

Address: See **FABER-CASTELL**.

History: c1939 to 1945 this was a (secret?) brand name for **FABER-CASTELL** used during the years of World War II.

HEMMLER (USA)

Address: Not known.

History: c1946 there was a Hemmler speed slide rule for use with tethered model aircraft to calculate scale speed. Not known if this was purely a specialist make, or made by someone else.

HERION (Germany)

Address: Herion Werke KG, Fabrik für Regel und Steuertechnik, 7 Stuttgart 1, Postfach 2970, Fernspr Stuttgart 566181 (5071)

History: Not known.

HEUER (Switzerland)

Address: Tag Heuer, Switzerland.

History: Produced a slide rule watch, the Calculator model, with a circular slide rule on a rotating bezel (C scale) round the edge of the watch, rotating against a fixed D scale. This has recently been reissued.

HICKS (GB)

Address: James J. Hicks, 8, 9, 10 Hatton Garden, London.

History: Established 1862.

Absorbed into **W.F. STANLEY** (with **HEATH** and **PERKEN**) in 1926.

Slide Rules:

1	2	3	4	5	6	7
Gunnery Slide Rule						

HOCO (USA)

Address: Not known - see **HUGHES-OWEN**

History: See **HUGHES-OWEN**.

HOFFMANN (Germany)
Address: Address unknown.
History: Not known.

Slide Rules:

1	2	3	4	5	6	7
-	5"L	CF	Pl	pl	K,A/B,CI,C/D,L//S,ST,T	

HOPE
Address: Hope, address unknown.
History: Not known.

Slide Rules:

1	2	3	4	5	6	7
No 51.	25L	CF	Pl	pl	S,K,A/B,CI,C/D,L,T	
No 54.	12.5L	CF	Pl	pl	A/B,CI,C/D,K//S,L,T	
No 57.	12.5L	CF	Pl	pl	cm\A/B,CI,C/D,K\ins//S,L,T	

HUBER (Switzerland)
Address: Jacob Huber, Zurich, Switzerland.
History: c1920, Huber developed the "Elementenschieber", the Data Slide, which was sold under the name **NORMUS**. These were sold in many countries under this name, and were ultimately developed by **IWA** in Germany, **OMARO** in France, and latterly by many other firms world-wide.

HUGHES-OWEN (USA)
Address: Hughes-Owen Co. Ltd. USA?.
History: c1951, a **HOCO** No 1777 slide rule was distributed with *The VERSALOG Slide Rule* by Feisenheiser et al. published by Frederick Post Co. These rules were made by **HEMMI** to Hughes-Owen specification.

HUNTER (USA)
Address: Hunter Calculators.
History: c1960 had a very comprehensive range of circular slide rules

HURTER & DRIFFIELD
Address: London.
History: Late C19, 1888 Patent range of Actinographs.

HUTTON (USA)
Address: Jos. H. Hutton, New York, USA.
History: sold a variety of slide rules which may have been of **NESTLER** origin.

Slide Rules:

1	2	3	4	5	6	7
Merchants slide rule					similar to Nestler 40	1913

HYDE (USA)
Address: Herbert, New York, USA.
History: c1950 made an optical slide rule.

HYDROMATH (USA)
Address: American Hydromath Company, USA.
History: Made specialist slide rules for ships loading.

Slide Rules:

1	2	3	4	5	6	7
Capacity Computer.	6"C	MD	Pl	pl	2 sides 1 cursor, 2 circular rules in set. Pat Pending	1945

INFO (USA)
Address: Info Inc. address not known.
History: c1967 known to have sold **OTIS-KING** calculators, **CONCISE** circular slide rules, **TYLER** slide rules etc.

IONIC (GB)
Address: Ionic Laboratories, Slough, Bucks, England.
History: c1944 known to have produced a series of multi-disc plastic circular slide rules for specialist applications.

Slide Rules:

1	2	3	4	5	6	7
"Flik-O'-Disk" Electrical slide rule	4"C	MD	Pl	pl	3D,1C; specialist electrical scales	

IWA (Germany)
Address: c1926,IWA Rechenschieberfabrik, F. Riehle GmbH; Calwerstraße 18, Stuttgart, Germany.
1929, Küferstraße 13, Eßlingen, Germany.; c1971: F. Riehle GmbH & Co. kG. D-7300 Eßlingen, Germany.
History: c1924, as Dipl. Ing. E. **WILLI** GmbH, licenced **NORMUS** data slides to be made in Germany, became IWA with various meanings to the abbreviation. Probably the best known European manufacturers of data slides and for their variety of advertising rules, circular and linear, for special calculations. Nearly 4000 master designs with 3 variants giving approximately 12,000 different designs since 1945.

Slide Rules:

1	2	3	4	5	6	7
IWA 0674 Regle a Calcul de Vannes	L	CF	Pl	pl	Special gas/liquid/vapour scales.	
IWA 1072 slide rule						
Exposure slide rule	L	CF	P	n/a	"Belichtungszeitschieber fur Agfacolour Kopierarbeiten"	
IWA 1638 ABC Calculator	C	MD	Pl	pl	System WERN, with spiral scales.	
IWA 1650 IWAMATIC Circular slide rule	C	??	??	??	??; patented "Planetsgetriebe" (Planetary drive) which drove an inner scale at half the rate of the outer.	
IWA 1660						
Vötsch slide rule	C	??	?/	??	version of 1650 with scales for climate calculations, patented.	
IWA 14127 %CO$_2$ in beer	L	??	??	n/a	Made for Heineken Breweries	©1978
IWA 14162 Vol CO$_2$ in beer	L	??	??	n/a	Made for Heineken Breweries	©1982
Fisher Controls Valve sizing rule	L	??	??	n/a		
IWA 5125	12.5L	CF	Pl	pl	K,A/B,CI,C/D,L,S,ST,T	
RS229 Moglijanski (steel mills)	xL	CF	Pl	pl	x=30x6; 2 slides, Walzwerks Rechenschieber	
522.21 Dr. Schwarze (steel mills)	xL	CF	Pl	pl	x=37x8; special steel milling scales	
IWAFLEX	L	CF	Pl	pl	Generic slide rules in many patterns	1962

Series 1, Machinery rules including:
516.1, 018, 023 and 027.
Series 2, Tool-making including:
701, 1411, 1412, 138, 157, 159, 160, 160G, 169, 170, 194, 51.1, 57.1, 526.1, 511.3.
Series 3, Worktime:
079.
Series 4, Welding technology including:
197, 59.3, 5912.
Series 5, Hardness testing:
188, (Cardboard)
Series 6, Weights of metals:
30, 517.1, 212, 218, 221, 222.
Series 7, Lifting equipment:
018, 049, 082.
Series 8, Layout:
082.
Series 9, Iron construction:
514.4.
Series 10, Building:
208, 209.
Series 11, Textile Industry:
52011.
Series 12, Electrotechniques:
5706.
Series 13, Miscellaneous technical:
514.8, 514.9, 248, 514.7, 514.8, 722, 726, 747, 748.
Series 14, Commercial calculations:
51205, Pocket Rietz slide rule.

JACKSON (GB)
Address: Jackson Bros. Ltd., Armley, Leeds, England.
History: Not known.

Slide Rules:

1	2	3	4	5	6	7
Wood slide rule	L	CF	W	m	A/B,C/D//L&C.	

JAKAR (Japan)
Address: Not known.;   English Offices: Jakar International Ltd., Hillside House, 2-6 Friern Park, London N12 9BX.
History: Slide rules appear to be generally made in Japan, sets of mathematical instruments have been seen which are made in Germany. History of the company not known.

Slide Rules:

1	2	3	4	5	6	7
No 523A	5"L	CF	Pl		13cm/A/B,CI,C/D,K\5inch//S,L,T	
No 523D	5"L	CF	Pl	?		
No 511	5"L	CF	Pl			
No 1005	10"L	CF	Pl			
No 1006D	10"L	CF	Pl	?	10inch/K,DF/CF,CIF,CI,C/D,A//T2,T1,L,S	
No 1011	10"L	CF	Pl	pl	LL3,LL2,LL1,A/B,BI,CI,C/D,DI,K,LL0//T2,T1,L,S	
No 1200	10"L	CF	Pl		Plastic Pocket rule	
No 1211	5"L	CF	Pl		Plastic Pocket rule	
No 22 Darmstadt	12.5L	CF	Pl	pl	LL2,LL3,A/B,CI,C/D,K,LL1//S,T,ST,L	
No 28 log-log						
No 29 Students log/log	10"L	CF	Pl	pl	//S,LL1,LL2,LL3,Blk,cos/cot, reverse slide	
No 29 Students log/log	10"L	CF	Pl	pl	//S,LL1,LL2,LL3,Red,cos/cot, reverse slide	
No 29 Students log/log	10"L	CF	Pl	pl	//S,LL2,LL3 reverse slide	
No 33 Duplex log/log	5"L	Dup	Pl	pl		

JEPPESEN (USA)
Address: Jeppesen & Co, Denver, Colarado, USA.
History: Parent of **WEEMS AERO**, a number of Aero Computers and other Navigation calculators were also sold under the Jeppesen name. 1986, part of Jeppesen Sanderson Inc., producing flight charts to date.

Slide Rules:

1	2	3	4	5	6	7
Computer Mod CR2	C	MD	Pl	?	Special scales	c1955
Computer Mod R2	C	MD	Pl	?	Special scales	c1955
3 way slide rule E-6B	C	MD	Pl	?	Special scales	
Flight Computer CSG-9	C	MD	Pl	?		
Jeppesen Sanderson Inc MU514200B	C	MD	Pl	?		

JIANG SHI (China)
Address: Jiang Shi, not known.
History: Produced a range of slide rules, including:

Slide Rules:

1	2	3	4	5	6	7
-	25L	CF	Pl	m/p	LL1,CU/B,S,SH,CI,C/D,LL2,LL3; pat 129.	
404 Rietz	12.5L	CF	Pl	pl	K,A/B,CI,C/D,L; joint Shanghai/Jianyi	
501	5"L	CF	W/C	pl	K,A,B,CI,C/D,L//S,ST,T	

JOHNSON (GB)
Address: Johnson of Hendon, London, England
History: Produced numerous Photographic exposure calculators:

Slide Rules:

1	2	3	4	5	6	7
Johnson Standard Exposure Calculator						1957
Johnson Exposure Calculator					Johnson Photo Year Book	1956
Johnson Standard Exposure Calculator (BSI/Scheiner)				?		
Johnson Colour & Cine Calculator						>1957

JORGENSON (Denmark)

Address: Ole Jorgenson Creative Workshop Inc. Copenhagen, Denmark.
History: Not known. Rules were generally a "design project" based on a plastic coloured beaker. Four models believed to exist.

Slide Rules:

1	2	3	4	5	6	7
Universal Circle Slide Rule	B	?	?	?	?	c1970
Pencil Cup slide rule	U	?	?	?	?	
International Circle Time Converter	U					

KANE (USA)

Address: Kane Aero Co, St Paul, Minnesota, USA.
History: c1957 Dead reckoning computer.

KATZWINKEL (Germany)

Address: Paul Katzwinkel, Zusenhausen bei Heidelberg, Germany.
History: Produced slide rules under the trade name **PYTHAGORAS.**

Slide Rules:

1	2	3	4	5	6	7
1/517C System Katzwinkel	25L	CF	W/C	pl	α°,cos,sin,10T,A/B,R*,CI,C/D,arc,c"//	c1954

KELVIN, BOTTOMLEY & BAIRD (GB)

Address: Kelvin, Bottomley & Baird Ltd., Cambridge St, Glasgow.
History: Identified as a slide rule maker in the 1922 *Dictionary of British Scientific Instruments*, but no slide rule types identified.

KELVIN HUGHES (GB)

Address: Kelvin Hughes Division (Smiths Instruments)
Not known.
History: Not known.

Slide Rules:

1	2	3	4	5	6	7
Slide rule for Trig calculations.	10"L	CF	Pl	pl	-	1961

KENT (Japan)

Address: Japan.
History: Comprehensive range of slide rules made or sold. Sold a variant of **ROYAL** possibly for the Dutch/Benelux market.

Slide Rules:

1	2	3	4	5	6	7
7153LL	25L	?	Pl	pl?	L,K,A/B,CI,C/D,S,ST,L//25cm/S,LL2,LL3/8inch	c1970, a.
7153	25L	?	Pl	pl	L,K,A/B,CI,C/D,S,ST,T//25cm/ - /8inch	
7300LL	25L	?	Pl	pl	T,K,A/B,CI,C/D,S,ST,L//LL01,LL02,LL03	b.
7903LL	25L	?	Pl	pl	T,K,A/B,CI,C/D,S,ST,L//LL01,LL02,LL03	
-	25L	?	Pl	pl	cm\K,A/B,CI,C/D,L//S,ST,T	
369 Student	25L	OF	Pl	plm	L,K,A/B,CI,C/D,S.ST,T	

Comments:
a.   Numerous variants of colour.
b.   Several variants in colour.

KERN & Cie (Switzerland)

Address: Aarau, Switzerland.
History: in existance from 1819 to 1988. Detailed history not known.

Slide Rules:

1	2	3	4	5	6	7
Kern Regle a Calcul pour la Stadia Topographique						1890
Matthe's The US Geographical Survey Topographic slide rule						1902

KEUFFEL & ESSER (USA)

Address: Keuffel & Esser, 127, Fulton Street, New York, N.Y.

History: Started business in June 1884 making mathematical instruments. Sold their first bought-in slide rules in 1886, including Dennert & Pape rules from Germany. K&E made their first slide rules in 1891, and went on to become the largest manufacturer of slide rules in America. Their last slide rule, manufactured in 1976, was donated to the Smithsonian Institute.

Slide Rules:

1	2	3	4	5	6	7
100, Demonstration Rule	7'L					1955
N105, Demonstration Rule	8'L					1955
1740, N, 1741, N-4012/3 Thacher rule	T					1892 - 1949
1742, 4015, Fuller calculator	T					1895 - 1921
1743, 4020, Charpentier circular,	6C	MD	M	m	C/D,√,//L,T,S; Mod 1743 to 1900	1895 - 1927
1743 1/2, Boucher watch calculator	W					1895 - 1906
1744 1/2, Duplex	5"L	Dup				1895 - 1906
1744, Duplex	10"L	Dup				1895 - 1906
4065/6, Duplex, Interchangeable slides	5"L	Dup				1900 - 1903
1744B, 4076, Duplex, Interch'able slides	10"L	Dup				1900 - 1903
4080, Duplex, Interchangeable slides	20"L	Dup				1900 - 1903
1745, Gunter's rule	10"	n/a	W	No cursor		1887 - 1899
4028, Gunter's slide rule	10"L	n/a	W	none		1900 - 1906
4030, etc. Mannheim	5"L					1887 - 1943
4035, Mannheim	8"L					1887 - 1943
4040, etc. Mannheim	10"L					1887 - 1943
4041 Mannheim	10"L	CF				
4045, Mannheim	16"L					1887 - 1943
4050, 4051, Mannheim	20"L					1887 - 1943
4058W, Beginners	10"L	CF	W			
also Mannheim Students, Beginners, Favorite, many numbers.						
1748-3, Favorite	10"L					1899
4100, Stadia	10"L			with cursor		1895 - 1955
1749, Stadia	20"L			without cursor		1895 - 1955
1749-1, Students, paper facings	10"L			cell		1897 - 1899
1749-2, Colby's Sewer Computer	20"L					1897 - 1906
1749-3, Colby's for Stadia	50"L					1897 - 1903
1749-4, Crane's Sewer Slide rule	10"L					1899 - 1927
4016, Sperry's watch calculator	W			geared differently		1906 - 1936
4017, Sperry's watch calculator	W			geared same		1906 - 1936
4018, K&E Circular watch calculator	W					1909 - 1921
4019, -A, Goodchild mathematical chart						1903 - 1906
4019-B, Triangular rule for Goodchild						1903 - 1906
4054, Mannheim, Favorite White	10"L					1900 - 1943
4056, Mannheim, Favorite Wood	10"L					1900 - 1943
4053, etc. Mannheim, Polyphase	8"L	CF	W/C pl		A/B,CI,C/D,K//S,L,T	1909 - 1955
4053-3, etc. Mannheim, Polyphase	10"L	CF	W/C pl		A/B,CI,C/D,K//S,L,T	1909 - 1955
4053-5, Mannheim, Polyphase	20"L	CF	W/C pl		A/B,CI,C/D,K//S,L,T	1909 - 1955
Improved Mannheim	13"L					c1943
4061-T, Duplex, Trig	5"L					1909 - 1915
4065-T, Duplex, Trig	8"L					1909 - 1915
4071-T, Duplex, Trig	10"L					1909 - 1915
4083-T, Duplex, Trig	16"L					1909 - 1915
4087-T, Duplex, Trig	20"L					1909 - 1915
4070-3, Polyphase Duplex Trig	10"L	Dup	W/C pl		DF/CF,CIF,CI,C/D,L K,A/B,T,ST,S/D,DI; Trig scales in degrees & mins	1943 - 1955
4071-3, Polyphase Duplex DeciTrig	10"L	Dup	W/C pl		DF/CF,CIF,CI,C/D,L K,A/B,T,ST,S/D,DI; Trig scales in degs & fractions	1943 - 1955
4080-3, Log Log Duplex Trig	10"L	Dup	W/C pl		L,LL1,DF/CF,CIF,CI,C/D,LL3,LL2 LL0,LL00,A/B,T,ST,S/D,DI,K; Trig $^\circ$/min	1943 - 1955
4080-5, Log Log Duplex Trig	20"L	Dup	W/C pl		L,LL1,DF/CF,CIF,CI,C/D,LL3,LL2 LL0,LL00,A/B,T,ST,S/D,DI,K; Trig $^\circ$/min	1943 - 1955
4081-3, Log Log Duplex Decitrig	10"L	Dup	W/C pl		L,LL1,DF/CF,CIF,CI,C/D,LL3,LL2 LL0,LL00,A/B,T,ST,S/D,DI,K; Trig $^\circ$/fractions	1943 - 1955
4081-5, Log Log Duplex Decitrig	20"L	Dup	W/C pl		L,LL1,DF/CF,CIF,CI,C/D,LL3,LL2 LL0,LL00,A/B,T,ST,S/D,DI,K; Trig $^\circ$/fractions	1943 - 1955
Log Log Duplex Decitrig (Braille)	10"L			to special order, scales as above.		
4083-3, Log Log Duplex Vector	10"L	Dup	W/C pl		L,LL1,DF/CF,CIF,CI,C/D,LL3,LL2 LL0,LL00,A/B,T,ST,S/D,Th,Sh2,Sh1; Trig $^\circ$/fractns	1943 - 1955
4083-5, Log Log Duplex Vector	20"L	Dup	W/C pl		L,LL1,DF/CF,CIF,CI,C/D,LL3,LL2 LL0,LL00,A/B,T,ST,S/D,Th,Sh2,Sh1; Trig $^\circ$/fractns	1943 - 1955

Model	Length				Notes	Dates
4088-1, Polyphase Duplex	5"L	Dup				1912 - 1936
4088-2, Polyphase Duplex	8"L	Dup				1912 - 1936
4088-3, Polyphase Duplex	10"L	Dup				1912 - 1936
4088-5, Polyphase Duplex	20"L	Dup				1912 - 1936
4090, Universal, 2 slides	10"L					1900 - 1909
4091, Universal, 2 slides	16"L					1900 - 1909
4090-3, Log Log Trig	10"L					1936
4091-3, Log Log Decitrig	10"L					1936
4092, Log Log Duplex	10"L	Dup				1909 - 1936
N-4092-5, Log Log Duplex	20"L	Dup				1909 - 1936
4093-3, Log Log Vector	10"L					1936
4093-5, Log Log Vector	20"L					1936
4094, Merchants, Mannheim	10"L					1936 - 1955
4095, Triangular metal	10"L					1900 - 1903
4095-1, Merchants Duplex	5"L	Dup				1915 - 1936
4095-3, Merchants Duplex	10"L	Dup				1915 - 1936
4096, Mannheim metal slide rule	10"L					1906 - 1912
4096M, Desk slide rule, Mannheim	20"L					1927 - 1955
4097B, "Ever There" Xylonite	5"L					1936 - 1949
4097D	6"L	CF	Pl			
4098, Duplex metal slide rule	10"L	Dup	M			1906 - 1912
4098A, K&E Pocket rule Xylonite	5"L					1936 - 1945
N-4100, Stadia, Mannheim	10"L					1927 - 1949
N-4101, Stadia, Mannheim	20"L					1927 - 1949
N-4102, Surveyors Duplex	20"L					1915 - 1949
4105, Stadia, Cylindrical, Webb's	15"T				No cursor	1903 - 1921
4128, Nordell Sewer Slide Rule	20"L				No cursor	1909 - 1921
4133, Roylance Electrical	8"L					1915 - 1921
N-4133, Roylance Electricians	8"L					1927 - 1949
N-4135, Power Computing Rule	5"L				No cursor	1915 - 1921
4135, Power Computing Rule	7.5"L				No cursor	1915 - 1921
4139, Cooke Radio Rule	10"L					1943 - 1955
4140, Hudson's Horsepower, cardboard	4.5"L		P		No cursor, also Xylonite & Boxwood	1900 - 1915
4142, Allan Friction Head Rule	20"L					1915 - 1921
4150, Pocket Slide rule	5"L					1955
4160, Chemist's Duplex	10"L	Dup				1915 - 1936
4161-3, Mannheim Plastic	10"L		Pl			1921
4165, Urea Index rule	10"L					1921
4175, Kurtz Psychometric	10"L					1936
9068, Doric plastic	5"L	Dup	Pl	pl	DF/CF,CI,C/D,L	1949
N-9081-3, Doric plastic	10"L	Dup	Pl	pl	DF/CF,CI,C/D,L	1949
Cox's Stadia slide rule	L	Dup?				1906
10,000	6"L					
Deci-Lon	10"L					
681617	10"L				military version of 4053-3	

Cursors:

Metal	1892 - 1912	
Frame	1897 - 1912	
Xylonite	1900 - 1945	
Framed	1900 - 1955	
Frameless	1915 - 1936	
Improved	1921 - 1955	Mannheim plastic
Improved	1936 - 1955	
Plain	1936 - 1955	

KINGSON (Hong Kong)

Address: Hong Kong

History: Known to have made Addiators, as well as metal slide rules with Addiators on the back. Addiators were sold under various names such as "Magic Brain" and others.

Slide Rules:

1	2	3	4	5	6	7
-	12.5L	CF	M	n/a	K,A/B,CI,C/D,L	

KLAWUN (Germany)

Address: Feinmessinstitut Jos. Klawun, 1924, Berlin-Charlottenburg 5, Germany. also, 1946, Feinmessinstitut Klawun, Hanover, Germany.
History: Founded 1924, moved to Hamburg after WW2 in 1946. Became part of the **CENTRIC** organisation after the death of Jos. Klawun, the founder, in 1978. Logo is a backward F with a K on the same stem, Klawun under the square shield. Apart from their own make of specialist slide rules, they also sold slide rules from **FABER-CASTELL, D&P (ARISTO), NESTLER** and **IWA**. Specialised in educational slide rules.

Slide Rules:

1	2	3	4	5	6	7
No1001/S Der Rechenstab fur Kaufmanns	25L	CF	W/C	m/p	KZ/T,%,T/Z//s-d,£; Berlin address.	
Mentor 300	30L	CF	Pl	?	L,DF/CF,CI,C/D,A,P,S,T; Hanover address.	
Faber-Castell Klawun 31C	14L	CF	Pl	?	K,A/B,CI,C/D,L//S,ST,T; Hanover address, "SMG Salzgitter".	

KNOTT (USA)

Address: L.E. Knott Apparatus Co, Boston, USA.

History: c1907 The L.E. Knott slide rule, history not known.

KOCH (Germany)

Address: Koch, Huxhold & Hannemann, Hamburg, Germany.
History: Made three types of slide rule including a Water Slide Rule under DRGM 1,214,848. See also **PFENNINGER.**

Slide Rules:

1	2	3	4	5	6	7
-	12.5L				A/B,C/D//S,T	
-	25L				A/B,C/D//S,T	
Water rule	25L				special scales	

KOJEVNIKOV (Russia)

Address: Russia.
History: Not known.

KOLESCH (USA)

Address: Kolesch & Co. New York, USA.
History: c1908 through to 1932 sold slide rules and drafting supplies.

Slide Rules:

1	2	3	4	5	6	7
"Midget" Slide Rule	L	?	?	?		1907
"Vest Pocket" Slide Rule	L	?	?	?		1907
"Triplex" Slide Rule	L	?	?	?		1907
-	10"L	?	W/C	?	A/B,C/D/S,L,T	

KONSAISO (Japan)

Address: K.K. Konsaiso. Tokyo, Japan.
History: Not known. Made the **FULLERTON** range of circular plastic slide rules sold by **DARGUE** and others.

KORTE (Germany)

Address: Friedrich Korte Ing. Braunschweig, Gutenberger Str 34.
History: Not known.

KOSINE (GB)

Address: Kosine Ltd., 1 Blenheim Grove, Peckham, London SE15, England.
History: Producers of specialist calculators for the trade. Copyright **OMARO** (England) Ltd.

Slide Rules:

1	2	3	4	5	6	7
Model P1					Steel and Iron, Plates, Squares, Sheets and Flats.	1940

KRAMER (USA)

Address: Samuel Kramer, Post Office Box 282, Fairfield, Ohio, USA.
History: Producer of weather condition calculators.

Slide Rules:

1	2	3	4	5	6	7
Weather Computer	14C	MD	Pl	n/a	Use in "problematic weather conditions".	

KRATSCHMER (Germany)

Address: Theofried Kratschmer Verlag, Bad Homburg, rdh. Germany.
History: Not known.

KRYGOWAJA (Russia)

Address: Not known.
History: In existence Oct 1917 to 1967. A watch type slide rule coded KL-1 is known.

K.S.L.A. (Holland)

Address: See **SHELL**.
History: Stands for Koninklijke Shell Laboratorium te Amsterdam.

KURVERS (Holland)

Address: Holland.
History: A Dutch wholesaler of slide rules.

KUTSUWA

Address: Not known.
History: Made a Basic Mannheim slide rule (FN801) and a Rietz (FN401) at some time.

LADISH (USA)

Address: Ladish Co., Cudahy, Wisconsin 53110, USA
History: In existence 1964, made flight computers.

LAFAYETTE (USA)

Address: Lafayette Radio Electronics Corporation. Long Island, New York, USA
History: c1950 produced a 99-7055 Log-Log, 99-71029 and a 666 slide rule of unknown function. Still in existence in 1960. Probably also sold **HEMMI** or other Japanese bamboo slide rules.

Slide Rules:

1	2	3	4	5	6	7
99-70559 Log-Log	25L	CF	B/C	pl?	Identical to **DIETZGEN** 1743 Log-Trig	
Electronics	L	?	B/C	?	impedance, reactance, decibels etc.	
F-428 Double Log Duplex Decitrig	10"L	?	?	?		

LANGE (Germany)

Address: H.C. Lange, Hamburg - Elmshorn, W. Germany.
History: Calculator system H.C. Lange - a range of slide rules.

LAWRENCE (USA)

Address: Lawrence Engineering Services, 1935-38: Wabash, Indiana, USA.; 1938-47: Peru, Indiana
History: Originally set up as Lawrence Slide Rule Company in Chicago in the late 1920s or early 1930's, and then moved to Wabash Indiana where the better known Lawrence Engineering Services was set up to make slide rules amongst other equipment. The move to Peru took place in 1938, the company changed its name to **ENGINEERING INSTRUMENTS** in 1947 where it continued in Peru, Indiana. Range of cheap all wood printed scale, slide rules, sold pre-war in Woolworths for 6d.

Slide Rules:

1	2	3	4	5	6	7
8-A	8"L	CF	Wp	m/p	A/B,C/D; Instructions printed on the back.	
8-B	7.5"L	CF	W	m/p	A/B,Cl,C/D,K, Crude, card box	1935

8-B	7.5"L	CF	W	m/p	reversed instructions on rear	1945?
Antenna Slide Rule	8.5"L	CF	W	m/p	Angle/Gain/dia. "Confidential", Box.	
					Radiation Laboratory, MIT, Cambridge, Mass.	1945
10-C Engravers & Photog. Prop. Calc.	10"L	CF	Wp	??	orig size/copy size, %size/%area8-L Volt Drop Calc. or Copper	
Wire sele.	8"L	CF	Wp	??		a.
10-B	10"L	CF	Wp	pl/plm	A/B,CI,C/D,K	b, c.
10-DO Lumber Calculator	10"L	CF	Wp	m/p	width, thickness, length, board feet.	
10-F Pricing and Inventory rule	10"L	CF	Wp	??	copyright 1946	
10-N Printers Proportion Rule	10"L	CF	Wp	??	ins/ins,pica/pica	
12-H Copyfitter Rule	12"L	CF	Wp	??	meas/lines,C.P.P/Chars; copyright O.T. Taylor, Dayton.	1946
250-BT	10"L	CF	Wp	??	A/B,CI,C/D,K//S,L,T; equivalents etc.	
250-V	10"L	CF	Wp	??	A/B,CI,C/D,K//S,L,T; equivalents etc	
Secret Code Maker	7.65"L	CF	Wp	none	arrays of letters and numbers, copyright 1939 and 1945.	
Dimensions of model trains	L	?	?	?		
9-K Music Transposer	9"L	CF	Wp	none	Blue and Red scales of chords, copyright	1946
Flashrule	15L	CF	Wp	??	Guide No etc.	1946
Slide Rule Model of 1917	18"L	CF	W/C	??	brass back, reversible slide, artillery rule	d.

Comments:

a. Variants with either advertising or instructions on the back.
b. Variants with and without magnifying cursor, some are all painted, others plain back, some with instructions, others
     with weights etc.
c. An example has Wabash address on back, and Peru address on front.
d. Made for **WEIL** Philadelphia.

LAWES RABJOHNS (GB)

Address: Lawes Rabjohns Ltd., Abbey House, Victoria Street, Westminster, London SW1.; also at (1952):; 91 Cornwall St, Birmingham;
Faraday House, Todd Street, Manchester;
Factory: Dacre Works, Brooklands Road, Weybridge, Surrey.
History: 1952: Major distributor of Mathematical instruments with a large range of slide rules, including **F-C, ARISTO, BRL,
THORNTON, DIWA, OTIS-KING, FOWLER, FULLER** etc. The Instruments are described in the appropriate makers section.

LEDDER (USA)

Address: G.G. Ledder, Boston, USA.
History: c1903, Hazen-Williams Hydraulic Slide Rule. Further history not known.

LEMAIRE (Belgium)

Address: Lemaire Frères, Rue Morinval 24, Liège, Belgium.
History: Not known.

LEMGO (Germany)

Address: Lemgo, Germany.
History: c1898, Piper's Logarithmische Skale, details and history not known.

LENZ (USA)

Address: Lenz Apparatus Co.
History: c1925 sold slide rules and drafting supplies.

LEWIS (USA)

Address: Lewis Institute, Chicago, USA.
History: c1909, a range of slide rules designed by Woodworth.

Slide Rules:

1	2	3	4	5	6	7
Woodworth's Slide Rule for Electrical Wireman						1909
Woodworth's Slide Rule for Calculations with Volts, Amperes, Ohms and Watts						1909

LEWIS & TAYLOR (GB)

Address: Lewis & Taylor Ltd. Address not known.
History: Not known.

Slide Rules:

1	2	3	4	5	6	7
Gripalcalculator slide rule (patent)						1929

LIETZ (USA)
Address: A. Lietz Co.
History: c1923, also known to exist in 1928, sold **GILSON's** rules. In 1950 sold the Gilson Atlas circular rule.

Slide Rules:

1	2	3	4	5	6	7
Dial Rule	4.2"C	MC	Al	pl	C,Cl,A,Af,L,Fa,LL1,LL2,DS,DT,M//C,S,ST,T; sold by **PICKETT**	a.

Comment:
a.   Af=fractional squares, Fa=fraction addition/subtraction, DS=Drill size, DT=Double depth of Threads, M=Metric conversion.

LILLEY (GB)
Address: Messrs John Lilley & Sons Ltd, see also Wilson & Gillie, The New Quay, North Shields.
History: Identified as a slide rule maker in the 1922 *Dictionary of British Scientific Instruments*, but no slide rule types identified.

LINDNER (Germany)
Address: Herbert Lindner, Berlin, Wittenam, Germany.
History: Not known. Plastic 20L slide rule for machining calculations.

LINEX (Denmark)
Address: Linex A/S, 7 HoersKaetten, DK-2630 Taastrup, Denmark; Phone: 45 42 52 71 11; Fax: 45 42 52 78 69
History: Agents in the United Kingdom are:
Pelltech Ltd, Station Lane, Witney, Oxon OX8 6YS.; Phone: 01993 77 64 51
also marked **BPM** or **BIPM** when the slide rules were made for Shell.

Slide Rules:

1	2	3	4	5	6	7
-	10"L	CF	Pl	pl	K,A/B,CI,C/D,L//S,ST,T; unusual tensioner. Marked BIPM.	
1253	5"L	CF	Pl	pl	5inch/A/B,CI,C/D,14cm	

LOEWE (GB)
Address: The Loewe Engineering Co. Ltd. London.
History: Slide rule developed by A. Magilansky.

LOGA (Switzerland)
Address: Loga Calculator A.G. Uster, Switzerland.
History: Cajori shows Daemen-Schmid's Rechenwalze as produced by "Orlikon" in 1908. They were "La Maison Daemen-Schmidt & Cie" c1913, and became Loga Calculator A.G at about this time. In existence c1918 producing circular slide rules. Heinrich Daemen-Schmidt is associated with the firm and a number of patents, particularly the famous and exceptional models of Loga-Calculator (Rechenwalz?). Ceased manufacture in c1980.

Slide Rules:

1	2	3	4	5	6	7
1m Cylinder	T	n/a	M	n/a	65mm dia, 210mm length.	
10m Cylinder	T	n/a	M	n/a	160mm dia, 510mm length.	
15m Cylinder (Rechenwalz)	T	n/a	M	n/a	60 lengths of scale each 50cm long, half overlapped	
2.4m Cylinder	T	n/a	M	n/a		

Circular slide rules for Commercial applications.

30sE	12.5C	MD	Pl	pl	C/D;	
30sR	12.5C	MD	Pl	pl	C/D,CI;	
30sT	12.5C	MD	Pl	pl	C/D,CI,A,K,L;	
30sRZ	12.5C	MD	Pl	pl	C/D,CI,%,Fractions (0,125-1.125).	
30Rtx	12.5C	MD	Pl	pl	C/D,CI, and weaving scales.	
30Eh	12.5C	MD	Pl	pl	C/D; as 30sE with turning lever.	
30Rtxh	12.5C	MD	Pl	pl	C/D,CI, and weaving scales; as 30Rtx with turning lever.	
30Rh	12.5C	MD	Pl	pl	C/D,CI; as 30sR with a turning lever.	

30Th	12.5C	MD Pl	pl	C/D,CI,A,K,L; as 30sT with a turning lever.	
30RZh	12.5C	MD Pl	pl	C/D,CI,%,Fractions (0,125-1.125); as 30sRZ with lever.	
30Tt	12.5C	Dup Pl	pl	C/D,CI,A,K,L	
				S,ST,T,LL1,LL2,LL3	
30RC	12.5C	MD Pl	pl	C/D,CI,%,Fractions (0,125-1.125).	
				£I,£II,£III/B,L,SC,W,D	a.
75E	29C	MD Pl	pl	C/D; for office use.	
75T	29C	MD Pl	pl	C/D,CI,A,K,L,S,T; for office use.	
75RZ Linear	29C	MD Pl	pl?	Patent Heinrich Daemen, Zurich. As 30sRZ for office use.	
Model Topo 6400	5"C	MD Al	pl	B,C/D,SC1,SC2; stand for desk use.	

Additional technical range of slide rules.

Comment:

a.   L,SC,W,D are scales with American and English weights and measures constants.

LOGAREX (Czechoslovakia)

Address: Not known.

History: There was no relationship with the other **"Logo/Loga"** products. The history is not known. Logo of 'Toison Dor' in laurel wreath.

Slide Rules:

1	2	3	4	5	6	7
27103 System Rietz	L	?	?	?	K,A/B,CI,C/D,L//S,ST,T	
27204-11, System Rietz	12.5L	CF	Pl	pl	K,A/B,CI,C/D,L//S,ST,T	
27205						
27403-11 System Darmstadt	25L	CF	Pl	pl	L,P,A/B,CI,C/D,S,T,ST//LL1,LL2,LL3; with and without cms/	
27602-11 Exponent	25L	Dup Pl	pl		L,P,K,A/B,CI,C/D,S,T,ST	
					LL01,LL02,LL03,DF/CF,CIF,CI,C/D,LL3,LL2,LL1	
27606	L				Ahrend; Växjö Sweden?	
27607-1 Studio	25L	Dup Pl	pl		ST,T1,T2,DF/CF,CIF,CI,C/D,P,S	
					LL01,LL02,LL03,A/B,L,K,C/D,LL3,LL2,LL1	

LOGOMAT (Germany)

Address: Logomat Rechengeräte GmbH, Germany.

History: c1970-83, Circular calculators/slide rules in various patterns. Used for advertising, made in various materials. c1976, Slide rule production reduced.

Slide Rules:

1	2	3	4	5	6	7
Fuelling Computer					© R. Barge	
Logomat 1300	C	MD M	pl		largest type	
Pfiffikus	C	MD M	pl		as 1300, but smaller	
Devis-o-mat						
P-disc	C	MD Pl	pl			
Mini-2000	C	MD Pl	pl		various versions, merchandise, technical, scientific	
Kaloriezähler	C	MD Pl	pl			

LOGOS (Switzerland)

Address: Not known.

History: Made cheap pocket sized slide rules in plastic. Logo is a "Bow" as in Bow and Arrow.

Slide Rules:

1	2	3	4	5	6	7
-	12.5L	CF	Pl	pl	A/B,C/D	

LONDON NAME PLATE MANUFACTURING (GB)

Address: London Name Plate Manufacturing Co. London and Brighton, England.

History: Made navigational and other flight computers, also other specialist calculating instruments. Mainly appear to be "make to print" instruments for the UK Armed Forces. History not known.

Slide Rules:

1	2	3	4	5	6	7
Computer Dead Reckoning Mk4A and 4B	4.25"C	MD M		No cursor	various plastic inserts and detail differences	
Range and Endurance Computer, various	5"C	MD M		?		
Height and Airspeed Comp. Mk I & II	4.25"C	MD M		n/a		

Computer Moving Film	6.5"C	MD	M	n/a	
Plate Protractor Navigation, various	U	n/a	Pl	n/a	grid conversions

LORD (GB?)
Address: Not known.
History: Not known. Lord's calculator was made by **ELLIOTT Bros** and others.

Slide Rules:

1	2	3	4	5	6	7
Lord's calculator	6W	n/a	M	n/a	metal scales, C/D plus others	c1910

MAHN (USA)
Address: Mahn & Co, address not known.
History: c1890 sold Colby's Stadia Slide Rule - made by **K&E**?

MAHO (Germany)
Address: Mayr Hörmann & Co, Southern Germany.
History: Known as MAHO, slide rules were made on a plastic covered metal base, and also used perspex as part of the slide.

Slide Rules:

1	2	3	4	5	6	7
No1540 System Rietz	10"L	CF	P/M	m/g	K,A/B,CI,C/D,L//S,ST,T. metal/plastic cursor	

MALLAT (France)
Address: France.
History: Sold a range of circular and linear slide rules, it is possible that they sold or were agents for **CONCISE** circular rules. The logo is a pair of stylised fish on either side of a square containing a pen through a piece of curved paper.

MANTISSA (Germany)
Address: See **V.E.B.**
History: The brand name or a series of slide rules made by **V.E.B.** originally from East Germany. May also have been the company name: VEB-MANTISSA.

MANUFRANCE (France)
Address: St Etienne (Loire), Paris, France.
History: The French equivalent of "Sears Roebuck" who included slide rules in its catalogue. A catalogue from the late 1920's shows three slide rules in boxwood. It also sold boxwood slide rules pre World War 1. It shares the same town as **BARBOTHEU** which might indicate that the rules sold via this catalogue were made by Barbotheu. Links to Barbotheu need confirmation or correction.

MARC (France)
Address: 41 Rue de Maubenge, Paris, France.
History: The Brand Name, or Logo, used on slide rules made by **UNIS** of France. Also used the brand name **MINERVA**.

Slide Rules:

1	2	3	4	5	6	7
-	12.5L	?	Pl	?	A/B,C/D	c1940
No 6	12.5L	?	Pl	?	K,A/B,CI,C/D,L//S,ST,T	c1950
Minerva Rietz No 6A	12.5L	CF	Pl	pl	cm\K,A/B,CI,C/D,L//S,ST,T	

A set of instructions for "Marc" pocket slide rules shows 6 models with the Rietz as No 6, thus by assumption:

No 1 Student	12.5L	CF	Pl	pl	?
No 2 Mannheim	12.5L	CF	Pl	pl	?
No 3 Beghin	12.5L	CF	Pl	pl	?
No 4 Sine-Tangent	12.5L	CF	Pl	pl	?
No 5 Electician	12.5L	CF	Pl	pl	?
No 6 Rietz	12.5L	CF	Pl	pl	?

MARCANTONI (Italy)
Address: Marcantoni, Milan, Italy.
History: Not known, but a major producer or seller of slide rules. Also Cassani-Marcantoni, or **CASSANI**. Possibly also **MARKANT**?

Slide Rules:

1	2	3	4	5	6	7
-	15L	?	Pl	?	?	
25R	25L	?	Pl	?	K,A/B,CI,C/D,L//S,ST,T	

MARINE & OVERSEAS SERVICES (GB)
Address: Not known.
History: Set up in 1939, known to be the proprietors of the **BRITISH SLIDE RULE COMPANY** in 1944.

MARKANT (Italy?)
Address: Not known.
History: See **MARCANTONI**.

MARTIN - St MARTIN (GB)
Address: St Martin's Scale Aids Ltd, London WC2.
History: Not known.

MARTINI (Italy)
Address: Vittorio Martini, Italy.
History: Not known.

Slide Rules:

1	2	3	4	5	6	7
3991, Rietz	12.5L	CF	Pl	pl	K,A/B,CI,C/D,L//S,ST,T	

MARTZ (Germany)
Address: A. Martz, Stuttgart, Germany.
History: c1903, Frank's Einscala-Rechenschieber. No other details known.

MASCHINOIMPORT (Russia)
Address: Maschinoimport, Address not known.
History: Probably only imported slide rules which were then "Badged", used in Russia and possibly re-exported.

Slide Rules:

1	2	3	4	5	6	7
Un-named	5"L	CF	Pl	pl	Plastic slide rule	1965

MASCOT (USA)
Address: Not known.
History: c1930 Mascot and Monitor were names associated with slide rules, Circular two disc plastic known.

MATTERN (USA)
Address: Jimmie Mattern Co. P.O. Box 281, Burbank, California, USA.
History: c1949/50 various navigational and STD instruments.

MAYER & MULLER (Germany)
Address: Mayer & Muller, Berlin.
History: c1902 produced Furle's Rechenbletter, details and history not known.

MEAR (GB)
Address: M.H. Mear & Co, 56 Nettleton Road, Dalton, Huddersfield.
History: In existence c1978.

Slide Rules:

1	2	3	4	5	6	7
Pipe Flow calculator						
M1 Comprehensive Conversion calculator						1978

MEGA (Holland)

Address: Megapromotions, Holland.
History: c1986 produced a patent Radio Netherlands calculator in cardboard, designed to find radio stations. Possibly commissioned from **ALRO.**

MEHMKE (Germany)

Address: R. Mehmke, Stuttgart, Germany.
History: c1898, Mehmke Rechenshieber fur Komplexe Grossen, otherwise no details.

MEIERHOFER (Swiss)

Address: Hans Meierhofer, Fabrique de plaques metaliques, Mellingen A.G. Switzerland.
History: SwissAir Navigators computer, also data slides made by them.

MEISSNER (E. Germany)

Address: Meissner, East Germany.
History: Not known.

Slide Rules:

1	2	3	4	5	6	7
30cm slide rule	30L					1950
Rietz	30L	?	Pl	?	K,A/B,CI,C/D,L//S,ST,T	
75/2b	30L	CF	W/C	?	K,A/B,CI,C/D,L//S,ST,T	
KG Rietz Novo 1	30L	CF	Pl	?	L,A,DF/CF,CI,C/D,S,ST,T; not a true Rietz	
Schulrechner 1172	30L	Dup	Pl	?	A/B,CI,C/D C/D,C/D	
Record 1731421	30L	Dup	Pl	?	L,K,A/B,CI,C/D,LL1,LL2,LL3 P,S,ST,DF/CF,CIF,CI,C/D,T1,T2,DI	
KG Rietz Arcus 1731543	30L	CF	Pl	?	L,A,DF/CF,CI,C/D,S,T,ST	
KG Arcus LL Rietz 1 1731543/1	30L	CF	Pl	?	L,A,DF/CF,CI,C/D,S,T,ST//LL1,LL2,LL3	
KG Electric 1731591	30L	CF	Pl	?	LL1,LL2.LL3,A/B,L,CI,C/D,P,S,T	

MERRIT (USA)

Address: Not known.
History: c1923 a slide rule sold by **WEBER.**

METTALOGRAPH (USA)

Address: Mettalograph Corp, New York, USA.
History: Made the Musketary Rule Model of 1918, 12"L Metal.

MICRONTA (USA)

Address: Not known.
History: c1960 produced an Electro slide rule. Micronta was the brand name of Radio Shack, known latterly for the production of electrical equipment including some of the earliest home computers.

MIGGETT (USA)

Address: W.L. Miggett, USA.
History: c1906 produced Miggett's Cost Estimating slide rule in linear and conical forms. No other details known. *Cajori* shows Younger's Logarithmic Chart for Computing Cost as the replacement for Miggett in 1907, but does not mention manufacturer.

MOLER (USA)
Address: Moler Rule Co?
History: c1935 a line loss and volt drop calculator was produced, whether this was made by Moler, or invented by Moler and made elsewhere is not known.

MONDIA (Switzerland)
Address: Mondia, Not known.
History: Produced a slide rule watch, the "Automatic 'Memory' Model with an in-built circular slide rule. This one is more unusual as it has the two scales within the watch crystal, the scales being driven from external buttons which give the initial impression that the watch has stop-watch functions. The two knobs rotate both scales, there is a hair line engraved on the crystal.

MONROE (USA)
Address: Not known.
History: Not known.

Slide Rules:

1	2	3	4	5	6	7
Pencil slide rule.	P	n/a	Pl	pl	A/B,CI,K,C/D, only slides right, not left.	

MORIN (France)
Address: Henri Morin, 11 Rue du Long, Paris.
History: Régle Henri Morin, unknown date, wood/celluloid 15L & 28L slide rules in various designs.

MUCHALL
Address: Muchall & Co, Address not known.
History: Not known.

Slide Rules:

1	2	3	4	5	6	7
Circular Metal slide rule	C		M			1974

NASHUA (USA)
Address: Nashua Corp, Los Angeles, California 90064, USA.
History: Approx 1970 took over **PERRYGRAF** which became the Perrygraf Division of Nashua Corp.

NATIONAL (Switzerland)
Address: National-Rechenmaschinen AG; also probably National-Rechenwalzen AG, Zurich, Switzerland.
History: Produced cylindrical slide rules, when and why the name changes occurred is not known.

NEGRETTI & ZAMBRA (GB)
Address: Messrs Negretti & Zambra Ltd, Holborn Circus, London EC1.
History: Established 1850, still making slide rules in the 1920's.

Slide Rules:

1	2	3	4	5	6	7
Boucher's Calculating Circle.	W					

NESTLER (Germany)
Address: Albert Nestler, Lahr, Germany.
History: Started design and manufacture of slide rules in 1878 and continued for 100 years to 1978 when slide rule production was discontinued. Nestler had the unfortunate habit of re-assigning numbers when a particular type was discontinued. Basic type names continued to 1955 when the two digit type number which accompanied the name was replaced with a four digit type number. Some slide rules were made in Switzerland, who owned this factory is not known.

Slide Rules:

1	2	3	4	5	6	7
Universal slide rule	L	?	W/C	?	also sold by K&E	1899
Rietz's Rechenschieber	L	?	W/C	?		1903
Peter's Universal Rechenschieber	L	?	W/C	?		1908
"Peter & Perry" slide rule	L	?	W/C	?		1908

Name					Description	Date
Precision slide rule	L	?	W/C	?		1908
Slide rule for wiring calculations	L	?	W/C	?		1909
Chemists No33 (Chemiker 33)	L	?	W/C	?	Ch/Ch',C/D	c1915-55
Commercial No 40 (Kaufmänn 40)	L	?	W/C	?	A/B,%/D//£sd, A,B,D = 1-10	c1915-55
Reinforced Concrete	L		W/C	?		c1935
Timber calcs No 31 (Holzhändler 31)	L	?	W/C	?		c1915-55
Electro 37 (Elektro 37)	L		W/C	?	V,A(PS)/B(KW),C/D,Speed/K,LL	c1915-55
Nestler Cylindrical	T	n/a	M	n/a	Large 18" cylindrical	c1935-45
Nestle No 30	L	L	W/C	?	E,C/D,C'/K,F; E&F are a 20" scale, C' is reversed C.	c1915-38
Nestler	L	?	W/C	?	As Nestle (note name change)	c1945-55
Hanaur	L	?	W/C	?	A,C/D,A,C'/D,K	c1938
Rietz No 23	L	?	W/C	?	K,A/B,C/D,L	1902-55
Reitz	L	?	W/C	?	As Rietz (note spelling change)	1945-55 2.
Precision	L	?		W/C	? All scales double length, spread on stock and slide.	1915-55
Universal	L	?	W/C	?	L,D/C,S,C/A/K//S,T	1915-55
Fix No 29	L	?	W/C	?	DF'/B,C/D	1915-31 1.
Baur	L	?	W/C	?	cm(L)A/B,C/D. Special cursor allows figs from cm to be transferred to the face of the rule.	c1938
Anido	L	?	W/C	?	L,A/B,C/D,L' (allows multiplication of negative numbers)	c1938
"Long"	L	?	W/C	?	Similar to Precision with one scale reversed.	c1915-55
Präzision No 27a,	58L	?	W/C	?	One metre C & D scales, in halves above each other.	
No 14	10"L	CF	W/C	m/g	A/B,C/D//S,L,T	1905?
No 23 Rietz	10"L	CF	W/C	Glass	K,A/B,C/D,L//S,ST,T	
No 11 ZO	5"L	CF	Pl	pl	A/B,C/D	3.
1/B?	10"L	CF	W/C	?	A/B,C/D//S,L,T. DRP173660	>1905
Nr 0208	10"L	CF	W/C	pl	A/B,C/D//S,L,T. DBP917215	>1957
No 0249 *Elemathic*	10"L	CF	Pl	pl	K,A/B,CI,C/D,L	
Nr 0250 Multimathic	10"L	CF	Pl	pl	K,A/B,CI,C/D,L//S,ST,T,LL1,LL2	
No 0251 Alpha	10"L	CF	Pl	pl	K,A/B,CI,C/D,L//S,ST,T	
Nr 23/52	10"L	CF	Pl	m/g	K,A/B,CI,C/D,L//S,ST,T	
-	5"L	CF	Pl	p/m	A/B,?,C/D	
System Rietz No 23	10"L	CF	W/C	m/g	K,A/B,CI,C/D,L//S,ST,T	
Rietz						
Darmstadt						
Duplex Log Log Decitrig	L		Dup			
Rechenwalze, 12.5m					cf. Loga	
Nr0123	12.5L	CF	Pl	pl	example in leather wallet	
Nr0290 Polymath Duplex	25L	Dup	Pl	pl		
System Landis & Gyr	25L	CF	W/C	m/g	K,A/B,CI,C/D,Watt//α1,L,α2	

<u>From a French book describing Nestler slide rules and their use, c1924 (DRP410,565).</u>

Name					Description
No 1c, Règle à calcul pour étudiants	25L	CF	W	pl	A/B,C/D
No 2c, Règle à calcul pour étudiants	25L	CF	W	pl	A/B,C/D//S,T
No 5, Règle à calcul bon marché	25L	CF	W/C	pl	A/B,C/D
No 11, Règle à calcul de poche	12.5L	CF	W/C	m/g	A/B,C/D
No 12, Règle à calcul de poche, acajou	12.5L	CF	W/C	m/g	A/B,C/D
No 12a, Règle à calcul de poche	12.5L	CF	W/C	m/g	cms\A/B,C/D
No 12c, Règle à calcul de poche	12.5L	CF	W/C	m/gm	cms\A/B,C/D
No 14, Marque "Nestler"	25L	CF	W/C	m/g	cms\A/B,C/D
No 22R, Système Rietz	15L	CF	W/C	m/g	cms\K,A/B,CI,C/D,L//S,T
No 23R, Système Rietz	25L	CF	W/C	m/g	cms\K,A/B,CI,C/D,L//S,T
No 23aR, Système Rietz	35L	CF	W/C	m/g	cms\K,A/B,CI,C/D,L//S,T
No 23/3, Système Rietz	25L	CF	W/C	m/g	cms\K,A/B,C/D,L//S,T
No 24R, Système Rietz	50L	CF	W/C	m/g	cms\K,A/B,CI,C/D,L//S,T
No 25, Système Perry	25L	CF	W/C	m/g	cms\LL,A/B,C/D,LL//S,T
No 27, Précision	25L	CF	W/C	m/g	cms\LL,C/D,CI,CF/DF,LL//S,T
No 27/1, Précision	15L	CF	W/C	m/g	cms\LL,C/D,CI,CF/DF,LL//S,T
No 27a, Précision	50L	CF	W/C	m/g	cms\LL,C/D,CI,CF/DF,LL//S,T
No 28, Universelle	25L	CF	W/C	m/g	cms\L,C/S,C/D,A\K,x//S,T	4.
No 28/1, Universelle	15L	CF	W/C	m/g	cms\L,C/S,C/D,A\K,x//S,T	4.
No 28a, Universelle	50L	CF	W/C	m/g	cms\L,C/S,C/D,A\K,x//S,T	4.
No 30, Système Nestle	25L	CF	W/C	m/g	cms\E,A/B,C/D,F//S,T
No 31, Marchand de Bois	20L	CF	W/C	n/a	A/B,-/D; D is 10 to 100
No 33, Chimiste	25L	CF	W/C	m/g	cms\Consts/Consts,C/D,L; Consts are Chemicals.
No 35, Système Peter	25L	CF	W/C	m/g	cms\LL,A/B,C/D,LL//S,T
No 37, Electro	25L	CF	W/C	m/g	cms\V,A/B,CI,C/D,U\x,y//S,T
No 37/1, Electro	15L	CF	W/C	m/g	cms\V,A/B,CI,C/D,U\x,y//S,T
No 37a, Electro	50L	CF	W/C	m/g	cms\V,A/B,CI,C/D,U\x,y//S,T

No 38, Perfekt	25L	CF	W/C m/g	cms\LL,A/B,CI,C/D,K//S,T	
No 39, Electro Système Besser	12.5L	CF	W/C m/g	cms\special electrical scales	
No 40, Pour Commerçants	25L	CF	W/C m/g	cms\Z/T,CI,D/K	
No 41, Liliput	10L	CF	W/C m/g	A/B,C/D	
No 43, Système Nestler	25L	CF	W/C m/g	cms\K,A/B,fe,C/D,L//S,T	

From a Nestler Catalogue, c1932/37:

No 1	25L	CF	W/C pl	A/B,C/D	
No 2	25L	CF	W/C pl	A/B,C/D//S,T	
No 3	25L	CF	W/C m/g	A/B,C/D	
No 4	25L	CF	W/C m/g	A/B,C/D//S,T	
No 5	12.5L	CF	W/C pl	A/B,C/D; cheaper student rule	
No 5	25L	CF	W/C pl	A/B,C/D; cheaper student rule	
No 7	25L	CF	W/C pl	q,Z/T,M,D,K,L; cheaper student rule, kaufmännischer.	
No 8, Rietz	25L	CF	W/C pl	K,A/B,C/D,L; cheaper rule	
No 9, Rietz	25L	CF	W/C pl	K,A/B,C/D,L//S,T; cheaper rule	
No 10	25L	CF	W/C m/g	cm\A/B,C/D//S,T	
No 11	12.5L	CF	W/C m/g	cm\A/B,C/D	
No 11 CO	12.5L	CF	W/C pl	cm\A/B,C/D; 5mm thick rule.	
No 11 CM	12.5L	CF	W/C pl	cm\A/B,C/D//S,T; 5mm thick rule.	
No 11 CR, Rietz	12.5L	CF	W/C pl	cm\K,A/B,CI,C/D,L//S,T; 5mm thick rule.	
No 11 ZO	12.5L	CF	Pl pl	cm\A/B,C/D; 3mm thick rule.	
No 11 ZM	12.5L	CF	Pl pl	cm\A/B,C/D//S,T; 3mm thick rule.	
No 11 ZR, Rietz	12.5L	CF	Pl pl	cm\K,A/B,CI,C/D,L//S,T; 3mm thick rule.	
No 11 ZE, Electro	12.5L	CF	Pl pl	cm\LL2,A/B,CI,C/D,LL3//S,T; 3mm thick rule. Elec scales in well	
No 11 ZK, Kaufmann	12.5L	CF	Pl pl	cm\q,Z/T,R,D/K,L; 3mm thick rule. £.s.d graduations.	
No 12	12.5L	CF	W/C m/g	A/B,C/D	
No 12a	12.5L	CF	W/C m/g	cm\A/B,C/D	
No 12b	12.5L	CF	W/C m/gm	cm\A/B,C/D; magnifying cursor	
No 12c	15L	CF	W/C m/g	cm\A/B,C/D	
No 12d	15L	CF	W/C m/gm	cm\A/B,C/D; magnifying cursor	
No 14/1	25L	CF	W/C pl/m	cm\A/B,CI,C/D//S,T; 1 line cursor	
No 14/3	25L	CF	W/C pl/m	cm\A/B,CI,C/D//S,T; 3 line cursor	
No 18/3	50L	CF	W/C pl/m	cm\A/B,CI,C/D//S,T; 3 line cursor	
No 20/1, Rietz	25L	CF	W/C m/g	cm\K,A/B,C/D,L//S,T; 1 line cursor	
No 22 R, Rietz	15L	CF	W/C pl/m	cm\K,A/B,CI,C/D,L//S,T; 3 line cursor	c1937
No 22 L, Rietz	15L	CF	W/C m/gm	cm\K,A/B,CI,C/D,L//S,T; 3 line cursor, magnifying	c1937
No 23/1, Rietz	25L	CF	W/C pl/m	cm\K,A/B,C/D,L//S,T; 1 line cursor	
No 23/3, Rietz	25L	CF	W/C pl/m	cm\K,A/B,C/D,L//S,T; 3 line cursor	
No 23 L, Rietz	25L	CF	W/C m/gm	cm\K,A/B,CI,C/D,L//S,T; 3 line cursor, magnifying	c1937
No 23 aL, Rietz	36L	CF	W/C m/gm	cm\K,A/B,CI,C/D,L//S,T; 3 line cursor, magnifying	c1937
No 23 R/1, Rietz	25L	CF	W/C pl/m	cm\K,A/B,CI,C/D,L//S,T; 1 line cursor	c1937
No 23 R/3, Rietz	25L	CF	W/C pl/m	cm\K,A/B,CI,C/D,L//S,T; 3 line cursor	c1937
No 23 aR/1, Rietz	36L	CF	W/C pl/m	cm\K,A/B,CI,C/D,L//S,T; 1 line cursor	c1937
No 24 L, Rietz	50L	CF	W/C m/gm	cm\K,A/B,CI,C/D,L//S,T; 3 line cursor, magnifying	c1937
No 24 R/3, Rietz	50L	CF	W/C pl/m	cm\K,A/B,CI,C/D,L//S,T; 3 line cursor	c1937
No 24 R/1, Rietz	100L	CF	W/C pl/m	cm\K,A/B,CI,C/D,L//S,T; 1 line cursor	c1937
No 25, Perry	25L	CF	W/C m/g	cm\LL2,A/B,CI,C/D,LL2; 3 line cursor	
No 26, Betriebs-Rechenschieber	28L	CF	W/C m/g	cm\x,A/B,x,C/D,x,x; x are special machinery scales	
No 27/15, Präzision	15L	CF	W/C pl/m	cm\A,C/D,CI,CIF,CF/DF,L\S,K; 3 line cursor	
No 27/25, Präzision	25L	CF	W/C pl/m	cm\A,C/D,CI,CIF,CF/DF,L\S,K; 3 line cursor	
No 27a/50, Präzision	50L	CF	W/C pl/m	cm\A,C/D,CI,CIF,CF/DF,L\S,K; 3 line cursor	
No 28/15, Universal	15L	CF	W/C pl/m	cm\L,D/cos^2,sin.cos,C/D,A\x,x; 3 line cursor	4.
No 28/25, Universal	25L	CF	W/C pl/m	cm\L,D/cos^2,sin.cos,C/D,A\x,x; 3 line cursor	4.
No 28/50, Universal	50L	CF	W/C pl/m	cm\L,D/cos^2,sin.cos,C/D,A\x,x; 3 line cursor	4.
No 29, Duplex	25L	Dup	W/C pl/m	DF/CF,CIF,C/D K,A/B,CI,C/D,L	
No 29a, Duplex	25L	Dup	W/C pl/m	LL0,A/B,S,T,C/LL3,LL2,LL1 K,A/B,CI,C/D,L	
No 30, Nestle	25L	CF	W/C m/g	cm\E,A/B,C/D,F	
No 31, Holzhändler-Rechenschieber	20L	CF	W/C m/g	C/L,consts/D	
No 33, Chemiker	25L	CF	W/C m/g	cm\chem consts/chem consts,CI,A/B	
No 34, Hannaur	25L	CF	W/C m/g	cm\O,A/D,q,R/B,C	a.
No 36, Elektro	15L	CF	W/C pl/m	cm\LL2,A/B,CI,C/D,LL3\\D,V	
No 36, Elektro	25L	CF	W/C pl/m	cm\LL2,A/B,CI,C/D,LL3\\D,V	
No 36a, Elektro	25L	CF	W/C pl/m	cm\LL2,A/B,CI,C/D,LL3\\D,V/K,x	
No 37/15, Elektro	15L	CF	W/C pl/m	cm\V,A/B,CI,C/D,U\LL2,K	
No 37/25, Elektro	25L	CF	W/C pl/m	cm\V,A/B,CI,C/D,U\LL2,K	
No 37a/50, Elektro	50L	CF	W/C pl/m	cm\V,A/B,CI,C/D,U\LL2,K	
No 38, Perfect	25L	CF	W/C pl/m	cm\LL2,A/B,CI,C/D,K	

No 40/15, kaufmännischer R-schieber	15L	CF	W/C	m/g	cm\q,Z/T,M,consts,D/K,L	b.
No 40/25, kaufmännischer R-schieber	25L	CF	W/C	m/g	cm\q,Z/T,M,consts,D/K,L	b.
No 40a/50, kaufmännischer R-schieber	50L	CF	W/C	m/g	cm\q,Z/T,M,consts,D/K,L	b.
No 41 Z, Liliput	10L	CF	Pl	pl	cm\A/B,C/D//S,T	
No 42 Z, Liliput	10L	CF	Pl	pl	cm\A/B,C/D	
No 43, Eisenbetton sys Nestler-Hoffmann	25L	CF	W/C	m/g	cm\K,A/B,fe,C/D,L	
No 44, Rechenwalze	25T	n/a	M	n/a	25 x 5.5cm dia; equivalent to 1.6m slide rule	
No 44a, Rechenwalze	33T	n/a	M	n/a	33 x 9cm dia; equivalent to 3.75m slide rule	
No 45, Rechenwalze	53T	n/a	M	n/a	53 x 16cm dia; equivalent to 12.5m slide rule	

Demonstration slide rules:

No 14D	100L	CF	W/p	m/g	A/B,C/D	
No 14D	200L	CF	W/p	m/g	A/B,C/D	
No 23D, Rietz	100L	CF	W/p	m/g	K,A/B,CI,C/D,L//S,T	
No 23D, Rietz	150L	CF	W/p	m/g	K,A/B,CI,C/D,L//S,T	
No 23D, Rietz	200L	CF	W/p	m/g	K,A/B,CI,C/D,L//S,T	
No 37D, Elektro	200L	CF	W/p	m/g	V,A/B,CI,C/D,U\LL2,K	
No 40D, Kaufmann	200L	CF	W/p	m/g	q,Z/T,M,consts,D/K,L	b.
No 5D	100L	CF	W/p	m/g	A/B,C/D	
No 7D	100L	CF	W/p	m/g	q,Z/T,M,D,K,L; kaufmann.	
No 8D, Rietz	100L	CF	W/p	m/g	K,A/B,C/D,L	

From a Nestler Catalogue, c1967:

Pocket slide rules:

0115, Techniker	12.5L	CF	Pl	pl	cm\A/B,C/D,//S,ST,T
0123, Rietz	12.5L	CF	Pl	pl	cm\K,A/B,CI,C/D,L//S,ST,T
0121, Darmstadt	12.5L	CF	Pl	pl	cm\P,K,A/B,CI,C/D,S,T//LL1,LL2,LL3
0129, Polymath-Duplex	12.5L	Dup	Pl	pl	LL01,LL02,LL03,A/B,L,K,C/D,LL3,LL2,LL1 T,ST,DF/CF,CIF,CI,C/D,P,S
0137, Elektro	12.5L	CF	Pl	pl	cm\V,A/B,CI,C/D,U,cos//S,ST,T
0149, Stahlbeton-Maarschalk	12.5L	CF	Pl	pl	cm\k$_z$,A/B,C/D,Tabelle//Tabele/S,ST,T/Tabelle
0127, International	12.5L	CF	Pl	pl	cm\ins(1/16),mm,A/B,CI,C/D,consts//^{0}F,^{0}C
0140, Kaufmann	12.5L	CF	Pl	pl	cm\DF/CF,CI,C/D//LL1,LL2,LL3

School slide rules:

0239, Schul-Rietz	25L	CF	Pl	pl	cm\K,A/B,CI,C/D,L//S,ST,T
0251, Alpha	25L	CF	Pl	pl	K,A/B,CI,C/D,L//Greek alphabet/cms,ins
0252, Beta	25L	Dup	Pl	pl	K,A/B,CI,C/D,L cm,DF/CF,CIF,CI,C/D,ins
0253, Gamma	25L	Dup	Pl	pl	K,A/B,CI,C/D,L T1,T2,DF/CF,CIF,CI,C/D,S,ST
0254, Delta	25L	Dup	Pl	pl	L,K,A/B,CI,C/D,LL3,LL2,LL1 T1,T2,DF/CF,CIF,CI,C/D,S,ST
0409, Schul-Merkur	25L	CF	Pl	pl	cm\DF/CF,CI,C/D//LL1,LL2,LL3
0232, Rietz	25L	CF	W/C	pl	cm\K,A/B,CI,C/D,L\ins//S,ST,T
0235, Rietz	50L	CF	W/C	pl	cm\K,A/B,CI,C/D,L\ins//S,ST,T
0238, Rietz	25L	CF	Pl	pl	cm\K,A/B,CI,C/D,L//S,ST,T
0210, Darmstadt	25L	CF	W/C	pl	cm,L\K,A/B,CI,C/D,P\S,T//LL1,LL2,LL3
0215, Darmstadt	50L	CF	W/C	pl	cm,L\K,A/B,CI,C/D,P\S,T//LL1,LL2,LL3
0218, Darmstadt	25L	CF	Pl	pl	cm\L,K,A/B,CI,C/D,P,S,T//LL1,LL2,LL3
0289, Rietz-Duplex	25L	Dup	Pl	pl	L,K,A/B,BI,CI,C/D,R1,R2 T1,T2,DF/CF,CIF,CI,C/D,P,S,ST
0291, Polymath-Duplex	25L	Dup	Pl	pl	LL01,LL02,LL03,A/B,L,K,C/D,LL3,LL2,LL1 T,ST,DF/CF,CIF,CI,C/D,P,S
0292, Multimath-Duplex	25L	Dup	Pl	pl	LL00,LL01,LL02,LL03,DF/CF,CIF,CI,C/D,LL3,LL2,LL1,LL0 P,S,T1,T2,A/B,K,CI,C/D,R1,R2,L,ST

Special slide rules:

0260, Mecanica	25L	CF	W/C	pl	special mechanical scales
0281, Geometer	25L	CF	W/C	pl	cm,L,D/sin,cos^2,C/D,A,K,BI//S,ST,T
0330, Chemiker	25L	CF	Pl	pl	special chemical scales
0370, Elektro	25L	CF	W/C	pl	cm\V,A/B,CI,C/D,U,LL,K//S,ST,T
0297, Electronic	25L	Dup	Pl	pl	special electronic scales
0440, Stahlbeton-Maarschalk	25L	Dup	Pl	pl	special electronic scales k$_z$,K,A/B,k$_h$,CI,C/D,k$_z$,L T1,T2,DF/CF,CIF,CI,C/D,P,S,ST
0401, Merkur	25L	CF	Pl	pl	cm\%,DF/CF,CI,C/D,Umr//LL1,LL2,LL3

Demonstration slide rules:

0203, Alpha	150L	CF	W/C pl	K,A/B,CI,C/D,L,S,ST,T
0206, Rietz	100L	CF	W/C pl	K,A/B,CI,C/D,L//S,ST,T
0207, Kaufmann	100L	CF	W/C pl	DF/CF,CI,C/D//LL1,LL2,LL3
0208, Delta	150L	Dup	W/C pl	L,K,A/B,CI,C/D,LL3,LL2,LL1
				T1,T2,DF/CF,CIF,CI,C/D,S,ST
0205, Darmstadt	100L	CF	W/C pl	L,K,A/B,CI,C/D,P,S,T//LL1,LL2,LL3
0204, Polymath-Duplex	150L	Dup	W/C pl	LL01,LL02,LL03,A/B,L,K,C/D,LL3,LL2,LL1
				T,ST,DF/CF,CIF,CI,C/D,P,S
0202, Multimath-Duplex	150L	Dup	W/C pl	LL00,LL01,LL02,LL03,DF/CF,CIF,CI,C/D,LL3,LL2,LL1,LL0
				P,S,T1,T2,A/B,K,CI,C/D,R1,R2,L,ST

From a Belgian Price List c1972:

Modèle de poche:

0149, Beton	12.5L	CF	Pl	pl
0121, Darmstadt	12.5L	CF	Pl	pl
0123, Rietz	12.5L	CF	Pl	pl
0130, Multimath-Duplex	12.5L	Dup	Pl	pl
0137, Electro	12.5L	CF	Pl	pl
0140, Commerçant -Handel	12.5L	CF	Pl	pl
0129, Polymath-Duplex	12.5L	Dup	Pl	pl
0113S	12.5L	CF	Pl	pl

Modèle scolaire:

0239, Rietz scolaire	25L	CF	Pl	pl
0252, Béta	25L	CF	Pl	pl
Elemath RS	25L	CF	Pl	pl
Elemath log-log	25L	CF	Pl	pl
Elemath Standard	25L	CF	Pl	pl
0409, Commerçant - Handel	25L	CF	Pl	pl

Règle à calculer de précision:

0218, Darmstadt	25L	CF	Pl	pl
0238, Rietz	25L	CF	Pl	pl
0289, Rietz Duplex	25L	Dup	Pl	pl
0292, Multimath Duplex étui cuir	25L	Dup	Pl	pl
0292, Multimath Duplex étui en plastique	25L	Dup	Pl	pl
0297, Electronic	25L	CF	Pl	pl
0401, Commerçant - Handel	25L	CF	Pl	pl
0440, Béton Maarschald	25L	CF	Pl	pl
Polymath	25L	CF	Pl	pl

Règle à calculer de précision en bois:

0210, Darmstadt	25L	CF	W/C pl	
0232, Rietz	25L	CF	W/C pl	
0260, Mécanica	25L	CF	W/C pl	
0281, Géomètre - Landmeter	25L	CF	W/C pl	
0370, Electro	25L	CF	W/C pl	
0215, Darmstadt	50L	CF	W/C pl	
0235, Rietz	50L	CF	W/C pl	

Règle à calculer de démonstration:

0201, Rietz	150L	CF	Pl?	pl
Elemath log-log	150L	CF	Pl?	pl
0209, Multimath Duplex	150L	CF	Pl?	pl
Elemath Standard	150L	CF	PL?	pl

Règle à calculer de projection:

Schul-Rietz	25L	n/a	C	n/a
Elemath Standard	25L	n/a	C	n/a
Elemath log-log	25L	n/a	C	n/a

Comments:

1. The Fix slide rule became the Doppelr. 29 in approx 1931. DF is folded at 7854 ($\pi/4$) for calculations involving circles.
2. The change of spelling from Rietz to Reitz may be a printing error in English books, or may be an "Anglicization" of the name.
3. The older two digit Nestler codes used all sorts of additional letter codes to indicate material, scales, length etc. The 11 ZO stood for "Zelluloid, Ohne (without) Sinus".
4. 360^0 or 400^0 variants available.
a. The scales are actually A,D/C,B,CI/A,K by function and as scales listing. Nomenclature above is slide rule markings.
b. The scales are actually %,C/D,CI,%/C,L by function and as scales listing. Nomenclature above is slide rule markings.

NICHOLL (GB)

Address: J.A. Nicholl & Co, London.
History: Not known.

Slide Rules:

1	2	3	4	5	6	7
Artillery slide rule.						1917
Fire (Artillery) slide rule	40L	?	W	?	?	1918
Observation of Fire slide rule	16"L	OF	Al	m	special scales	

NICHOLLSON (GB)

Address: J.A. Nichollson & Co, London?
History: Not known. See also **FOUNTAIN & SEENEY.**

Slide Rules:

1	2	3	4	5	6	7
Applied observation of shot slide rule	24"L	OF	Al	m	special scales	1917

NICKEL (USA)

Address: E.F. Nickel, New Jersey, USA.
History: c1906 Nickel's Runnerless slide rule, otherwise history not known.

NICOLET (USA)

Address: Not known.
History: c1930 in existence, products not known.

NORDEN (Sweden)

Address: Sweden
History: May have had some link with A.B. Bofors, 2/48 System Darmstadt slide rule known.

NORMA (Germany)

Address: B. Norma, Germany.
History: Known for its **GRAFIA** series of circular slide rules with their very unusual design of cursor which appears to "hang" in the air on the outside of the rule. A number of the rules are labelled DBGM - the German registered trade mark notation. Also marked on some slide rules: Norma Marque Depose - System Huber, Cervin Swiss, Nederlandsee maatschaapij....

Slide Rules:

1	2	3	4	5	6	7
Grafia 100	10C	?	Pl	?	C/D; Paper formats;	
Grafia 190	19C	?	M	?	Packaging/Paper scales;	
Gambrinus 100	10C	?	Pl	?	A/B,Brewing (x2),cos; For Brewing & Malting;	
Mercuria 190	19C	?	M	?	C/D,CI,%,%;	
190	19C	?	M	?	C/D;	

NORMUS (Switzerland)

Address: See **HUBER.**
History: The brand name of the original slide charts as would ultimately be made by many manufacturers to deal with the specialist needs of many specific industries. See also **HUBER, OMARO, IWA etc.**

NORTON & GREGORY (GB)

Address: Norton & Gregory Ltd., 1917: Castle Lane, Buckingham Gate, Westminster, London S.W.1.
also 34 Robertson St, Glasgow, Scotland; and 71 Queen Street, Glasgow. (Unknown date, but at the same time as Castle Lane, London.)
History: Sold slide rules for **UNIQUE** in their early days, also other mathematical instrument makers slide rules and general mathematical instruments.

Slide Rules:

1	2	3	4	5	6	7
Cardboard slide rule	25L	CF	P	c	A/B,C/D; a very simple cardboard slide rule.	

NOVOTINI (USA)
Address: Not known.
History: John J. Novotini produced a number of slide rule designs starting in 1920, a very popular design was the Novotini Weight Rule still being sold in 1960. There was a tie with **WARREN KNIGHT** who also sold the Novotini as well as other weight rules.

NYSTROM (USA)
Address: J.W. Nystrom, 31 Union Street, Philadelphia, USA.
History: 1851, designed Nystrom's Circular Calculating Rule, he made the earliest (prototypes?), quantity production (probably no more than 100) was made by **YOUNG** and **THORSTED**.

OMARO (France)
Address: 13 Rue de la Nation, Paris, France; also 13 Rue de Sofia, 75018, Paris (n.b. Rue de Sofia used to be Rue de la Nation).
History: Founded 1928 to distribute the Swiss **NORMUS** data slides, still in business today. Produced various slide rules and slide charts. See also **KOSINE** who used the Omaro name as a trade mark on their calculators during the same period. The Omaro P1 is a slide chart associated with steel sections.

Slide Rules:

1	2	3	4	5	6	7

An undated catlogue sheet lists approximately 36 slide charts, some examples:
C8 Conversions
F1 Metric Nuts and Bolts
H5 Hydraulic circulating conductors
M1 Expressions Trigonometriques
P19 Metal sections - beams etc.
R3 Spring calculations
U3 Tolerances

OTIS-KING (GB)
Address: Various, see **CARBIC**.
History: Otis-King Calculators were made by **CARBIC** for many years, full details under **CARBIC**.

P.I.C. (GB)
Address: Precision Instrument Company, address not known.
History: 1901. taken over by **A.G. THORNTON**. P.I.C. name continued to be used by Thornton to c1970.

PARK
Address: Park Instrument Co, address not known.
History: Not known.

Slide Rules:

1	2	3	4	5	6	7
-	12.5L	?	?	?	A/B,CI,C/D,K//S,L,T	

PEASE (USA)
Address: Not known.
History: c1928, sold slide rules and drafting supplies. There was an association with **US BLUEPRINT**.

PENOL
Address: Not known.
History: Not known. Use an anchor with the letters C & O around it as a logo.

Slide Rules:

1	2	3	4	5	6	7
102A	25L	?	Pl	?	A/B,CI,C/D,K	
528A	15L	?	Pl	?	A/B,CI,C/D,K	
1005A	25L	?	Pl	?	S,K,A/B,CI,C/D,L,T	

PERILOG (France?)

Address: Not known.
History: Not known what if any link there was with **PERRYGRAPH**, range of slide rules produced or sold. May have been French from a "Modele Depose" plastic circular example.

PERRAY (France)

Address: Le Perray, Yvelines, France.
History: Known to have made a 13cm Metal circular slide rule.

PFENNINGER (Switzerland)

Address: Zurich, Switzerland.
History: Distributer under their own brand name of **KOCH** slide rules.

PERRYGRAPH (USA)

Address: 1962: Perrygraf, Maywood, Illinois, USA.; 1971: Perrygraf Division, Nashua Corp, Los Angeles, California 90064, USA.
History: Produced numerous specialist calculators and slide rules, in cardboard and plastic, for sophisticated equipment suppliers. Still making slide charts to date, may be some link with **SLIDE CHARTS**. Examples include:

Slide Rules:

1	2	3	4	5	6	7
Antenna Calculator (Gabriel)					Specialist scales	1959
Noise Figure Slide Rule (Cutler Hammer)					Specialist scales	1964
RF Filter Performance Chart, (Telonic Engineering)					Special scales	1965
Capacitance, Inductance, Reactance Calculator, (Hewlett Packard)						1971
Signal Distribution System Performance Calculator (Norlin Comms)						1978
Microwave/Power Amp calculator (Varian)						1986
Valve sizing calculator						1962

PICKETT & ECKEL (USA)

Address: Pickett & Eckel, Chicago 3, Illinois, USA.
also, Pickett Inc, Santa Barbara, California, USA.; and R.C. Pickett, Alhambra, California, USA. (c1949, P&E Inc);
and Pickett A.F. Eckell (c1949)
History: Not known what the relationship between the various addresses and companies was. P&E were in business in 1947 and to 1960 and beyond, made approx 100 models of slide rule, early models have tongue on slide, intermediate models have tension springs fitted, latest models have groove on slide. Well known for its Aluminium slide rules, but also made some Plastic slide rules.
Model numbers have suffix ES for "Eye-saver yellow scales", or, T for Traditional White, or N for Nylon cursor (>1950).

Slide Rules:

1	2	3	4	5	6	7
Demonstration slide rule	4'L				√,K,A/B,S,ST,T,CI,C/D,?,?	
N.3-T Powerlog exponential	10"L	Dup	Al	pl	Aluminium Duplex.	1960
Deci-Point	L	Dup				
Model 1	L	Dup				
Model 2	L	Dup				
Model 3	L	Dup				
Model 4	L	Dup				
Model 6	L	Dup			Quality Control slide rule	1946
200 Series	6"L					
300 Series	6"L					
600 Series	6"L					
Model 901	L	CF	Al	?	A/B,CI,C/D,K//instructions	
Model 902	L					
Mannheim	L	CF?	Al	?	sheet metal	
Linear	L	??	Pl	pl		
Circular	C					
Fisher Control Valve sizing rule	L	??	Pl	?	made for the Fisher Governor Co, Marshaltown, Iowa.	
Lietz Dial rule	4.2"C				See **LIETZ**	
N4-ES Vector Log-Log dual base speed	10"L	Dup	Al	pl	A,B,C,D,CI,CF,DF,CIF,DI,CI,T,ST,S,[3]√,√, DF/M,CF/M,TH,SH,Ln,L,LL1,LL2,LL3,LL4,LL01,LL02,LL03,LL04	1959,a.
N-200-ES Pocket Trig	5"L	Dup	Al	pl	K,A/B,CI,C/D,L S,ST,T,C/D	
N-500-ES Hi Log Speed rule	10"L	Dup	Al	pl	LL1,LL02,A/B,T,ST,S,C/D,DI,K,[3]√,√, LL2,LL02,DF/CF,CIF,L,CI,C/D,LL3,LL03	1962

N-515-T Electronics	10"L	OF	Al	pl	A,B,C,D,CI,L,electrical scales; electrical formulae on back.	a.
N-531-ES Capitol Radio Engineer	10"L	?	Al?	?	?	
N-600-T Log-log	5"L	Dup	Al	pl	Aluminium Duplex.	
N-800-T Log-Log Speed rule	10"L	Dup	Al	pl	LL1,LL01,A/B,S,ST,T,C/D,DI	
					LL2,LL02,DF/CF,CIF,L,CI,C/D,LL3,LL03	
N-902-ES Simplex Trig	10"L	OF	Al	pl	K,A/B,CI,S,T,C/D,L; also in - T	
800-ES Log-Log Synchro scale	10"L	Dup	Al	pl	A,B,C,D,DI,CF,CIF,CI,DF,K,L	
					LL1,LL2,LL3,LL01,LL02,LL03,S,ST,T	a.
803T Dual Base Log-Log	10"L	??	Al	?	?	
1010 Power Trig	10"L	Dup	Al	pl	DF/D,ST,T,S,C/D,L	
					A/B,CI,CIF,C/D,DI,K	1948,b.
Microline 120-ES	10"L	OF	Pl	pl	?	
Microline 121 Simplex	10"L	OF	Pl	pl	K,A/B,S,T,CI,C/D,L	
Microline 140 Duplex	10"L	Dup	Pl	pl	A,B,C,D,CF,DF,CI,K	
					LL0,LL00,LL1,LL2,LL3,S,ST,T	a.
ED-26, Projection rule	10"L	n/a	Pl	pl	A/B,CI,C/D; made by Projection Optics Company, Inc.	
Trainer	10"L	CF	Pl	pl	S,K,A/B,CI,C/D,L,T; equivs on back	
No 111						
B1						
120T Student	10"L	?	Pl	?	?	
140-ES Log-Log	10"L	?	Pl	?	?	
140-T Microline Log-Log	10"L	?	?	?	?	
160-CT Pocket	6"L	?	Pl	?	?	
101C	4.5"C					
104C Profit Calculator	8"C					
105C Profit Calculator	4"C					
106C Proportional Calculator	6"C					

Comment:
a.   scales listed, disposition not certain.
b.   Pickett, A.F. Eckel

PICOLET (USA)
Address: L.E. Picolet, Philadelphia, USA.
History: Not known.

Slide Rules:

1	2	3	4	5	6	7
Picolet Circular Slide Rule	C	MD	M	n/a	2 disc 1 cursor	c1915-45

PICKWORTH GB)
Address: C.N. Pickworth, c1908: Manchester, England; later: 17 Griffith Drive, Southport, UK
History: Pickworth had a design of slide rule made for him by **FABER-CASTELL**, and is also the author of one of the standard works on slide rules from c1900 to 1954.

Slide Rules:

1	2	3	4	5	6	7
Pickworth's Power Computer	L					1908
Engine Power Computer	L				special scales for engine power calcs	c1915-55
Faber - Pickworth slide rule	10"L	CF	W/C	m/g	special back for access to the rear scales on slide.	c1917

PIGNONE (Italy)
Address: Pignone Sud, Italy
History: Known to have produced a range of circular Control Valve Sizing rules. Some marked W.E. Kreiss?

PILOT
Address: Not known.
History: Not known, but a Rietz pattern plastic slide rule with this name is known, made in Japan.

PLASTOGRAPH (Italy)
Address: Como, Italy.
History: Not known.

Slide Rules:

1	2	3	4	5	6	7
Circular	10C	MD	Pl	pl	A/B; cursor is double ended.	

PLATH (Germany)

Address: C. Plath, Hamburg, Germany.
History: Not known.

Slide Rules:

1	2	3	4	5	6	7
Dreieckrechner (flight computer)	15C	MD	Al	?	Types 1.1 and 2,2	c1930

POST (USA)

Address: Frederick Post Co. Chicago, Illinois, USA.
History: Established c1921, sold **GILSON, HEMMI** as well as its own slide rules. Various associations with other companies: c1936, **POST-HEMMI**; c1959, **POST-DIETRICH**; c1970, **POST-TELEDYNE**. What the nature of the early associations was is not known, however Teledyne bought out Post c1970. Also have had known slide rules "made in Occupied Japan".

Slide Rules:

1	2	3	4	5	6	7
Hemmi Log-Log Vector	10"L	Dup	W/C	m/p	L,K,A/B,CI,C/D,T,GΦ	
					Φ,RΦ,P/Q,Q',C/LL3,LL2,LL1	1947
Versalog	10"L	Dup	W/C	m/g	LL0,LL01,K,DF/CF,CIF,CI,C/D,R1,R2,L	
					LL01,LL02,LL03/T,ST,S,C/D,LL3,LL2,LL1	1955
No 111	C	?	?	?	?	
Gilson Atlas	C	?	?	?	see GILSON	1950
1444K	5"L	CF	Pl	m/g	A/B,CI,C/D,K//S,L,T	a.
1447	10"L	?	?	?	?	a.
Versatrig 1450	10"L	Dup	B/C	m/p	S,DF/CF,CIF,S,T,ST,C/D,R1,R2	
					L,T,SecT,Cos,S,K	a.
Versalog 1460	10"L	Dup	B/C	m/p	LL0,LL01,K,DF/CF,CIF,CI,C/D,R1,R2,L	
					LL01,LL02,LL03/T,ST,S,C/D,LL3,LL2,LL1,Sec,Cos	a.

Comment:
a.   Made by Hemmi, disposition of scales uncertain.

PRECISION (USA)

Address: Not known.
History: PRECISION was the brand name of a range of cheap plastic slide rules and calculators made by **STERLING Plastics**.

Slide Rules:

1	2	3	4	5	6	7
Precision.		10"L	CF	Pl	?	Plastic rule.	

PRESTO

Address: Presto Calculators.
History: Not known. Possibly made **PRESTOLOG** circular slide rules, e.g. Profitmeter 5"C c1935.

PRESTOLOG

Address: Presto Calculators.
History: Not known. See **PRESTO**.

PRESTON (GB)

Address: E. Preston, Birmingham, England.
History: A well known UK rule maker, bought out by **RABONE** in 1932. Particularly known for Two foot, two fold rules of the Coggeshall variety.

PROELL (Germany)

Address: R. Proell, Dresden, Germany.
History: c1902 made Proell's Rechentafel, also known as Proell's Pocket Calculator, sold by **GRIFFIN**. Otherwise history not known.

PULLER (Germany)
Address: E. Puller, Saarbrucken, Germany.
History: c1900, Puller's Rechenschieber, otherwise no details known.

PYTHAGORAS (Germany)
Address: Paul Katzwinkel,Zusenhausen bei Heidelberg, Germany.
History: PYTHAGORAS is the trade name for slide rules made by **KATZWINKEL**.

QUEEN (USA)
Address: James W. Queen, The Union, 91 Walnut Street, Philadelphia, USA.
History: Established c1850, still in business 1897, known to have sold **FABER** slide rules c 1902. See also **THORSTED** and **YOUNG** who made/sold **NYSTROM**'s calculator.

Slide Rules:

1	2	3	4	5	6	7
Nystrom's Calculator.						1852

RABONE (GB)
Address: James Rabone & Sons, Birmingham, England; also offices in the USA, address unknown.

History: Long established in the UK, known to have done business in the USA c1880. Bought out **E. PRESTON** in 1932. Still in existence making ordinary measuring rules, it had stopped making slide rules by 1963.

RSC (USA?)
Address: Not known.
History: may have been **LAWRENCE** - shown on antenna slide rule.   Also shows Radiation Laboratory, MIT, Cambridge, Mass, USA.

REED (Australia)
Address: Sydney, Australia.
History: Known from one slide rule, history not known.

Slide Rules:

1	2	3	4	5	6	7
Service Electronic Engineers	10"L	?	W/C	?	K,A/B,CI,C/D//D/L	

REGAL (Japan)
Address: Not known.
History: These may have been the Regal Series of the Helix (Universal) Co Ltd of Stourbridge, Worcestershire included F106 for Engineers, Technician Colleges, University etc. See also **HELIX**.

Slide Rules:

1	2	3	4	5	6	7
No F106	10"L	CF	Pl	pl	K,DF/CF,CIF,CL,C/D//TI2,TI1,L,S1	
A50LL	13L	?	Pl	?	LL2,A/B,C/D,LL3	
-	29L	?	Pl	?	K,DF/CF,CIF,CI,C/D,L	

7 types of Rule, 5 off 10", 2 off 5".

REISS (Germany)
Address: Reisszeugfabrik, Nurnberg, Germany.
History: 1923 patent. Still making a range of slide rules as late as 1970, the later identified slide rules are from the *Concise Encyclopedia of Mathematics* from Van Nostrand Rheinhold c1971.

Slide Rules:

1	2	3	4	5	6	7
15cm aluminium rule.	15L		M			
3212	5"L	CF	Pl	pl	cm\K,A/B,CI,C/D,L	
3214 Darmstadt Record	10"L	Dup	Pl	m/p	P,S,ST,DF,CF,CIF,CI,C/D,T1,T2,CI	
					L,K,A/B,CI,C/D,LL1,LL2,LL3	
-	10"L	CF	Pl	??	K,A/B,CI,C/D,L//??	
-	10"L	CF	W/C	pl	L,?/?,CI,C/D,£.s.d	
Rietz?	10"L	CF	W/C	pl	K,A/B,CI,C/D,L//??	

Darmstadt	10"L	Dup Al	pl	K,A/B,CI,C/D,L	
				Cos,S/T/LL1,LL2,LL3/Sin,Tan	
Cylindrical	T				
3223 Progress	25L	Dup Al	pl	A,DF/CF,CI,C/D,L	
				T1,T2/LL1,LL2,LL3/D,S	

RELAY (Japan)

Address: Relay Keisanjaku, Japan; also Relay Industrial Co, Japan, and San-Ai Keikico Ltd.
History: Established 1951, later taken over by **RICOH**, became Ricoh Keisanjaku. Relay slide rules were sold in the USA in 1955, an Electric Vector and an Electro model are known.

Slide Rules:

1	2	3	4	5	6	7
151	10"L	Dup	B/C	?	LL00,L,K,A/B,S,ST,T,C/D,DI,P,LL0	
					LL01,LL02,LL03,DF/CF,CIF,CI,C/D,LL3,LL2,LL1	
80DK1	L				marked Relay Industrial Co	
DT1015	L					
514	12.5L	?	B/C	?	11 scales	c1962
#82 Skyline	?					

RELIABILITY (USA)

Address: Reliability & Confidence Slide Rules, address unknown.
History: c 1960 were selling **AGA**, **AERO and ASTRO** Corporation slide rules.

RELIANCE (GB)

Address: Not known..
History: Not known, probably British, may have been a retailer only. One example known is a Zinc mesh framework with plastic covering. Logo is a head in a helmet.

Slide Rules:

1	2	3	4	5	6	7
Rule-Slide 10"	10"L	CF	W/C	pl	A/B,C/D//S,L,T; Zinc mesh frame?	c1950

REVERE (USA)

Address: Revere Corp, Wallingford, Connecticut, USA.
History: Made the **FELSENTHAL** range of Flight Computers and its own Load Adjusters. See also **COX & STEVENS**.

RICHARDSON (USA)

Address: George Washington Richardson. Richardson Slide Rule Co, Chicago, Illinois.
History: Made slide rules between 1910 and approx 1940. Provisional Patents 1-31-11, and 3-26-12. Used "Owl" Logo as **GILSON**.

Slide Rules:

1	2	3	4	5	6	7
The Pyramid Slide Rule No 1898	5"L	CF	W?	pl	special format of scales to give 200" equiv length. $8.00	1917
Direct Reading Slide Rule No 812	10"L	CF	M	pl/m	A/B,C/D; folded light gauge metal, printed pl scales, $2.50	1917
Adding & Subtracting rule No 1812	10"L	CF	M	pl/m	add/subtract,mult/div	1917
Polymetric Slide Rule No 1776	10"L	CF	M	pl/m	A/B,CI,C/D; $3.50	1917
Logometric Slide Rule No 1860-LL	10"L	CF	M	pl/m	A/B,LL1,LL2,LL3,C/D; $4.00	1917
Binary Polymetric Slide	n/a	n/a	M	n/a	slide for use with above;/C,CI,inv. Binary fractions,integers 1-10.	
Business Mans rule	L					
Pantocrat Slide Rule	L				$10.00 with leather case	
Educator	L	CF	M	pl/m	$1.50	
Naval Gunnery rule	17"L	OF	W/M		A/B,C/D//B,C	1917
International Correspondence Schools	10"L	CF	M	pl	K,A/B,CI,C/D	1939
Also some Gilson circular rules						

RICOH (Japan)

Address: Ricoh Keisanjaku, Japan. Also a United Kingdom Office.
History: See **RELAY**.

Slide Rules:

1	2	3	4	5	6	7
Darmstadt Slide Rule	10"L		Pl	pl		
No 105 (Rietz)	10"L	CF	B/C	?	K,A/B,CI,C/D,L//S,ST,T	
No 252 Duplex	10"L	Dup	B/C	pl	DF/CF,CIF,CI,C/D,L	
					K,A/B,T1,T2,ST,S/D,DI	
No 403	10"L	CF	B/C	?	A/B,CI,C/D//S,L,T	
No 514	10"L	CF	B/C	?	K,DF/CF,CI,C/D,A//T1,T2,L,S	a.

Comment.
a.   With advertising information.

RIEGER (Germany)
Address: Prof. J. Rieger, Germany.
History: Not known.

Slide Rules:

1	2	3	4	5	6	7
Slide rule.						1920

RINGPLAN (GB)
Address: See **BOOTS**.
History: RINGPLAN was the Brand name of slide rules and other mathematical instruments sold by **BOOTS**.

RISTA (Denmark?)
Address: Not known.
History: Not known, but made or sold a standard range of slide rules. There is a possible link with **UTO**, and indeed it may have been UTO sold under the Rista name (an example has UTO mentioned in the instructions).

Slide Rules:

1	2	3	4	5	6	7
Rietz 1	27L	?	Pl	pl	27cm\K,A/B,CI,C/D,L//S,ST,T;	
Electro 2	27L	?	Pl	pl	27cm\Motor,K,A/B,CI,C/D,Volt,T,S//LL1,LL2,LL3;	
Darmstadt 3	27L	?	Pl	pl	27cm\L,K,A/B,CI,C/D,P,T,S//LL1,LL2,LL3;	
Poly-e4	27L	Dup	Pl	pl	L,K,A/B,T,ST,S/D,DI,P;	
					LL01,LL02,LL03,DF/CF,CIF,CI,C/D,LL3,LL2,LL1;	
Rietz 11	13L	?	Pl	pl	13cm\K,A/B,CI,C/D,L//S,ST,T;	
Electro 22	13L	?	Pl	pl	13cm\Dyn.Motor,Volt,A/B,CI,C/D,cos.cos,L;	
Darmstadt 33	13L	?	Pl	pl	13cm\L,K,A/B,CI,C/D,vi-z2,S,T//LL1,LL2,LL3;	

ROBERTS (USA)
Address: E.P.Roberts, Pittsburg, USA.
History: c1906, Robert's slide rule for wiring calculations, otherwise no details known.

ROOS (USA)
Address: The Roos Co, USA.
History: History not known, but known to have had a link with **CHARVOZ**.

Slide Rules:

1	2	3	4	5	6	7
-	10"L	?	W	?	A/B,CI,C/D,K//S,L,T	

ROSS (GB)
Address: Ross Ltd, Clapham Common, London SW4.
History: Better known as Optical Instrument makers, famous for lenses, early cameras and binoculars.

Slide Rules:

1	2	3	4	5	6	7
Anderson Slide Rule	10"L	CF	W/C	pl	scale in 4 sections.	
Fuller's Calculating Slide Scale.	T					

ROSS (USA)

Address: San Francisco, California, USA.

History: In existence 1910, and in some association or selling **LEITZ** rules in 1923. Not known if there was any history with **ROSS** of UK (Optical Instrument makers).

ROSSIER (France)

Address: J. Rossier, 49 Rue Tapis-Vert, Marseille, France.

History: Made the **ROULETTE** "X" Système Fabien Carlo circular slide rules (24cm).

ROTARULE (USA)

Address: see **DEMPSTER**.

History: Both spellings, Rotarule and Rotorule, have been seen. History not known. Established 1963?

Slide rules:

A range of sophisticated circular slide rules made by **DEMPSTER**.

ROTHER (Germany)

Address: Rother, Bayern, Germany.

History: c1900, Rother's Rechenschieber, otherwise no details known.

ROULEAU (Switzerland)

Address: Rouleau Calculator National SA., Zurich, Switzerland.

History: This may have been another name for **NATIONAL**.

Slide Rules:

1	2	3	4	5	6	7
Cylindrical Grid-iron slide rule, No 16	T	n/a	Al	Al	18" long, 6" dia.	1920

ROULETTE (France)

Address: See **ROSSIER** France.

History: Brand name or series of slide rules made by **ROSSIER**.

ROYAL (Japan)

Address: Ushida Yoko Co Ltd., Japan.

History: The same manufacturer may have made slide rules for both Royal and **T.S.**, Royal is the more common name in Benelux countries.

Slide Rules:

1	2	3	4	5	6	7
-	33L	?	Pl	?	L,K,A/B,CI,C/D,S,ST,T//25cm,DF/CF,C/D,9inch;	
No 1002A	10"L	CF	Pl	pl	Simplex open frame	
No 1002	10"L	CF	Pl	pl	S,K,A/B,CI,C/D,L,T;	
No 1002	10"L	CF	Pl	pl	S,K,A/B,CI,C/D,L,T; Diff font, no Japan.	
1051	33L	?	Pl	pl	25cm,DF/CF,CIF,CI,C/D,A,K;	
1053	10"L	OF	Pl	?	L,K,A/B,CI,C/D,S,ST,T//25cm/S,LL2,LL3/8ins;	c1972
1081	10"L	Dup	Pl	pl	T1,T2,A/B,BI,CI,C/D,P,S,ST LL1,LL2,LL3,DF/CF,CIF,,CI,C/D,L,K	c1970; b
1091	36L	Dup	PL	pl	LL01,LL02,LL03,A/B,L,K,C/D,LL3,LL2,LL1 T2,T1,ST,DF/CF,CIF,CI,C/D,P,S	
1053 Ecola	10"L	OF	Pl	?	L,K,A/B,CI,C/D,S,ST,T//25cm/S,LL2,LL3/8ins;	c1972
1081 Technica	10"L	Dup	Pl	pl	T1,T2,A/B,BI,CI,C/D,P,S,ST LL1,LL2,LL3,DF/CF,CIF,,CI,C/D,L,K	c1970
1091 Technicum	36L	Dup	PL	pl	LL01,LL02,LL03,A/B,L,K,C/D,LL3,LL2,LL1 T2,T1,ST,DF/CF,CIF,CI,C/D,P,S	

From Royal 5" slide rule leaflet for Engineers and Students

1	2	3	4	5	6	7
511	5"L	CF	Pl	pl	cm\LL3,LL2,LL1,A/B,BI,CI,C/D,DI,K,LL0//T2,T1,L,S	1.
525A	5"L	CF	Pl	pl	A/B,CI,C/D,K	
525D	5"L	CF	Pl	pl	K,DF/CF,CI,C/D,A	
522A	5"L	CF	Pl	pl	A/B,CI,C/D,K//S,L,T	
522D	5"L	CF	Pl	pl	K,DF/CF,CI,C/D,A//T2,T1,L,S	

523A	5"L	CF	Pl	pl	cm\A/B,CI,C/D,K//S,L,T\ins	
523D	5"L	CF	Pl	pl	cm\K,DF/CF,CI,C/D,A//T2,T1,L,S\ins	
56A	5"L	CF	Pl	pl	A/B,CI,C/D,K//S,L,T	a.
56D	5"L	CF	Pl	pl	K,DF/CF,CI,C/D,A//T2,T1,L,S	a.
57AI	5"L	CF	Pl	pl	ins\A/B,CI,C/D,K//S,L,T	a.
57AM	5"L	CF	Pl	pl	cm\A/B,CI,C/D,K//S,L,T	a.
57DI	5"L	CF	Pl	pl	ins\K,DF/CF,CI,C/D,A//T2,T1,L,S	a.
57DM	5"L	CF	Pl	pl	cm\K,DF/CF,CI,C/D,A//T2,T1,L,S	a.

Comment.
1.   Variant illustrated is LL3,LL2,A/B,CI,C/D,K,LL1
a.   All have CI scale in Red.
b.   Has a T.S. Emblem and "Techlog" on the slide rule.

RUXTON (USA)
Address: Not known.
History: 1928 known to have produced a MultiVider slide rule - assumed to be a linear slide rule. Also the Pencil slide rule at the same time.

Slide Rules:

1	2	3	4	5	6	7
Pencil slide rule.	P	n/a	Pl	n/a	C/D, duplicated scales, only slides apart.	1928
MultiVider slide rule						1928?

RYNJA (Holland)
Address: P. Rynja, Amsterdam, Holland.
History: Issued a series of cardboard and plastic slide rules for calculating the size and number of wooden ceiling and floor beams required for the dimensions, weight, type of construction and so on of a building.

Slide Rules:

1	2	3	4	5	6	7
-	14C	MD	P	?	building calculation scales.	
-	26C	MD	Pl	?	building calculation scales.	

SALLERON (France)
Address: Salleron, Paris, France.
History: became **DUJARDIN**.

SANDERS (GB)
Address: Sanders and Sons, London.
History: They made a Proof Rule 31 cm long in wood, not known when.

SANTOK / SANTOKU (Japan)
Address: Not known - Japan.
History: Not known.

SAMA AND ETANI (USA)
Address: Sama & Etani Inc, Groton, Massachusetts, 01450, USA
History: 1960 was selling slide rules, also designed slide rules for **CONCISE**.

SCHACHT & WESTERICH (Germany)
Address: Schacht & Westerich, Technischer Zeichenbedarf, Hamburg, Germany.
History: Still in business today making drawing instruments and selling writing materials.

Slide Rules:

1	2	3	4	5	6	7
System Rietz	10"L	CF	W/C	m/p		1924
Circular watch calculator	2.25"W		n/a	n/a	two sided, C/D//4 scales	

SCHUPPISSER & BILLETER (Switzerland)
Address: See **BILLETER**.
History: See **BILLETER**.

SCHUTS (Germany?)
Address: Not known.
History: Known to have made a range of circular slide rules.

SCIENTIFIC INSTRUMENT COMPANY (SIC) (USA)
Address: Berkeley, California, USA.
History: Not known. Slide rules made in Japan, bamboo, possibly **RICOH**. Also circular plastic rules which were probably **CONCISE**, and metal circular slide rules from an unknown manufacturer.

Slide Rules:

1	2	3	4	5	6	7
5500	12.5L	Dup	B/C	pl	A/B,S,T,C/D,K DF/CF,CI,C/D,L	1961
5510	12.5L	Dup	B/C	pl	A,B,C,D,CI,DI,CF,CIF,DF,LL01,LL02,LL03,LL0 LL0,LL1,LL2,LL3,K,S,ST,T	1961,a.

Comments:
a.   Disposition of scales uncertain.

SCIENTIFIC PUBLISHING CO. (GB)
Address: 53 New Bailey Street, Manchester, England.
History: Established pre 1898 (When Fowlers Calculators were set up) and best known for the publication of technical pocket books. The 1904 *Mechanical Engineers Pocket Book* shows a watch type calculator. Owned by **FOWLERS**.

Slide Rules:

1	2	3	4	5	6	7
Watch type calc, one knob, one central button						1915
Watch Type calculator, two knobs						1904

SEEHASE (Germany)
Address: c1923, Dr. Ing. Seehase, Waltman Delft, Berlin, Germany
History: Seehase was an engineer who produced slide rules to his patents. He is known to have been in Berlin 1923, Waltman Delft - not known what the relationship was. A number of Swiss patents are known.

Slide Rules:

1	2	3	4	5	6	7
-	14L	?	P	?	A/B,C/D	
-	14L	?	W/C	?	A/B,C/D	
-	12.5L	?	P/C	pl	A/B,C/D//5inch,13cm	

SEGER (Sweden)
Address: Not known.
History: Not known, but there is a plastic log-log slide rule known.

SELBY (Australia)
Address: B.H. Selby & Co Pty. Ltd., Melbourne, Australia.
History: 1904-17: H.B. Silberg; 1917-63 and probably later, B.H. Selby. Scientific instrument dealers who sold the Fuller and Fowler range of calculators.

SELECTRON (USA?)
Address: Selectron, Not known.
History: Maker of a slide rule watch, similar to other models from more expensive makers.

SERVO (Denmark)
Address: Servo Calculator Co, Denmark.
History: 1960, Frequency Response slide rule made for Boonshaft & Fuchs.

SLIDE RULE MANUFACTURERS AND RETAILERS

SHANGHAI (China)
Address: Shanghai Calculation Rule Factory, Shanghai, China.
History: Shanghai was also in a joint venture with Jianyi Instruments factory.

Slide Rules:

1	2	3	4	5	6	7
1018 Electricians slide rule	25L	Dup	Pl	pl/m	LL01,LL0,LL1,K,A/Q,QI,L,I/D,DI,LL02,LL2 Φ,RΦ,P',P/B,T,S,C/J,J',T,GΦ	
404 Rietz	12.5L	CF	Pl	?	K,A/B,CI,C/D,L; joint Shanghai/Jianyi	
57 Calculation rule	12.5L	Dup	Pl	pl	K,A/B,BI,CI,C/D,P,S,T// formulae.	a.

Comment:
a.   A variant with the same type number has scales K,A/S,ST,T,C/D,DI,L.

SHELL (Holland)
Address: Shell Labs, Amsterdam.
History: Numerous specialist slide rules in various formats were commissioned by Shell and produced under the Shell name. These were made by a range of manufacturers. Generally Shell would have stamped rules bought by them for general use stamped **K.S.L.A.** (Koninklijke Shell Laboratorium te Amsterdam) with an identity number.

SIDA (China)
Address: Not known.
History: Duplex slide rules, like Hemmi. Logo is a strange curled 'tick' mark.

Slide Rules:

1	2	3	4	5	6	7
-	25L	Dup	Pl	m/p	L,LL1,DF/CF,CIF,CI,C/D,LL3,LL2 LL0,LL00,A/B,T,ST,S/D,Th,Sh2,Sh1.	
1015	25L	Dup	Pl	m/p	LL01,LL02,LL03,DF/CF,CIF,H,CI,C/D,LL3,LL2,LL1 SH1,SH2,K,A/B,T,ST,S,C/D,DI,L,Th.	
1083	25L	Dup	Pl?	m/g	L,LL1,DF/CF,CIF,CI,C/D,LL3,LL2 LL0,LL00,A/B,T,ST,S/D,Th,Sh2,Sh1.	

SIEMENS (Germany)
Address: Not known.
History: A range of specialist electrical and component calculators, whether made by Siemens or for them is not known.

SILLCOCKS-MILLER (USA)
Address: Sillcocks-Miller Co., Special Devices Division, USA.
History: Various flight computer and other navigational calculators for the US armed forces.

SIMI (Sweden)
Address: Simi, Malmö, Sweden.
History: 154/47 International, Universal slide rule - System Darmstadt.

SIMPLIFIED (USA)
Address: Simplified Flight Calculator Co, Beddell Bldg, San Antonio, Texas, USA.
History: c1942, John E. Clemons Patent 1,969,939 (1932) produced Flight Calculators.

SIMPLON (GB)
Address: See **DARGUE**.
History: SIMPLON was the brand name of slide rules initially sold by **DARGUE** possibly made by **HEMMI** and later manufactured by themselves for a short period.

SIVAD (GB)
Address: England.
History: Slide rule in existence.

SKALA (Poland)

Address: Spóldzialnia Pracy, Warsaw, Poland.

History: Founded as a "collective" for the manufacture of mathematical instruments in 1948, and started making slide rules c1955. Ceased manufacture of rules in 1976. Their Logo was a "Globe" with a stem through, and Skala across the Globe.

Slide Rules:

1	2	3	4	5	6	7
-	12.5L	CF	Pl	?	K,A/B,CI,C/D,L//S,ST,T	
SLPP	33L	?	Pl	?	L,P,K,A/B,CI,C/D,S,T,ST	
-	12.5L	CF	Pl	pl	K,A/B,CI,C/D,L//??	c1960
-	5"L	?	W/C	?	K,A/B,CI,C/D,L; marked DOL'MEL WROCK A.W.	

SLIDE CHARTS (GB)

Address: Slide Charts (Britain) Limited, 13 Burcot Park, Burcot, Abbingdon OX14 3DH, England.

History: Associated with **AMERICAN SLIDE CHART CORP.** Suppliers of numerous specialist slide charts as technical support for sophisticated equipment manufacturers, charts available to this day. Examples as follows:

Slide Rules:

1	2	3	4	5	6	7
Trox Acoustic Vibration Control selection Guide						1990
Date Forecaster						1991
GEC Alsthom HRC Fuse Link selector						1992
Molsiv Adsorbents						1992
Northern Telecom Digital μwave Radios						1992
AEG Modicon Product Reference Guide						1993
Pirelli Cables Accessory Selector						

SLIDE RULE CLUB (USA)

Address: Slide Rule Club, New York, USA.

History: J.L. Hall's structural slide rule, a Column slide rule, Beam slide rule and Mannheim slide rule all in one, otherwise history not known.

SMALL (USA)

Address: Small & Co., Waltham. Mass., G. Small, Boston, Mass.

History: SMALL circular calculator, beautifully made, approx 2" dia with scales on edge and faces and an ingenious mechanism for allowing the two halves to work together.

Slide Rules:

1	2	3	4	5	6	7
"Small" Pocket Calculator - as patented	2"U	u	M	m	C/D on edge only.	1900
"Small" Pocket Calculator	2"U	u	M	m	C/D on edge,S,T,"//A,S1,T1,L.	

SMITH. G (USA)

Address: George Smith, Address unknown.

History: 1940, The slide rule of Weights Models 5 and 10. No details.

SMITH. J.W. (GB)

Address: J. W. Smith, Coventry

History: Seen illustrated in various books approx 1945 only. (*Handbook of Mechanical Calculations* by Chapman)

SMITH. W.H. (GB)

Address: W.H. Smith,

History: Stationers. Sold **BLUNDELL-HARLING** rules with their own name on, also **FABER-CASTELL** rules were sold with W.H Smiths name on.

Slide Rules:

1	2	3	4	5	6	7
Master Duplex A505	10"L	Dup	Pl	pl	Plastic tube.	
Log-log trig	10"L	OF	Pl	pl	Plastic case	
Log-log trig	10"L	OF	Pl	pl	diff font	

Simplified Rietz	5"L	OF	Pl	pl	plastic wallet	
Simplified Rietz	5"L	OF	Pl	pl	diff font	
Simplified Rietz	10"L	OF	Pl	pl	Pl wallet	
Simplified Rietz	10"L	OF	Pl	pl	diff font	
Rietz	10"L	OF	Pl	pl	Pl tube	1980

SMITHS CALCULATORS (GB)
Address: Smiths Calculators, Beaminster, England.
History: Producers of specialist calculators and slide charts, as **SLIDE CHART LIMITED, PERRYGRAF** etc.

Slide Rules:

1	2	3	4	5	6	7
Metric/Imperial Concrete Calculator (Mixamate)						
Determination of Pipe Thermal Expansion						

SOLTMANN (USA)
Address: E.G. Soltmannn, New York, USA.
History: W.H. Breithaupt's Reaction scale and general slide rule, c1894, otherwise not known.

SPENCER (GB)
Address: John Spencer & Co (Publishers) Ltd, 131 Brackenbury Road, London, W6.

Slide Rules:

1	2	3	4	5	6	7
"METRIDISC" Metric Converter No1						

SPHINX (USA)
Address: Not known.
History: c1923 a slide rule sold by **WEBER**.

STAEDLER (Germany)
Address: Staedler, Nürnberg, Germany.
History: Staedler (UK) Ltd, based in Wales, took on the Aristo agency from TS. Slide rules may have been made by Staedler, or badged from another manufacturer e.g. **ROYAL** or possibly **ARISTO**. Known as STAEDLER MARS.

Slide Rules:

1	2	3	4	5	6	7
544 DLL	10"L					
544.52 Noris Log Log	10"L					
544.55 Noris Rietz	10"L					
54.500 Commercial	5"L					
544A	10"L	?	Pl	pl	K,DF/CF,CIF,CI,C/D,A	
544LL	10"L	OF	Pl	pl	L,K,A/B,CI,C/D,S,ST,T//S,LL1,LL2,LL3	
544S	10"L	?	Pl	pl	T1,T2,K,A/B,CI,C/D,S,ST,L	
54103 Rietz	10"L	?	Pl	pl	K,A/B,CI,C/D,L//S,ST,T	
54106 Electro	27L	?	Pl	pl	V,A/B,CI,C/D,U//LL3,K	
54401 Darmstadt	27L	?	Pl	pl	L,K,A/B,CI,C/D,P,S,T//LL1,LL2,LL3	1.
54403 Rietz	27L	?	Pl	pl	K,A/B,CI,C/D,L//S,ST,T	
54428 Duplex	27L	Dup	Pl	pl	P,S,T1,T2,A/B,K,CI,C/D,R1,R2,L,ST	
					LL00,LL01,LL02,LL03,DF/CF,CIF,CI,C/D,LL3,LL2,LL1,LL0	
Noris LL 54452	27L	Dup	Pl	pl	K,A/B,CI,C/D,L	
					S,ST,T,LL1,LL2,LL3	
Noris Rietz 54455						
Rietz 54503	13L	?	Pl	pl	K,A/B,CI,C/D,L//S,ST,T	

Comment:
1.   Many variants of detail.

STANDARDGRAPH (Germany)
Address: Not known.
History: Not known, made circular and linear slide rules.

Slide Rules:

1	2	3	4	5	6	7
900	4 7/8"C	MD	Pl	pl	%,A/B,B,consts; 2D,1C	
905	12C	MD	Pl	pl	C/D	
910 Disco Rietz	5"C		MD	Pl	pl	L,%,A/B,S,T,√x; 2D,1C	
No 9982 Stano Log	25L	Dup	Pl	pl	L,K,A/B,BI,CI,C/D,LL1,LL2,LL3	
					T1,T2,DF/CF,CIF,S',C/D,S,ST,P	

STANLEY (GB)

Address: W.F. Stanley & Co Ltd. 1927: 286 High Holborn, London W.C.1. Head Showrooms: 79-80 High Holborn, London WC1.
also at: 13 Railway Approach, London SE1, 8 Hatton Gardens, London EC1, (from 1951 - Heath, Hicks & Perken)
52 Bothwell St, Glasgow. 1967: New Eltham, London SE9. (Head Office and Main Works). Glasgow: 52 Bothwell St. C2.
Factory: 1951 - 1958: Head Office & Main Works: New Eltham. London. Possibly ex **HEATH** Factory. Also later.
History: One of the oldest firms in the mathematical instrument making business, see C.J.A. Allens *History of Stanley*. They
bought/absorbed many instrument making companies.

Slide Rules:

1	2	3	4	5	6	7
Honeysett's Hydraulic slide rule	L					1902
Hale's slide rule for indicator diagrams	L					c1902
Froude's Displacement rule	L					c1902
Shepherds cubing and squaring slide rule	L				quantity surveyors.	1902
Hudson's Horse-Power scale	?					c1902
Hudson's Shaft, Beam and Girder scale	?					c1902
Hudson's Pump Scale	?					c1902
Hudson's Photo-Exposure Scale	?					c1902
Anthony's Improved Circular Hydraulic calc	C					1905
Fuller-Bakewell Calculating slide Scale	T					1905
Stanley-Boucher Calculator	W					1905
Gravet's Tacheometer Slide Rule	L?					1905
Interest Computing rule						
Bouchers Calculating Circle.	C					
Wireless.	L					
Davis Stokes Field Gunnery Slide Rule.	L					
Fullers Calculating Slide Scale.	T					
Gravette - an earlier form.	L					
Multiplex (cubes / cube roots)	L					
Coopers 100"					20 parallel lines 5" long with movable frame	c1935-45
Maitland Hydraulic					solves Kutters formula with n=0.013	c1915

From the Stanley A catalogue - 1958.
The following range of slide rules and other calculators were shown in the catalogue.
The Fuller Calculator.

Model No 1.	Log scale 500" long on cylinder, with table of useful data on inner cylinder.
Model No 2.	as No 1, with log and sine scales on inner cylinder instead of data table. Two versions:
Model No 2A	1907: Instructions with sine scales instead of data tables.
Model No 2B	1940: 4 place log scales as well as sine scale.
Model No 3.	Fuller-Bakewell, sin^2 and sin.cos scales on inner cylinder.
Model No 4.	The 1879 catalogue mentions a miniature version with 200" scale lengths instead of the usual 500" scale length. No examples are known.
Other specials	One example has special scales for Vector calculations. (Complex Number Slide Rule).

Some 14,000 Fuller calculators were made between 1879 and 1975.
Versions made after 1938 have plastic (bakelite) handles rather than wood, the papier maché tubes carrying the scales replaced with celluloid (linenised bakelite) tubes, and the brass fixed and moveable pointers are replaced by Perspex pointers for special models.

"OTIS KING" Calculator.
Model K
Model L

"FOWLER" Pocket Calculators

Universal	3.25"C
Magnum	4.25"C
Jubilee Magnum	4.25"C

Lawes Rabjohns Technical Catalogue - 1952
Fullers Calculating Instruments
Fullers Calculating Rule No 1

Fullers Calculating Rule No 2
Fullers Calculating Rule No 3 - Fuller-Bakewell

-		10"L	CF	W/C m/p	A/B,C/D//S,L,T; wood/cell rule.	1936
-		10"L	CF	W/C ?	25cm/A/B,C/D\10inch//S,L,T	

STANLEY MANUFACTURING CO. LTD (Canada)

Address: Stanley Manufacturing Co Ltd, Toronto, Canada.
History: Not known whether there was any relationship with either Stanley (USA) or Stanley (UK).
Known to have made an Aircraft Navigation Computer of standard design.

STANLEY RULE & LEVEL CO. (USA)

Address: Stanley Rule & Level Co. USA.
History: In business 1880. Made Routledge's rule pre World War 1. Not known what if any association with the UK W.F. Stanley Co, or with W. Stanley of Peru, Indiana was (guess that this was an earlier name). Later address in Philadelphia PA.

Slide Rules:

1	2	3	4	5	6	7
Barnard's Coordinate calculator						1931

STEPHENS (USA)

Address: Stephens Co, address not known.
History: In business 1881. Related to **Chapin-Stephens** ?

Slide Rules:

1	2	3	4	5	6	7
Supermatic slide rule.						

STERLING PLASTICS (USA)

Address: Mountainside, New Jersey, USA.
History: Founded in 1938, made a range of cheap plastic slide rules from 1961. Brand name is **PRECISION** for one range. Made the **ACCU-MATH** range of Duplex slide rules.

Slide Rules:

1	2	3	4	5	6	7
Pocket Metric Converter	15L	?	Pl	?		
Decimal Trig LL	30L	Dup	Pl	?		
Precision	30L	OF	Pl	?	A/B,CI,C/D,K//S,L,T; reversible slide	a.
Precision	30L	OF	Pl	?	ins\A/B,CI,C/D,K\ins//S,L,T; reversible slide	b.
584	30L	OF	Pl	?	A/B,CI,C/D,K//S,L,T	
587 (Rietz)	30L	OF	Pl	?	K,A/B,CI,C/D,L//S,ST,T	
587	5"L	?	Pl	?	A/B,CI,C/D,K//S,L,T	
Decimal Trig Log Log Duplex	30L	Dup	Pl	pl	22 scales.	
594	?L	?	?	?	Log-log scales	

Comment:
a.   An identical slide rule with "Japan" but nmo other names is also known. Version with all red scales on slide.
b.   Slide tinted green, inches by 1/8 on top and .050 bottom.

STERLING SLIDE RULE CO (USA)

Address: Mountainside, New Jersey, USA. (??)
History: Associated with **STERLING PLASTICS**, probably the same company, cheap plastic rules.

STEVENS

Address: Not known.
History: Not known, made a Rally Indicator Mod 25, being a Speed/Time/Distance (STD) calculator.

STEWARD (GB)

Address: J.H. Steward, 1945, 406 Strand, London WC2.
History: Established 1852.

Slide Rules:

1	2	3	4	5	6	7
Halls Nautical Slide Rule	L				Two slides, 8 scales	c1902-15
R.H.S. Calculator	T	n/a	M	n/a	Prof R.H. Smith Tubular calculator	c1906-15
Macleod Field Artillery	L					c1938
TOG	L				TOG Angles/B,C/TGO Angles	c1938

SUN (Japan)
Address: See **HEMMI**.
History: See **HEMMI** which was the parent company. The "Sun" logo was first registered in 1917 as part of the Hemmi company.

T.S. (GB)
Address: See **TECHNICAL SALES**.
History: As **TECHNICAL SALES**.

TAD (USA)
Address: TAD Calculators, unknown address.
History: 1965, Sold slide rules.

TAKEDA (Japan)
Address: Takeda Drawing Instruments Mfg Co Ltd, Japan.
History: Agents for **CONCISE** circular slide rules.

TAMAYA (Japan)
Address: Tamaya & Co, Tokyo, Japan.
History: Not known, but produced a Rietz slide rule with the SUN Japanese patent No 22129 shown, also:

Slide Rules:

1	2	3	4	5	6	7
-	5"L	CF	B/C	?	A/B,C/D//S,L,T; pat 22129.	

TARQUIN (GB)
Address: Tarquin Publications, Stradbrooke, Diss, Norfolk, IP21 5JP. Telephone: 01379 384 218. Facsimile: 01379 384 289
History: A small publishing house that made a limited range of self assembly educational slide rules in the early 1970's and are still in business producing educational books to date. Approx 100,000 linear card rules produced, a smaller number of plastic linear, and about 2000 circular. In 1973/4 the card rule was sold for 6.5 pence, the plastic linear for 13.5 pence. This had increased to 11p and 26p respectively by 1977/8. The plastic circular sold for 65 pence in 1975/6, and 77p in 1977/8. Tarquin also sold the **BLUNDELL** Academy 803 Rietz slide rule for £1.50 each for the "advanced user".

Slide Rules:

1	2	3	4	5	6	7
Tarquin Slide Rule (Card)	25.5L	CF	P	pvc	A/B,R,C/D; Paper kit for self assembly and learning.	c1971-78
Tarquin Slide Rule (Plastic)	25.5L	CF	Pl	pvc	A/B,R,C/D; Plastic kit slide rule as above.	1973-78
"Simplicity" Circular Slide Rule	11.5C	MD	Pl	pl?	A,C,K,CI; Plastic kit slide rule in plastic wallet.	1975-78

TARRANT (USA)
Address: Not known.

Slide Rules:

1	2	3	4	5	6	7
Hydraulic slide rule						

TARRANT (Australia)
Address: Tarrant Motor Co, Russel Street, Melbourne, Australia.
History: A motor dealer who sold a slide rule they appeared to make themselves, and a circular which may have been imported.

Slide Rules:

1	2	3	4	5	6	7
(Speed slide rule)	?L	?	M/P	n/a	Dist(Yds)/Time(mins&secs)/Speed(mph); Tinplate with card slide	c1897-1911
Blackwell Speed Calculator	?C	?	?	?	Speed calcs, may have been UK made.	

TAVELLA (USA)
Address: Tavella Sales Co. (USA)
History: 1940, sold **GILSON** rules (midget slide rule).

TAVERNIER-GRAVET (France)
Address: 3 Rue Mayet, Paris 4-20, France. also Rue Hayer 49, Paris, and 30 Rue de Babylone, Paris; but dates not known.
History: Stemmed from the firm **GRAVET**, and then Gravet et Lenoir in 1850. Made over 300 models.

Slide Rules:

1	2	3	4	5	6	7
Lenoir slide rule	L	?	?	?	original slide rule design from the firm	1850
Mannheim slide rule (original)	L	CF	W	n/a	first design followed by the design with cursor	1851
Mannheim slide rule	25L	CF	W	m	from 1856 to approx 1945	1856
Moinot, Regle Logaritmique pour la Tacheometrie	?	W	?	?	?	1868
Goulier, regle pour les Levers de Tacheometriques	?	?	?	?	?	1873
De Montrichard's, Regle pour le Cubage des Bois	?	?	?	?	?	1876
Cherapashinski's slide rule	25L	?	W	?	?	1883
Lebrun, regle pour les calculs de Terrasements	?	?	?	?	?	1886
Sanguet, Regle pour les Levers Tacheometriques	?	?	?	?	?	1888
Bosramier, Regle pour les Levers Tacheometriques	?	?	?	?	?	1892
Beghin slide rule, Regle des Ecoles.	25L	CF	W/C	m	made to 1945 at least	1894
Gallice. Regle pour les calculs nautique	L?	?	?	?		1897
Leven's Regle pour les Reports de Bourse	L	?	?	?	commercial slide rule	1903
Barrière slide rule	25L	CF	W/C	m		1942
2184 Electro	25L	?	Pl	?	LL2,A/B,CI,C/D,LL3//Dyn,Mot,Volt	
13 bis	25L	?	W/C	?	7 scales	c1915

TECHNICAL SALES - T.S.
Address: Technical Sales (London) Ltd. 9 - 10 Rivers Reach, Gartons Way, London SW11 3SX.
History: Sold slide rules that were made to its design in Japan, from 1948 to approx 1970. Prior to this it was the UK agent for **ARISTO**. It is still agent for **ECOBRA** mathematical instruments and **UNIQUE** instruments. It may also have had a link with **ROYAL**. The "Emblem" series consisted of 6 all-plastic slide rule types, all 10", as follows:

Slide Rules:

1	2	3	4	5	6	7
1051 - Tutor	10"L	CF	Pl	pl	DF/CF,CIF,CI,C/D,A	
1052 - Olympic	10"L	CF	Pl	pl	L,K,A/B,CI,C/D,S,ST,T	cm,ins
1053 - Olympic LL	10"L	CF	Pl	pl	L,K,A/B,CI,C/D,S,ST,T//cm/S,LL2,LL3/ins	
1071 - Commercial	10"L	CF	Pl	pl	F1/F2,CI,C/D//Decimal £ to LSD	
1081 - Techlog	10"L	Dup	Pl	pl	LL1,LL2,LL3,DF/CF,CIF,CI,C/D,L,K //T1,T2,A/B,CI,C/D,P,S,ST	
1091 - Executive	10"L	Dup	Pl	pl	LL01,LL02,LL03,A/B,L,K,C/D,LL3,LL2,LL1 //T2,T1,ST,DF,CF,DF/CF,CIF,CI,C/D,PS	

TECHNICAL SUPPLY COMPANY (GB)
Address: Technical Supply Co. Carden Avenue, Brighton, Sussex.
History: Made slide rules for **UNIQUE**, and may have been an earlier incarnation of Unique. TYSR shows some slide rules made by the Technical Supply Co.

Slide Rules:

1	2	3	4	5	6	7
Log-Log	5"L	CF	W/C	pl	LU,A/B,C/D/LL	
Electrical	10"L	CF	W/C	pl		c1971
Dualistic	10"L	CF	W/C	pl	//LL3,LL2,LL1	c1971
Brighton	10"L	CF	W/C	pl		c1958

TECNOSTYL (Italy)

Address: Tecnostyl, Milano, Italy.

History: Not known, but produced a range of circular and linear slide rules. Used a shaped *TS* for their logo.

Slide Rules:

1	2	3	4	5	6	7
Linear	L	?	?	?	S,K,A/B,CI,C/D,L,T	
40/R Rietz	27L	?	?	?	K,A/B,CI,C/D,L//S,ST,T	
2197	L	?	?	?	similar logo to **TS**.	
Circular	C	?	?	?	?	
41/C Concrete	12.5L	CF	Pl	m/g	A/B,Concrete,C/D	c1955
41/C4 Concrete	12.5L	CF	Pl	pl	A/B,Concrete1/2Concrete3/4,C/D	c1960
41/D Mathematical	12.5L	CF	Pl	pl	LL,K,A/B,CI,C/D,T,S	c1955
41/E Electrical	12.5L	CF	Pl	m/g	cos,A/B,CI,C/D,cos	c1955

TEKNOR (Sweden?)

Address: Not known.

History: Not known.

Slide Rules:

1	2	3	4	5	6	7
525	L	?	?	?	A/B,CI,C/D,K	
No 1051 Grundskola	25L	CF	Pl	?	DF/CF,CIF,CI,C/D,A	
Gymnasium 1081	25L	Dup	Pl	pl	T1,T2,A/B,BI,CI,C/D,P,S,ST LL1,LL2,LL3,DF/CF,CIF,CI,C/D,L,K	

TELEDYNE POST (USA)

Address: Not known.

History: c1970 Teledyne bought POST after which many original POST rules were sold under the TELEDYNE-POST name. Also sold **GILSON** rules. See **POST**.

Slide Rules:

1	2	3	4	5	6	7
Midget	4"C	MD	W/P	pl	see Gilson, painted Green	1940 - 60
Hemmi VersaTrig 44CA-500	L	Dup	B/P	?	see **Post**.	
44CA-600 Versalog II	L					

TELEX (USA)

Address: Telex Communications Inc, USA.

History: c1974 produced Flight Computers and navigational instruments, covered by US, Canadian and Belgian Patents.

THOMSON, SKINNER & HAMILTON (GB?)

Address: Not known.

History: c1914 sold Barnard's Coordinate calculating rule, still being sold by **STANLEY** in 1931.

THORNTON (GB)

Address: A.G. Thornton Ltd.,

1878: 39 Great Cheetham St, Manchester. In partnership with Halden.

1879: 109 Deansgate, Manchester. Enlarged 1890.

1895: South King Street, Manchester.

1897: 4 St. Marys Street, Manchester.

1904: Paragon Works, 2 King St. West, Manchester. Main office and warehouse.

History: Originally set up in 1878, limited company in 1940, and became **British Thornton** in 1967. He was originally in partnership with Joseph Halden, though this did not last long, from 1878 to 1880. Started to manufacture slide rules after 1900, final slide rules manufactured c1975, and the company name finally changed to Education and Science Furniture in 1992. Closed in 1995.

Factories:

>1880: Workshop at John Dalton Street, Manchester.

1897: Workshop at Bridge Street, Manchester.

1907: Minerva Works, Sydney Street, Salford, Manchester.

1920s: 41 King Street West, Manchester.

1925: Rusholme, Manchester; and North George Street, Salford.

>1945: Derby Street Works, Openshaw, Manchester. (slide rules only)

Slide Rules:

1	2	3	4	5	6	7
Faber's slide rule	L				sold by Thornton	1895
Routledge's slide rule	L				sold by Thornton	1895
Hawthorn's slide rule	L				sold by Thornton	1895
Carett's slide rule	L				sold by Thornton	1895
Calculex	W				sold by Thornton	c1905?
Boucher's Calculating Circle.	W	n/a	M	n/a		
Chemists slide rule.	L	?	W/C	?		
Fuller's Calculating Slide Scale.	T	n/a	W	n/a		
Gravette - an earlier form.	10"L	CF	W	m		
Perry's New slide rule	L	Cf	W/C	?		1908
Perry's log-log	L	CF	W/C	m/g		1902, 1915-45
Thornton Rietz	L	CF	W/C	m/g	K,A/B,CI,C/D,L	c1945
PIC with Differential Scales	L	CF	W/C	m/g		1938-45
PIC AC Electrical	L	CF	W/C	m/g		1938-45
Comprehensive slide rule	L	CF	W/C	?		1965
Students model	L					1965
Engineers new pattern	L				additional LL scales with C/D	c1945
Electrical	L	CF	W/C	m/g	additional LL scales, also V & D scales	c1945-55

From "Bonds" Catalogue, 1935 - Thornton's P.I.C. slide rules.

1	2	3	4	5	6
No1, Junior	10"L	CF	W/C	c	A/B,C/D; 6/6d
No2, Junior	10"L	CF	W/C	c	A/B,C/D//S,L,T; 10/-
No3654, PIC Standard	10"L	CF	W/C	c?	21/-
No4865, PIC Engineers Commercial	10"L	CF	W/C	c	32/6d

From the Lawes Rabjohns Technical Catalogue - 1952

PIC Slide Rules

1	2	3	4	5	6	7
PIC - Standard Pattern cursor	5"L	CF	W/C	m/g	A/B,CI,C/D//S,T,L;	1 line
PIC - Standard Pattern	10"L	CF	W/C	m/g	A/B,CI,C/D//S,T,L; 1 line cursor	
PIC - Standard Pattern	20"L	CF	W/C	m/g	A/B,CI,C/D//S,T,L; 1 line cursor	
PIC - Rietz Pattern	5"L	CF	W/C	m/g	K,A/B,CI,C/D,L//S,T,L; 1 line cursor	
PIC - Rietz Pattern	10"L	CF	W/C	m/g	K,A/B,CI,C/D,L//S,T,L; 1 line cursor	
PIC - Rietz Pattern	20"L	CF	W/C	m/g	K,A/B,CI,C/D,L//S,T,L; 1 line cursor	
PIC - Elec & Mech Engrs	5"L	CF	W/C	m/g	LL1,A/B,C/D,LL2\\V,D//S,T,L; 1 line cursor	
PIC - Elec & Mech Engrs	10"L	CF	W/C	m/g	LL1,A/B,C/D,LL2\\V,D//S,T,L; 1 line cursor	
PIC - Elec & Mech Engrs	20"L	CF	W/C	m/g	LL1,A/B,CI,C/D,LL2\\V,D//S,T,L; 1 line cursor	
PIC - Elec & Mech Engrs	5"L	CF	W/C	m/g	LL1,A/B,CI,C/D,LL2//S,T,L; 1 line cursor	
PIC - Elec & Mech Engrs	10"L	CF	W/C	m/g	LL1,A/B,CI,C/D,LL2//S,T,L; 1 line cursor	
PIC - Engineer-Commerce Pattern	5"L	CF	W/C	m/g	LL1,A/B,CI,C/D,LL2,D&ILL//S,T,L; 1 line cursor	
PIC - Engineer-Commerce Pattern	10"L	CF	W/C	m/g	LL1,A/B,CI,C/D,LL2,D&ILL//S,T,L; 1 line cursor	
PIC - Engineer-Commerce Pattern	20"L	CF	W/C	m/g	LL1,A/B,CI,C/D,LL2,D&ILL//S,T,L; 1 line cursor	
PIC - Electrical	5"L	CF	W/C	m/g	LL1,A/B,CI,C/D,LL2,D&ILL\\V,D//S,T,L; 1 line cursor	
PIC - Electrical	10"L	CF	W/C	m/g	LL1,A/B,CI,C/D,LL2,D&ILL\\V,D//S,T,L; 1 line cursor	
PIC - Electrical	20"L	CF	W/C	m/g	LL1,A/B,CI,C/D,LL2,D&ILL\\V,D//S,T,L; 1 line cursor	

all above with 3 line cursor at extra cost.

1	2	3	4	5	6	7
PIC - Student's	10"L	CF	W/C	pl	A/B,C/D	
PIC - Student's	10"L	CF	W/C	pl	A/B,C/D//S,T,L	
PIC - Junior Students	10"L	CF	W/C	m/g	A/B.C/D	
PIC - Junior Students	10"L	CF	W/C	m/g	A/B,C/D//S,T,L	
PIC - AC Electrical - Mod F	10"L	CF	W/C	m/g	S,C,A/B,Cl,Z,T⁰,C/D,T¬⁰,L\\LL,C; 1 line cursor	
PIC - AC Electrical - Mod F	10"L	CF	W/C	m/g	S,C,A/B,Cl,Z,T⁰,C/D,T¬⁰,L\\; 1 line cursor	
PIC - Standard - Mod A	5"L	CF	W/C	m/g	A/B,Dif,C/D,L; 1 line cursor	
PIC - Standard - Mod A	10"L	CF	W/C	m/g	A/B,Dif,C/D,L	
Pattern 131 - Comprehensive L-L	10"L,	CF,	W/C	m/g,	mahogany; electrical engineering	
Pattern 132 - Comprehensive L-L	20"L,	CF,	W/C	m/g	mahogany; electrical engineering	
Pattern 4880 - Estimator-Finance	10"L	CF	W/C	m/g	mahogany	
Pattern 4882 - Estimator-Finance	20"L	CF	W/C	m/g	mahogany	

Others:

1	2	3	4	5	6	7
No 131	10"L	CF	W/C	pl	Pat No 411,090; Electrical extension.	1938
No 3649	10"L	CF	W/C	m/g		1948
-	5"L	CF	W/C	pl	plain back	1935
-	10"L	CF	W/C	m/g		1945

- (Standard)	10"L	CF	W/C	m/gF	A/B,C/D//S,L,T; Inst Book.(J.Davis.)	1946
-	10"L	CF	W/C	m/g	A/B,C/D; "British", plain back, Al/glass cursor	1935
Junior	10"L	CF	W/C	m/g?	A/B,C/D	1938
Rule Slide Royal Artillery.	10"L	CF	W/C	m/g	A/B,C/D//S,ST,T; 2 cursors	1941
No 121	10"L	CF	W/C	m/g	LL1,LL2,A/B.CI,C/D,K+; pat 411,090	1944
No 144 AC Electrical	10"L	CF	W/C	m/g	S,?,A/B,CI,Z,C/D,T//LL1,2,3,C; reversible slide	1954
Thornton Mirless PIC	10"L	?	?	?	shaft hp,diam,rpm; Prov pat 3470/52	1952

Comment:

a.   Thornton slide rules normally have a date code stamped into the wood on the back of the rule.

THORNTON (BRITISH THORNTON) (GB)

Address: British Thornton PIC. P.O.Box 3, Wythenshawe, Manchester 22.

History: See also **A.G.Thornton**, British Thornton was formed in 1967. The firm finally went out of business in 1993, having ceased making slide rules some years earlier.

Won a British Design centre award in c1975. Best known for the range of Thornton's Plastic Slide Rule types as follows:

Slide Rules:

1	2	3	4	5	6	7

Type Ai. Labelled in red (on both sides for Dup) of the stock on the top left, straight bridge.

PIC No 221	25L	Dup Pl		m/p	LL01,LL02,LL03,DF/CF,CI,Isd-Td,C/D,LL3,LL2,LL1 S,ST,T,A/B,L,K,C/D,DI,Ps,Pt
PIC no 241	25L	OF Pl		m/p	
PIC No 251	25L	OF Pl		m/p	
PIC No 271	25L	OF Pl		m/p	LL2,LL3,L,A/B,CI,Sd-Td,C/D,S,ST,T
PIC No 281	25L	OF Pl		m/p	

Type Aii. pre 1962? Plastic tube? Labelled in Red on one side, top left of the stock. Straight bridge

Thornton PIC No 221- Comprehensive	25L	Dup Pl	pl	
Thornton PIC No 241 - Log-Log	25L	OF Pl	pl	
Thornton PIC No 251 - Standard	25L	Dup Pl	pl	cm,K,DF/CF,CI,C/D,A,ins//A/B,CI,C/D
Thornton PIC No 271 - Student Log-Log	25L	OF Pl	pl	LL2,LL3,L,A/B,CI,Sd-Td,C/D,S,ST,T
Thornton PIC No 281- Standard Student	25L	OF Pl	pl	cms,ins,DF/CF,CI,C/D,A,ins

Type Aiii. pre 1962? Plastic tube? Labelled in Black on one side, top left of the stock. Shaped bridge

Thornton PIC No 221- Comprehensive	25L	Dup Pl	pl	
Thornton PIC No 241 - Log-Log	25L	OF Pl	pl	
Thornton PIC No 251 - Standard	25L	Dup Pl	pl	cm,K,DF/CF,CI,C/D,A,ins//A/B,CI,C/D
Thornton PIC No 271 - Student Log-Log	25L	OF Pl	pl	LL2,LL3,L,A/B,CI,Sd-Td,C/D,S,ST,T
Thornton PIC No 281- Standard Student	25L	OF Pl	pl	

Type Bi. using the same model numbers as Aii above, but preceded with a "P" (for PIC?) and a name. Plastic tube or plastic "wallet".

Thornton No P221 - Comprehensive	25L	Dup Pl	pl	LL01,LL02,LL03,DF/CF,CI,Isd-Td,C/D,LL3,LL2,LL1 S,ST,T,A/B,L,K,C/D,DI,Ps,Pt
Thornton P241 - Log-Log	25L	OF Pl	pl	
Thornton P251 - Standard	25L	Dup Pl	pl	
Thornton No P271 - Log-log	25L	OF Pl	pl	LL2,LL3,L,A/B,CI,C/D,S,ST,T
Thornton P281 - Standard Student	25L	OF Pl	pl	

Type Bii. Called British Thornton and using the Pxxx numbers plus name. Pre 1971? Plastic wallet?

British Thornton No P221 - Comp'sive	25L	Dup Pl	pl	LL01,LL02,LL03,DF/CF,CI,Isd-Td,C/D,LL3,LL2,LL1 S,ST,T,A/B,L,K,C/D,DI,Ps,Pt
British Thornton No P241 - Log-Log	25L	Dup Pl	pl	LL2,LL3,L,A/B,CI,Isd-Td,C/D,S,ST,T cm,K,DF/CF,CI,C/D,A,ins
British Thornton No P251 -	25L	OF Pl	pl	
British Thornton No P271 - Log-log	25L	OF Pl	pl	LL2,LL3,L,A/B,CI,ISd-Td,C/D,S,ST,T
British Thornton No P281 - Standard	25L	OF Pl	pl	cm,K,DF/CF,CI,C/D,A,ins

Type Biii. Called British Thornton with a new series of model numbers. Plastic tube.

British Thornton AA010 -Comprehensive	25L	Dup Pl	pl	LL01,LL02,LL03,DF/CF,CI,Isd-Td,C/D,LL3,LL2,LL1 S,ST,T,A/B,L,K,C/D,DI,Ps,Pt
British Thornton AD050 - Log-log	25L	OF Pl	pl	LL2,LL3,L,A/B,CI,ISd-Td,C/D,S,ST,T
British Thornton AD060 - Standard	25L	OF Pl	pl	cm,K,DF/CF,CI,C/D,A,ins
British Thornton AD070 - Modern-maths	25L	OF Pl	pl	

Type Biv. As Biii, but model numbers only used. Plastic tube

British Thornton AA010.	25L	Dup Pl	pl	

British Thornton AD050.	25L	OF Pl	pl	
British Thornton AD060.	25L	OF Pl	pl	
British Thornton AD070.	25L	OF Pl	pl	K,A/B,CI,C/D,L

<u>Type C</u> which is physically identical to the type B, but has the British Thornton name and type designation on the right hand end of the slide rather than the top left hand side of the stock. This is also known as the Mark 2 by Thornton. pre 1978? Plastic tube.

British Thornton AA010 - Comprehensive	25L	Dup Pl	pl	
British Thornton AD070 - Modern Maths	25L	OF Pl	pl	K,A/B,CI,C/D,L
British Thornton AD150 - Advanced	25L	OF Pl	pl	LL1,LL2,LL3,A/B,CI,K,C/D,S,ST,T
British Thornton AD160 - Standard	25L	OF Pl	pl	K,DF/CF,CI,C/D,A

<u>Type Ci</u> As type C, but with model numbers only used. Plastic tube.

British Thornton AA010.	25L	Dup Pl	pl	
British Thornton AD070.	25L	OF Pl	pl	
British Thornton AD150.	25L	OF Pl	pl	LL1,LL2,LL3,A/B,CI,K,C/D,S,ST,T
British Thornton AD160.	25L	OF Pl	pl	

<u>Type D</u> Special function slide rules, shaped bridge. Plastic tube or wallet.

F5300, Pipe sizing rule water	25L	OF Pl	pl	E,A,B,BI,C/D,F (act,nom/wg,wg,m/t,t); Saunders pattern

<u>From 1960 **Dinsdale** catalogue, P.I.C.</u> Slide rules as follows:

F3654 Standard	10"L	CF W/C	m/g	A/B,CI,C/D//S,L,T; 50/-
VP3656 Standard	5"L	CF Pl	m/g	A/B,CI,C/D//S,L,T; 33/-
F3756 Standard	20"L	CF W/C	m/g	A/B,CI,C/D//S,L,T; £7/17/6d
F4458 Rietz	10"L	CF W/C	m/g	K,A/B,CI,C/D,L//S,ST,T; 60/-
VP114 Rietz	5"L	CF Pl	m/g	K,A/B,CI,C/D,L//S,ST,T; 30/-
F4459 Rietz	20"L	CF W/C	m/g	K,A/B,CI,C/D,L//S,ST,T; £8/15/0d
No131 Comprehensive Log-Log	10"L	CF W/C	m/g	LL1,LL2,A/B,CI,Is,Vs,C/D,P//S,ST,T\\Volt,Dyn; 72/6d
VP134 Comprehensive Log-Log	5"L	CF Pl	m/g	LL1,LL2,A/B,CI,Is,Vs,C/D,P//S,ST,T\\Volt,Dyn; 39/-
No132 Comprehensive Log-Log	20"L	CF W/C	m/g	LL1,LL2,A/B,CI,Is,Vs,C/D,P//S,ST,T\\Volt,Dyn; £9/5/0d
No121 Comprehensive Log-Log	10"L	CF W/C	m/g	LL1,LL2,A/B,CI,Is,Vs,C/D,P//S,ST,T\\Vector; 60/-
VP124 Comprehensive Log-Log	5"L	CF Pl	m/g	LL1,LL2,A/B,CI,Is,Vs,C/D,P//S,ST,T\\Volt,Dyn; 32/6d
No122 Comprehensive Log-Log	20"L	CF W/C	m/g	LL1,LL2,A/B,CI,Is,Vs,C/D,P//S,ST,T\\Volt,Dyn; £8/10/0d
P170 Standard Student	10"L	CF Pl	m/g	A/B,CI,C/D,L; 19/3d
P171 Rietz Student	10"L	CF Pl	m/g	K,A/B,CI,C/D,L//S,ST,T; 23/9d
P172 Log-Log Student (Deferential)	10"L	CF Pl	m/g	K,A/B,CI,C/D,L//S,ST,T; 36/-
P172E Log-Log Student (Electrical)	10"L	CF Pl	m/g	K,A/B,CI,C/D,L//S,ST,T; 36/-

<u>From the "Drawing Office Materials Manufacturers and Dealers Association catalogue, 1964.</u>

221 Comprehensive	10"L	Dup Pl	plm	LL01,ll02,ll03,DF/CF,CI,Sd-Td-ISd-ITd-C/D,LL3,LL2,LL1 S,ST,T,A/B,L,K,C/D,DI,Ps,Pt
241 Standard Log Log	10"L	OF Pl	plm	LL2,LL3,L,A/B,CI,Diff Trig,C/D,S,ST,T
251 Standard	10"L	OF Pl	plm	cms,DF/CF,CI,C/D,A,ins
271 Student	10"L	OF Pl	plm	LL2,LL3,L,A/B,CI,Diff Trig,C/D,S,ST,T
VP150 Standard	5"L	CF Pl	pl?	A/B,CI,C/D
VP114 Trigonometrical	5"L	CF Pl	pl?	K,A/B,CI,Diff Trig,C/D,L
VP124 Comprehensive Log Log	5"L	CF Pl	pl?	LL,A/B,CI,C/D,dB,Ps

THORSTED (USA)
Address: George Thorsted, Manufacturer, 23 Hammersly St, New York.
History: Not known.

Slide Rules:

1	2	3	4	5	6	7
Made **NYSTROM's** calculator.						1854

THREESCALE (Germany?)
Address: Not known.
History: Assumed to be German as the example known is covered by a DRGM. This may also have been a brand name for another manufacturer or a translation of the "Drei-skal" slide rule from various German manufacturers.

Slide Rules:

1	2	3	4	5	6	7
-	L	??	?	?	K,A/B,C/D,L//S,T	

TIANJAN (China)
Address: Tianjan, not known.
History: Not known.

Slide Rules:

1	2	3	4	5	6	7
501 (Rietz)	12.5L	CF	Pl	pl	K,A/B,CI,C/D,L//S,ST,T	

TIGER (GB)
Address: Tiger Toys, Petersfield, Hampshire.
History: Not known.

Slide Rules:

1	2	3	4	5	6	7
4 foot classroom slide rule.	4'L	CF	Wp	w/p	A/B,CI,C/D	
6 foot classroom slide rule.	6'L	CF	Wp	w/p	A/B,CI,C/D	

TREYSIT (Germany)
Address: Heinrich Ermel, D3578, Treysa, W. Germany.
History: Produced a Bakelite circular slide rule under the name Treysit.

TRÖGER (Germany)
Address: K. Emil Tröger, Mylam - Vogtl, Germany. Also Hans Tröger, Kirchenthumbach, Oberpfalz, Germany.
Also at Reichenbach and Mylau, Germany.
History: Not known, nor is the relationship between the two Trögers known.

Slide Rules:

1	2	3	4	5	6	7
Mod 1. Circular slide rule.	30C				C/D; Hans Tröger.	1910
Model 1:	30C				circular slide rule	
Cylindrical	T					1910

TROUGHTON (GB)
Address: Troughton & Simms Ltd, 340 Woolwich Rd, Charlton, London SE7.
History: Well known Mathematical and Optical Instrument makers.

Slide Rules:

1	2	3	4	5	6	7
Fullers Calculating Slide Scale.						

TUCK & BLAKEMORE (GB)
Address: Tuck & Blakemore Ltd, Coventry, England.
History: c1921, wholesale distributers of slide rules, particularly **OTIS-KING** in their early days in Coventry.

TYLER (USA)
Address: Not known.
History: c1967 slide rule sold by **INFO INC**.

UNIQUE (GB)
Address: Unique Slide Rule Co of Brighton Ltd., Telscombe Cliffs, Brighton, Sussex.
History: 1920, Unique Slide Rule Company founded by Burns Snodgrass in Moulescombe, Brighton. 1921, first slide rules delivered, made in the family home in Moulescombe.
c1928, moved to Carden Avenue, Brighton
1936/7, moved to a purpose built factory at London Road, Old Patcham, Brighton.
1951, becomes the Unique Slide Rule Company Limited
1954, Burns Snodgrass dies aged 68
1956, move to Buckhurst Road, Telscombe Cliffs, Brighton
1975, manufacture of slide rules stopped.
1993, Donald Snodgrass dies, he is the son of the founder who was Managing Director of the company following his fathers death to c1975, and ultimately c1980 when Donald's wife Elaine took over as Managing Director.

The most popular and common UK slide rules were made by Unique who also made other mathematical instruments. All instruments were exported throughout the Commonwealth and the rest of the world.
1994, Unique closes.

Slide Rules:

1	2	3	4	5	6	7
From a 1958 advertisement for these slide rules, we can identify:						
U1, Universal I	10"L	CF	W/C	pl		
Log-log	10"L	CF	W/C	pl		
Log-log	5"L	CF	W/C	pl		
Navigational	10"L	CF	W/C	pl		
Commercial	10"L	CF	W/C	pl		
Monetary	10"L	CF	W/C	pl		
Ten-Twenty	10"L	CF	W/C	pl		
Electrical	10"L	CF	W/C	pl		
U3, Dualistic	10"L	CF	W/C	pl		
Brighton	10"L	CF	W/C	pl		1

Comment:

(1)   This carried Darmstadt arrangement scales.

The information says that there were 25 slide rules, it is not known if this included the 10" and 50" dial calculators, and whether there were other 5" versions supplied. The accompanying photograph shows an all plastic cursor.

The advert for 1968 uses the same words, but the illustration is changed showing the same slide rule with a different plastic cursor. This later version has shaped and serrated plastic end pieces on the cursor.

A similar advert from 1971 shows:						
S500, Study 500	10"L	OF	Pl	pl		
Study 900 Reitz	10"L	OF	Pl	pl		
J181, Junior	10"L	CF	Pl	pl		
Darmstadt	10"L	?	Pl?	pl	Plastic or Wood?	
Universal One	10"L	CF	W/C	pl		
Log-log	10"L	CF	W/C	pl		
Electrical	10"L	CF	W/C	pl		1
Dualistic	10"L	CF	W/C	pl		1
Chemical	10"L	CF	W/C	pl		

Comment:

1.   These are shown as made by the **TECHNICAL SUPPLY Co.**

The information illustrates a plastic open construction rule, and says that rules were available in traditional wooden designs as well. The dial calculators are covered in the literature.

Slide Rules:

1	2	3	4	5	6	7
Comprehensive list of Unique Slide Rule Types.						
(Unmarked)	10"L	CF	W/C	w/pl	type 1, Manufactured in Moulscoomb.	
The "Unique" Slide Rule	10"L	CF	W/C	w/pl	type 2, A/B,C/D. Typed label. Scales early font?	
The "Unique" Slide Rule	10"L	CF	W/C	w/pl	type 2, A/B,C/D. Diff font (as later). Printed label	
Universal	10"L	CF	W/C	m/pl	LU,S,A/B,CI,C/D,T,LL	
The Unique Universal I Slide Rule	10"L	CF	W/C	m/pl	type 3	
Universal I, U1	10"L	CF	W/C	m/pl	LU,S,A/B,CI,C/D,T,LL	
Universal I, U1/2	6"L	CF	W/C	m/pl	LU,S,A/B,CI,C/D,T,LL	
Universal I, U1/1	10"L	CF	W/C	var	Additional low reading log-log scale	
Universal One, U1	10"L	CF	W/C	var	Plastic End Caps	
Universal One, U1	10"L	CF	W/C	var	LU,S,A/B,CL,C/D,T,LL	
Universal One, U1/1?	6"L	CF	W/C	var		
Universal II, U2	10"L	CF	W/C	var	LU,A/B,S,T,C/D,LL	
Universal Two, U2	10"L	CF	W/C	var	LU/A/B,S,T,C/D,LL	
Universal Three	10"L	CF	W/C	var		
The Unique Log-log Slide Rule	10"L	CF	W/C	var	small "Made in England", normal MiE. No CI	
The Unique Log-log Slide Rule, 5L/L	6"L	CF	W/C	var	with CI scale, B = Technical Supply Co.	
Log-log, 10L/L	10"L	CF	W/C	var	Without CI scale	
Log-log, 10L/L?	10"L	CF	W/C	var	With CI scale	
Log-log, 10L/L	10"L	CF	W/C	var	LU,A/B,C/D,LL - no CI scale.	
Log-log, 5L/L	6"L	CF	W/C	var	Without CI scale	
Log-log, 5L/L?	6"L	CF	W/C	var	With CI scale	
Log-log,	6"L	CF	W/C	var	Without CI scale	
Log-log, 5L/L	6"L	CF	W/C	var	White Plastic with Celluloid scales. Experimental?	
Log-log	7"L	CF	W/C	var		

Legible Rule, 10G	10"L	CF	Pl	var	Stock & Slide with metal inserts, experimental?
Legible Rule, 5G	5"L	CF	W/C	var	
The Unique 10/20 Precision	10"L	CF	W/C	var	
10/20 Precision, 10/20	10"L	CF	W/C	var	D,s/s,s/s,A; s = double length scales
The Unique 5/10 Precision	5"L	CF	W/C	var	D,s/s,s/s,A; s = double length scales
5/10 Precision, 5/10	5"L	CF	W/C	var	D,s/s,s/s,A; s = double length scales
Commercial, C	10"L	CF	W/C	var	H,K,M/N,R,C/D,U,V
Electrical, E	10"L	CF	W/C	var	LU,F,DF/CF,Cl,C/D,C,L
Navigational, N	10"L	CF	W/C	var	S1,A/B,S2,T2,C/D,T1
Dualistic, D1	5"L	CF	W/C	var	
Dualistic, D2	5"L	CF	W/C	var	
Dualistic, D3	10"L	CF	W/C	var	also Pat 583,637
Dualistic, D4	10"L	CF	W/C	var	P1/DF/CF,Q1,Q2,C/D,P2//LL1,LL2,LL3
Dualistic(?), U3(?)	10"L	CF	W/C	var	
Monetary, M	10"L	CF	W/C	var	M2,M1/B,C/D,M3
Brighton, B	10"L	CF	W/C	pl	K,A/B,Cl,C/D,L,($\sqrt{1-x^2}$//LL1,2,3//S,T
Brighton, B	10"L	CF	W/C	pl	as above, painted white.
Thin pocket (as 5L/L), T1	5"L	CF	W/C	pl	
Thin pocket (as 5/10), T2	5"L	CF	W/C	pl	
Thin pocket (as 5/G), T3	5"L	CF	W/C	pl	
Thin pocket (as U1/2), T4	5"L	CF	W/C	pl	LU,S,/A/B,Cl,C/D,Y,LL
Thin pocket (as D2), T5	5"L	CF	W/C	pl	
Study 500, S500	10"L	OF	Pl	pl	K,A/B,Cl,C/D,ST
Study 600, S600	10"L	OF	Pl	pl	seven scales
Study 700, S700	10"L	OF	Pl	pl	LU,S,A/B,Cl,C/D,T,LL
Study 900 Reitz, S900	10"L	OF	Pl	pl	K,L,A/B,Cl,C/D,S,T,ST. Also, "Study 900" in Red.
Darmstadt	10"L	CF	Pl	pl	
Chemical	10"L	CF	??	pl	wood/cell?
Chemical, O	10"L	CF	??	pl	LU,A/B,Cl,Ch,C/D,LL. Atomic weights under slide
Legible, J180	8"L	CF	Pl	pl	Short version of S900. K,A/B,Cl,C/D,S,T.STR
Junior, J181	8"L	CF	Pl	pl	A/B,C/D
Legible, J182	8"L	CF	Pl	pl	LU,S,B,Cl,C/D,T,LL. Logo in Red.
Junior, J	6"L	CF	W/C	None	Unusual folded scale, no cursor groove.
Jiffy	4"L	CF	W/C	pl	A/B,CI,C/D,K. No "Unique" name on rule.
Florida, F	10"L	CF	W/C	pl	
Florida, F5	5"L	CF	W/C	pl	
Area Calculator	6"L	CF	W/C	none	No cursor groove.
Double H Slide Rule	10"L	Dup	W/C	pl	Aluminium ends. Duplex. Experimental?
Craft Slide Rule	6"L	CF	W/C	none	
Mini-Mannheim	5"L	CF	W/C	pl	
Interval Calculator	10"L	CF	W/C	pl	
Pioneer Long Scale	10"L	CF	W/C	pl	
Unmarked	9"L	CF	W/C	pl	A/B,CI,C/D,K; Possible prototype.
Concrete Volume Computer	6"L	CF	W/C	none	No cursor groove.
Costing	10"L	CF	W/C	pl	
Development Piece	6"L	CF	W/C	none	dB scale, Marked REDIFFUSION
Tube Making	10"L	CF	W/C	pl	For Tube Investments.
Ratcliffe's Spring Calculator	10"L	CF	W/C	pl	2 slides. D* = D modified
Rolling Mill Calculator	10"L	CF	W/C	pl	2 slides, prototype?
Wilkinson's Sugar Rule	10"L	CF	W/C	pl	For W. Wilkinson. © 1953
Harfield Weight Calculator	10"L	CF	W/C	pl	For Parson & Crossland. Metric.
Harfield Weight Calculator	10"L	CF	W/C	pl	For Harrison & Whitfield. Imperial.
Friel-Sturdy	9"L	CF	W/C	pl	For Norfolk Labs. Prov Pat 29187/45
Caterers and Confectioners	10"L	CF	W/C	pl	Special scales.
T.E.R. 1	10"L	CF	W/C	pl	For G.P.O.
T.E.R. 2	10"L	CF	W/C	pl	For G.P.O.
T.E.R. 3	10"L	CF	W/C	pl	For G.P.O.
T.E.R. 4	10"L	CF	W/C	pl	For G.P.O.
T.E.R. 5	10"L	CF	W/C	pl	For G.P.O.
Circular Calculator 10"	5"C	MD	Pl	pl	a.
Circular Calculator 50"	5"C	MD	Pl	pl	a.
Study 500, S500	10"L	OF	W/C	pl	wood stock, plastic ends. Study 500 on slide.
Universal I, U1/3?	6"L	CF	W/C	pl	low reading log-log scale on rear of slide.

Comment:

a.   Advertised in *Teach Yourself the Slide Rule* but can not be confirmed as ever manufactured.

UNIS (France)
Address: 41 Rue de Maubeuge, Paris, France.
History: Not known, UNIS made **MARC** slide rules. Logo: "UNIS, FRANCE" in an ellipse. Believed to have been in business for many years, from c1919?

Slide Rules:

1	2	3	4	5	6	7
-	13L	?	Pl	pl	13cm\A/B,C/D	
-	13L	?	Pl	pl	13cm\K,A/B,CI,C/D,L//S,ST,T	
-	14L	?	Pl	pl	14cm\A,B/C,D,E/f,L//5inch/S,ST,T	
-	14L	?	Pl	pl	14cm\kW,A/B,C/D,Dyn-Mot	
Electricien No 5	13L	?	Pl	pl	13cm\kW,A/B,C/D,Dyn-Mot	
6	13L	?	Pl	pl	13cm\K,A/B,CI,C/D,L//S,ST,T	1919?
Rietz 6A	13L	?	Pl	pl	14cm\K,A/B,CI,C/D,L//S,ST,T	1919?

US BLUEPRINT (USA)
Address: Not known.
History: 1939 selling Sexton's Omnimeter, **HEMMI** and **HALDEN** slide rules, also **ARISTO**. Some relationship with **PEASE**. Assumed to be the same as **AMERICAN BLUEPRINT**.

Slide Rules:

1	2	3	4	5	6	7
Vapor Pressure slide rule						1946

UTO (Denmark)
Address: Fabrikken UTO. DK-7173 Vonge, Denmark. Telephone/Fax: +45 75 80 36 90
History: Still in business to date, slide rule manufacturing history not known. UTO may have been one of the manufacturers who made and issued thousands of slide rules with no makers name or other notation.

Slide Rules:

1	2	3	4	5	6	7
Type 601	5"L	CF	Pl	pl	cm,K,A/B,CI,C/D,L. No. on front. Made In Denmark.	
Type 601	5"L	CF	Pl	pl	ins,K,A/B,CI,C/D,L,. No. on rear, No Made In Denmark.	
601U	5"L	CF	Pl	pl	K,A/B,CI,C/D,L//S,ST,T; labelled Hellerman.	
930						

From UTO Instruction sheet for 10" and 5", 27cm and 13cm linear slide rules.						
Rietz	5"L	CF	Pl	pl	ins\K,A/B,CI,C/D,L//S,ST,T	a.
Rietz	10"L	CF	Pl	pl	ins\K,A/B,CI,C/D,L//S,ST,T	a.
Darmstadt/Duplolog	5"L	CF	Pl	pl	L,K,A/B,CI,C/D,P,T,S//LL1,LL2,LL3	a.
Darmstadt/Duplolog	10"L	CF	Pl	pl	L,K,A/B,CI,C/D,P,T,S//LL1,LL2,LL3	a.
Electro	5"L	CF	Pl	pl	motor,volt,A/B,CI,C/D,cos,T,S//LL1,LL2,LL3	a.
Electro	10"L	CF	Pl	pl	motor,K,A/B,CI,C/D,volt,T,S//LL1,LL2,LL3	a.

Comment.
a.   Probably versions also available in 27 cm. and 13 cm. versions with the inches scale replaced by a cms scale.

VARI-VUE (USA)
Address: Not known.
History: Produced a range of Decimal equivalents converters under US Patents, 2,799,938 and 2,815,310.

VEB-MANTISSA (Germany)
Address: Volks Eigens Betrieb, Deutshes Demokratik Republik.
History: Not known, but produced a comprehensive suite of slide rules. Possible brand name or range of slide rules known as **MANTISSA**. The Company may have been VEB-MANTISSA.

Slide Rules:

1	2	3	4	5	6	7
13L Duplex	L	?	?	?	ST,A,DF/CF,CIF,CI,C/D,P,S,T /LL1,LL2,LL3,L/	
Mantissa - Multi	13L	?	Pl	pl	ST,A,DF/CF,CIF,CI,C/D,P,S,T//LL1,LL2,LL3,L	
Rietz	25L	?	Pl	?	K,A/B,CI,C/D,L//S,ST,T; 1731511	
Mono Rietz	25L	?	Pl	?	L,K,A/B,CI,C/D,S,T,ST; 1731541	

Darmstadt II	13L	?	Pl	?	ST,A,DF/CF,CIF,CI,C/D,P,S,T//LL1,LL2,LL3,L;
Darmstadt Duplo	25L	Dup	Pl	pl	K,A/B,CI,C/D,L; 1731523/4
					P,ST/LL1,LL2,LL3/D,T,S

VOGELSANG (Germany)
Address: Carl Vogelsang, Hähne, Bielfeld, Ventile, Germany.
History: Not known. Logo "CV" on a R185 circular metal slide rule with C/D scales.

VÖGTLI (Switzerland)
Address: Olten, Switzerland.
History: Not known. Circular 12 cm plastic slide rule with advertising information.

VYNCKIER (Belgium)
Address: Gand 51, Nieuwe Vaart, Gent, Belgium.
History: Not known.

W&G (Australia)
Address: cWWII: White & Gillespie Pty Ltd, 17-19 Radford Toad, Reservoir, Victoria, Australia.
Later: White & Gillespie, Record Processing Co Pty Ltd. Melbourne, Australia.
History: Probably the premier Australian slide rule manufacturer. Applied for a patent, not known if granted. See also **REED** and **CAL** who sold and may also have manufactured slide rules in Australia. Logo >W&G<.

Slide Rules:

1	2	3	4	5	6	7
During World War II they made specialist rules:						
Mixing Ration Calculator						
Meteorological Bureau Circular Slide Rule ?C						
Australian Army Protractor						
Radiosonde Temperature Evaluator Slide Rule						
R.A.C. Plotter						
Circular Slide Rule						
RAAF Map Scale						
Howse Navigational Computer for Commercial Aircraft						
Slide Rule Model 421	12"L	CF	?	?	?	
150mm Rectangular Army Protractor						

The range in the 40's or 50's included, according to adverts in Instruction manuals, the following:
Model 432 Dualface Engineers slide rule, with Volt Drop and Dynamo scales. Also known as Comprehensive.
Model 443 Dualface Artillery slide rule.
Model 454 Dualface Survey slide rule.
Model 476 Dualface Trigonometrical slide rule.

Models 432 and 476 are known from examples, the others may have been "vapourware".

WAGNER (USA)
Address: Walter Wagner Corp, New Jersey, USA.
History: c1923 a slide rule sold by **WEBER**. No other details.

WAKCO (USA?)
Address: Not known.
History: Not known, see **WARREN KNIGHT**.

WARNER (USA)
Address: Warner Instruments, Chicago 56, USA.
History: c1951, a Plotter computer, patent 2,545,935, and c1964 Patent (L.A. Warner) Flight Computer - 3,131,858 (USA).

WARNER & TRASEY (USA)
Address: Warner & Trasey, USA.
History: c1906 Burnham's Circular slide rule. No further details available.

WARREN KNIGHT (USA)
Address: Not known.
History: c1950 sold the WAKCO weight slide rule, c1951 sold the **NOVOTINI** Weight Slide Rule.

WATSON (GB)
Address: W. Watson & Sons Ltd, 313 High Holborn, London WC1.
History: Not known.

Slide Rules:

1	2	3	4	5	6	7
Fullers Calculating Slide Scale.	T					
Multiplex (cubes / cube roots)	L					

WATTS (GB)
Address: Messrs E.R. Watts & Son, 123 Camberwell Rd, London SE5.
History: Not known.

Slide Rules:

1	2	3	4	5	6	7
Atmospheric -	L				Gunnery, corrections for temp and pressure.	
Bouchers Calculating Circle.	W					
Fullers Calculating Slide Scale.	T					
Course and Speed calculator.						

WEBER (USA)
Address: Not known.
History: 1923 were selling **WAGNER, SPHINX, CHEMIC, DWARF** and **MERRIT** slide rules, also other drafting materials.

WEBER (Germany)
Address: R. Webber, Aschaffenburg, Germany.
History: c1872, Weber's Rechenkreisen, this is where the Aschaffenburg comes from, not known if the later examples are the same address.
Slide Rules:

1	2	3	4	5	6	7
Weber's Rechenkreisen	?					1872
No 4285N	29.5L	?	W	?	K,A/B,CI,C/D,L//29.5cm/S,ST,T/11.5inch.	

WEBER (Switzerland)
Address: Wilhelm G.G. Weber, Zurich, Switzerland.
History: System Hofer and Broennimann slide rules. There are examples in the Kern collection in the City Museum in Aarau, Switzerland.

WEEMS (USA)
Address: Denver, Colorado. also Annapolis, Maryland, USA.
History: A division of **JEPPESEN & Co**, various Flight Computers under several Ray Lahr patents c 1955. Also a Nautical Slide Rule, assumed to be part of another range of instruments. Weems System of Navigation.

Slide Rules:

1	2	3	4	5	6	7
Dalton Dead Reckoning Computer E-6B	C	MD	Pl	?	Pat 2,097,116	
Dunlap Low Altitude Sun Azimuth Comp.	C	MD	Pl	?		c1962
Tyler slide rule	21U	MD	Pl	pl	special scales, U=21cm square with 19cm circular overlay	
Nautical Slide Rule FNS-3	C	MD	Pl	?		

WEIL (USA)
Address: Not known.
History: c1930 selling slide rules and drafting supplies.

WEST (GB)
Address: West and Partners, 15/17 Petty France, Westminster, London, England.
History: Not known, only example is a "British Made" slide rule. Known to have sold **DAVIS** rules under the West name, or had some relationship with Davis.

WHITE & GILLESPIE (Australia)

Address: Melbourne, Australia.
History: See **W&G** and **DUALFACE**.

WHITTON (GB)

Address: Whitton Precision Limited, Bridge Works, Durnsford Road, Wimbledon London, SW19.
History: >1962, Sold Otis-King Calculators. Not known what the relationship with Carbic was.

WICHMANN (Germany)

Address: Gebr. Wichmann mbH, c1900/10, Karl-Strasse 13, Berlin NW.6, Germany.
c1938, Karl-Strasse 13-14, and Marienstrasse 19-20, Berlin NW. 7, Germany.
History: Established 1873, sold other manufacturers rules, though it may also have made its own simple slide rules. In 1970 in the USA was selling its own as well as **ARISTO, D&P, HALDEN, SCHACT u. WESSERICH** and **FABER-CASTELL** slide rules.

Slide Rules:

1	2	3	4	5	6	7
From a c1910 Catalogue:						
No 458, Taschen-Rechenstab	15L	CF	W	plm	A/B,C/D//S,T	
No 474	L	?	?	?		
No 1389	L	?	?	?		
No 1399, Halden's Calculex	W	n/a	M	n/a		
No 1418, System Rietz	L	?	?	?	K,A/B,CI,C/D,L//S,T	
From a c1938 Catalogue:						
No 1318, Präzisions-Arbeitszeit	8.4C	MD	M	n/a	3D; System Thomas. Described as a "Rechenuhr".	
No 1319, Rechenschiebe Hansa	17.5C	MD	M	pl	3D,1C; A/B,consts.	
No 1320, Recehschiebe Fix	8.5C	MD	M	n/a	2D; converter.	
No 1396, Kreiss-Rechenschieber	5.8C	MD	M	m	2D,1C; similar scales to Calculex.	
No 1397, Kreiss-Rechenschieber	6.5C	MD	M	pl	2D,1C; single sided version of 1399	
No 1399, Kreiss-Rechenschieber	6.5C	MD	M	pl	2D,1C; double sided, similar to Calculex.	
No 1400, Watch type	5.5W	n/a	M	pl	A/B,CI; possibly Schact u. Wesserich.	
No 4488, Measuring tape slide rule	U	n/a	M	pl	C/D; 50cm scales, bakelite case.	
Various Others:						
Wichmann's Rechenschieber	L	CF?	W/P	?	Paper scales on wood	1895
R.R.	5"L	CF	W/C	m/g	A/B,C/D.ST wood/cell rule	
Nr 4430	5"L	CF	W	pl	A/B,C/D painted wood, card case	
4477 Darmstadt	?	CF	?	?	K,A/B,CI,C/D,P\S,T//LL1,LL2,LL3; made by F-C.	
-	?	?	?	?	A/B,C/D//15cm/S,ST,T/5inch; made by D&P.	
466	?	?	C	?	A/B,C/D	
1918	50L	?	W/C	?	A/B,CI,C/D//S,L,T; made by F-C; 4/98	
1327	25L	?	W/C	?	K,A/B,C/D,L	

WIESENTHAL (Switzerland)

Address: Wiesenthal & Cie, Switzerland.
History: Not known.

Slide Rules:

1	2	3	4	5	6	7
Hartmann's circular slide rule	C					1972

WIESENTHAL (Germany)

Address: Wiesenthal & Cie, Aachen, Germany.
History: c1875/7, Hermann's Recenknecht, a circular slide rule; otherwise no details. It is not known if there is a relationship with the Swiss company of the same name, it seems likely.

WILLI (Germany)

Address: Dipl. Ing. E. Willi GmbH, Calwerstrasse 18, Stuttgart, Germany.
History: 1924, set up the company that would become **IWA**. See also **NORMA** and **OMARO**.

WILLIAMS (USA)
Address: Chas E. Williams, USA.
History: c1968, a Volume and Velocity Computer.

WILLIS (USA)
Address: Division of Smiths International Inc.
History: Valve size and flow rule.

WILSON (USA)
Address: K.J. Sonney Wilson, Computers, 213 River Road, Little Hocking, Ohio, USA.
History: c1966 to 76, produced a range of Orifice Computers, small and large.

WILSON & GILLIE (GB)
Address: see also Messrs John Lilley & Sons Ltd, The New Quay, North Shields.
History: Identified as a slide rule maker in the 1922 *Dictionary of British Scientific Instruments*, but no slide rule types identified.

WINTERTHUR (Germany)
Address: Reiter Winterthur? Germany?
History: Not known.

WOLTERS-NOORDHOFF (Holland)
Address: Holland.
History: Sold Fuji slide rules, unlikely to have ever manufactured their own rules.

Slide Rules:

1	2	3	4	5	6	7
No112	25L	CF	Pl	pl	L,DF/CF,CIF,CI,C/D,A,K	
NoFJ102	25L	CF	Pl	pl	L,DF/CF,CIF,CI,C/D,A,K//S,ST,T1,T2	
324	L				made in Germany	
JE650	25L	Dup	Pl	pl	A,DF/CF,CIF,CI,C/D,K; made in Germany L,T1,T2/S,CI,C/D,S,ST	

WOOD (USA)
Address: R. Wood, New York, USA.
History: c1907, Slide Rule for Calculating sag in wires, otherwise not known.

WORTHINGTON (USA)
Address: Worthington Corporation, USA.
History: Worthington Controls Valve slide rule, c1965.

WURTH (Belgium)
Address: Wurth-Micha, Liege, Belgium.
History: c1904, Regle pour le calcul des Distributions de Vapeur, otherwise no details.

WYNNE-ROBERTS (USA?)
Address: R.O. Wynne-Roberts, address not known.
History: c1908, Wynee-Roberts' pocket hydraulic calculator. History not known.

YOUNG (USA)
Address: William J. Young, also Wm. J. Young & Sons or Young & Sons, Philadelphia, USA.
History: 1864, made and sold **NYSTROM**'s Calculating rule

MISCELLANEOUS SLIDE RULES:
It is interesting to note that in 1921, nine of the manufacturers in the *Dictionary of Scientific Instrument Makers* claimed to make the Fuller Calculators, and seven claimed to have made the Boucher Calculator. Both would be deemed as expensive calculating instruments, and in all probability these companies were only retailers. It should be noted that a number of these manufacturers were also better known in the

photographic world, not surprising for an optical manufacturers' publication.

There are a number of designs whose manufacturer is not known:

1807	Scale for calculating cargo weights
1899	F.J. Vaes, Regle pour le traction des Locomotives, Holland, see also 1905.
1900	Lallemand Regle a Calcul, from Leipzig.
1900	Herrgott, Regle a Deux Reglettes
1901	Muller's Hydraulischer Rechenschieber, Austria.
1901	Riebel's Geodetischer Rechenschieber, Austria.
c1901	Schweth's Rechenschieber.
1901	Pierre Weiss, Regle a Calcul.
1902	Cox's High Pressure Fluid Discharge Computer.
1903	Knowles' Calculating Scale.
1904	Baines' Slide Rule for solving equations.
1904	Mougnie Regle pour le Calcul des Conduites d'eau.
1904	M. Scnitzel Logarithmischer Kubizierungsmass-stab.
1904	Derivry, Carte a calcul.
1905	Toulon, Regle pour les Calculs de Terrassements.
1905	F.J. Vaes, Echelles Binaires, from Holland.
1905	Moehlembruck's Instrument.
1905	Masera's Rechenschieber, Zurich.
1905	Malassis Cercles Logarithmiques.
1905	Viaris, Ruban Calculateur.
1906	Niehan's Metalltachymeterschieber.
1906	Fearnley's Universal Calculator (By Royal Letters Patent)
1908	Niethammer's Prazisions-Schulstab, Germany.
1908	Koppe's Barometerhohenrechenschieber
1908	Vogler's Barometerhohenrechenschieber
1908	Hammer's Barometerhohenrechenschieber
1909	Schumacher's Rechenschieber mit Teilung in gleiche intervalle, Munich.
c1915	Best's simplified duo decimal, upper scales duodecimally divided.
c1915	Princeps Artillery Field slide rule, specifically for sighting, deflection
c1915	Baines Slide rule, invented by Mr H.M. Baines of Lahore, for the solution of Flamants formula, uses 4 slides connected with a form of Pantograph, or Jointed parallelogram.
c1915	Cultriss's Calculating Disc, similar to Charpentier Calculator (K&E)
1920	Slide rule for calculations concerning weight of fish
1920?	MacFarlane's calculating cylinder
1920?	Glasser's slide rule for cloth calculations
1920?	Pollitt's pipe calculator
??	Lloyd's Cotton cloth coster incorporated in a writing box
1923	Gladstone's Cross Gauge, 425" scale length, grid iron as Everett
??	Barth slide rule - see Bethlehem Steel
??	Moko Lightening Calculator. 4" dia watch type calculator, made in Germany.
c1910	Calcu-Rotor. Double-sided circular 2 disc, maybe made in Australia.

also:

Slide Rules:

1	2	3	4	5	6	7
Uniface	10"L	CF	W/C	m/g	S,A/B,L,C/D,T//??	
Bentley Long-Short rule	10"L	CF	W/C	m/g	L,C/D,X1,X2/Y2,Y1; X & Y are double scales as C/D but 20"	
"C.E.M"	10"L	CF	W/C	?	Civil, Electrical & Mechanical Engineers rule, extra V,D and eff scales	
Omnes rule	10"L	CF	M	pl	ins,A/R,C/D,ins//S,T; second slide: L,Dynamo,Volt	

OTHER MANUFACTURERS:

Various other "manufacturers" are known from the Dutch Circle of Slide Rule Collectors and a number of other sources. There is insufficient information to add them to the main listing of manufacturers; however there is a fine line between some of the manufacturers in the main listing and those listed here. In some cases it is probable that these were actually slide rules made by a major manufacturer which were marked with the name of an advertiser for whom they were made. In other cases they may have been small manufacturers who produced slide rules in small quantities and more information may yet be forthcoming. There are small numbers of these slide rules (generally in single figures) and they are listed here for completeness, with comments where appropriate.

A.

Abrams Instruments Corp - a photographic computer.

Ace - the ACE wood/celluloid 20" rule.

Acrow - known for building materials, this is a concrete volume computer.

Addimult

Aniba - Arnhem; a circular disc calculator.

Anoxal - Swiss manufacturer of data rules and slides.

Antica Fabbrica - Italy. Slide rule No 25 known.

Arman & Pillmeier - Cassel, Germany. Scherer's Logarithmic calculating tables.

Aster - possibly Italy.

AT&T Co. - Guy Rule, circular slide rule.

Atlas Copco - known for earth moving machinery. Atlas Copco AB, Sweden. Slide rules for air pressure etc; made for them?

Auxitrol - Dutch company, produced conversion calculators.

B.

Becel - slide chart maker.

Begemann - disc and linear slide rules for special applications. Dutch company.

Bell. Arthur L. Bell - Bell's Stadia Slide Rule, c1906.

Below - plastic slide rule for food calculations.

Blau Knox Ltd - card/plastic rule; A/B,C/D,E. USA.

Bock - Julius Bock KG, JUBO, Hauenstein (Pfalz), Kromberg (Taunus), Germany. Metal disc calculator.

Brigon - combustion calculator, plastic slide chart.

Brown Boveri - metal circular electric-motor calculator, Swiss?

Buss - 20" wood gaugers rule c 1890, Ivory 5" gaugers rule from earlier. 48 Hatton Garden, London.

C.

Caldwell - Caldwell Banker & Co, Los Angeles, California - Return on Investment circular slide rule, one off?

Cartesius. Known from a Rietz 1A.

Ca-Su - Spain. Regla de Cálculo. See CASU.

J. Chadwick - Chadwick's Improved Slide Rule, c1905.

Chrysler Corp - Detroit, Michigan. Psychrometric slide rule (Humidity). 20L, plastic.

Civic - marked on a Hemmi 269 slide rule.

Clarkson - Holland. Slide Charts.

Connecticut Spring Corp., Hartford, CT. 12"L, Metal, for spring calculations.

CTIP - c1960-9, slide charts.

CTM - Russian, disc calculators.

Culmann - Swiss.

CWI Breda - military disc calculator.

D.

Dalton - USA, possibly the Dalton Flight Computer made by many manufacturers.

dam - brand name for Damien?

Daniel Industries

Degussa - Hannau, Germany.

Denspa - plastic 5"L slide rule.

DEWE - Germany.

Diefenbach

Dikema en Chabot - slide charts, Holland?

Director - Director Instruments, 525 Grand Ave., New Haven, Conn. Course finder/plotter.

Disney - Walt Disney Productions, converters.

Draper, B. - made a Routledge slide rule.

Drayton - Drayton Regulator & Instrument Co. Ltd. West Drayton, Middlesex. Valve sizing circular calculator.

Droogtechniek Rotterdam - specialist printers?

Dupan - Type #CD14, Japan?

E.

Eastman - Eastman Oil Well Survey Co. Houston, Denver, Long Beach. "Ouija" deflecting tool setting slide chart.

Emeloid - type name?

EMF - Dortmund? Retailer of plastic circular slide rules.

Engelhard Industries - UK? One-off Flight Computer.

Engineering Services, USA - Dyna-Slide Kit, circular DIY slide rule.

Environmental Dynamics Inc. - USA?

ESAR - c1945, 2' metal gunnery rule.

Eschmann-Wild - Swiss, mentioned in Cajori.

Esso - probably made for the petrol company by a major manufacturer - e.g. Perrygraph.

Exxon Aviation Products - plastic circular flight calculator.

F.

Farrow

Favor-Ruhl & Co - W/C 5"L

Festus Mfg. Co. MO. USA. - Sold the ACU-RULE.

Filotecnica - Italy.

Flynn - Circular protractor/calculator.

Fogra - Germany? Plastic circulr slide rule.

Foxboro - valve sizing rule, The Foxboro Co, Foxboro, Mass, USA. Various valve and flow calculators.

G.

Goldwell - circular calculator.

GOST - SP - Russian State Factory, coded by year of production.

Graham, H.H. - plastic circular flow calculators.

Graphic Calculator Co. Chicago, Ill. - Oil industry?
Green, Alex, E.S. - c1945 Flight Engineers Computer - one off?
Grizzle

H.
Hacquebard - Dutch printing company. Card circular slide rule.
Hammaburg - possibly Germany.
Hansa - beer?
HCW Leiden
Healy, H.W. - c1923 made a slide rule for Vickers-Spearing Boiler Co, 20 Kingsway, London WC2, sole licensee and mfr.
Heinevetters - Stabrechner, 20L plastic with special scales.
Henzi - Swiss manufacturer of calculating discs.
Higgison - UK manufacturer. Birmingham. Made carpenter's rules.
Hiltpold, Walter - Zurich, Switzerland.
Hock, Dr A. F&M Lautenschläger GmbH, Munchen 2. Plastic slide rule.
Hoff - van t'Hoff & Jongepier N.V. Shiedam, Holland. Financial circular calculator.
Holly Hobby - USA?
Hono - known from a Rietz plastic slide rule.

I.
IDO/CWI Breda - military slide rules.
Ilford - photographic slide rule, made for the film manufacturer?
Imac - slide charts.
Imco
Itab - Germany, made various calculators also for Siemens.

J.
Jason - possibly Japanese.
Jensen Bros - USA. 6" specialist pump calculator.
Johnston - slide charts.
Jokl

K.
Kaspar - plastic & W/C slide rules, "System Kaspar".
Kelco - slide charts & special (e.g. horse racing) calculators.
Kemperman - Holland?
Kisseleff - Ingenieurburo Kisseleff AG, Küesnacht, ZH, Switzerland. System Kisseleff plastic/metal circular for hydraulic calculations.
Klix - Germany.
Koh-i-Noor - Czechoslovakia. Perhaps better known as a pencil maker.
Koh-i-Noor-Graphoplex - France.
Koh-i-Noor Hardtmuth - Italy.
Krama
Krebs - Zollikofen, BE, Switzerland.
Kröplin-Bützow - metal circular calculator for opticians.
Kvama - 25L W/C slide rule.
Kyd - Swiss

L.
Lancaster - Swiss. One linear slide rule in German silver.
Lautenschläger F&M. G.m.b.H. Munchen. Metal circular for technical calculation.
Lios - possibly Germany.
Löbker & Co KG. Dresden, Germany. W/C slide rule.
Löhr - Germany.
Losztalyu - Hungary. W/C identical to GAMMA.
Lummert - plastic slide rule.
Lutz - slide rule made in Japan.
Lyth - Antwerp, metal slide rule.

M.
Marabu - Germany.
Masera - Swiss.
Matsku - Russian watch type rule.
Matterson Ltd. - Shawclough, Rochdale. Card circular rule for gear calculations.
Matthijssen - Holland.
M.D.S. London - slide rule data slips.
Meiwa - makers of decimal handy calculator with Japanese patents, 41-16935, 758,588; and US pat 3,312,395.
Memmi - Japanese. Is this a variant of Hemmi?
Mercator - Flight computer, Alkmaar, Holland, Bergen?
Mercura - French circular slide rule for commercial calculations.
Mignon - for Lang & Co NV, Amsterdam, maker or advertiser?
Mimo - Swiss manufacturer of watches with slide rules.

Minerva - Rietz slide rule.
Mobil - circular calculators, maybe made for them?
Montana
MS - France.

N.

Nabla - Studio slide rule?
Nehim, Amsterdam.
Neolt - 12.5L plastic.
Nickel E.F. - Nickel's Runnerless slide rule, c1906.
Nijgh & Van Ditmar - Holland.
Nilsson - Sweden.

O.

Ohico - possibly Dutch, W/C Electro slide rule.
Ohmite - USA.
Orlikon - *Cajori* shows them (Oerlikon?) as making **DAEMEN-SCHMID's** Rechenwalze in 1908.

P.

Pacific Cable Board, Queen Annes Chambers, London SW1, Wood Cylindrical Telegraph Tariff Calculator.
Pattern - British, may be a logo or brand name.
Peerless - USA. Circular cardboard Capacity, Horsepower Efficiency etc. calc. c1959.
Pestalozzi - Swiss.
Philips - many disc and slide charts, maybe made for them?
Piechotta - Germany.
Pipeline - Service Pipeline Co, c1953, distributed by L.L. Ridgeway.
Planosec - medical calculator.
PMO
Polyflo - Flow computer, c1945,45 and 66.
Pratt & Whitney - makers of jet engines, c1970 made a Turbine Engine Performance calculator.
Produx - makers of Addiators.
Prolixan
Prometheus - Russian.
PTT - Radiology Circular slide rule.

R.

Racine - Swiss manufacturer of watch slide rules.
Ralo - Loewenthal-Raydt.
Ramenskoe - 10"L W/C.
Rani - Pontarlier, France. 5"L plastic slide rule.
Raytheon - probably made for them, Perrygraph?
Reciloga - Germany.
Rheinhold - Circular plastic rules.
Rhieta - 12.5L W/C.
Richter
Ridgeway, L.L. Inc. Distributers.
Roka - simple children's calculators.
Rotoflow - Pump manufacturers?
Rotor - patent Russian circular calculators.
Russisch - generic term for Russian slide rules? 19 examples.
Ruth - Austria.
Rynja - P. Rynja, Amsterdam. Distributers?

S.

Sabena - Airline, someone made a money converter for them?
Sajuz - Russian.
Samia - Renens-Lausanne VD, Switzerland. System Rognon slide rules, exhibited at the Technorama, Winterthur.
Sans & Striffe - possibly Japanese.
San Tech Inc - Flight Computer.
Schaur - pencil slide rule.
Schmacht - Germany.
Schneider - Germany.
Schuppisser & Billeter - Swiss.
Schurmanns - plastic circular slide rule.
Sdelaho - Leningrad.
SIG - possibly Japanese?
Sillocks-Miller - USA.
Simplex - brand name or type? Possibly Dutch.
Stambach - Swiss.

Standard Rule Co. - 2 fold 2' wooden rule. USA?
Stewart - 5"L military vehicles movement rule.
Stevenson, P. - Early Edinburgh maker.
Stücki - Germany.
Swaay - air conditioning slide rule.
Sweco - circular special calculators.

T.
Tata - E. Tata, plastic 5"L.
Teknolisk Institut - "Regnestok fur Maskinbearbejdning".
Tesa - Tesa SA, Renens VD, Switzerland. Exhibited at the Technorama in Winterthur.
H. Tilsley - Tilsley's slide rule for strength of gear teeth, c1907.
Tijdzone
TKF - cable slide rule.
TRD/GTI - energy slide rule.
Tretford
Trilux - c1962 Präzisions Rechenschiebe Licht Rechner (Light Meter).

U.
Uster - Zellweger Uster, Switzerland. Produced disc calculators for textile purposes.
Universum - Logo H.H.

V.
Ventura - Ventura Design on Time. Still produces slide rule watches.

W.
Waagschaal
Wasserwerke der Stadt, Wien, Austria.
Welch - 48" Instruction slide rule.
Wizard - Addiator makers?
Wolthekker - R.E.P. Wolthekker, Valeriansplein 6, 8916CS, Leeuwarden, Netherlands. 1961 Wood slide rule.
Wüest - Lucerne, Switzerland. Known from one example of linear civil engineers rule.
R.O. Wynne-Roberts - Pocket Hydraulic Calculator, c1908.

Y
Yoder - USA.

Z
Zeilschip

Appendix 1.
Glossary and Biographies.

Words in **Bold** are definitions of some of the terms used in the book; those in ***Bold italics*** cover the main characters in the development of the slide rule. The meanings of the terms are specifically as used in this book; unfortunately, not all manufacturers or authors were consistent in their usage of a particular term.

A

Adams: George Adams (1707 - 1773) was instrument maker to the Prince of Wales and later George III. In 1748 he produced a 12 inch brass circular slide rule with a spiral scale with ten windings.

Addiator: A simple Pescaline mechanical calculator capable of adding and subtracting. It requires a stylus to operate, and is usually fitted to the back of some slide rules, particularly Faber-Castell.

Allen: Two Allens were involved in the manufacture of slide rules. Elias (1580 - 1653) was a well known mathematical instrument maker who worked from near St. Clement's Church in the Strand in London. Elias was the father of the slide rule, as he made Oughtred's first slide rule, and indeed advised Oughtred that it would be easier to mark a circular slide rule with opening cursors, rather than the two disc design that Oughtred first designed. John Allen (1610 - 1670) may have been Elias Allen's son or nephew. John made Delamain's Mathematical Ring in either silver or brass.

B

Beghin: Auguste Beghin produced his Règle des Ecoles in France in 1907, this had a displaced scale by $\sqrt{10}$ and, with the Mannheim rule, became another standard used by a number of manufacturers. He also produced his Règle à Calculs in 1898.

Bion: Nicholas Bion published his *Construction and Principal uses of Mathematical Instruments* in France in 1723. This was translated into English by Edmund Stone and was published in several editions adding to the information available to instruments makers in France and England at that time.

Bissaker: In or about 1654 Robert Bissaker introduced the double faced slide. He is also apparently responsible for the slide/fixed stock principle.

Bones,

   Napier's   Napier's bones, invented by John Napier of logarithm fame in 1617, were several rods marked with numbers from an ancient Arabic series which could be used to perform complex multiplication. They remained in use for approximately 150 years.

Boucher   The Boucher calculator of 1876 was the seminal design of a circular pocket-watch type slide rule produced by W.F. Stanley and many other manufacturers in the UK, Europe and the USA.

Boulton: Mathew Boulton (1728 - 1809), also sometimes spelled Bolton, with James Watt, improved the accuracy of the graduations on slide rules, and in 1779 produced what became the standard "Soho" Engineers slide rule for work on their steam engine designs.

Briggs: Henry Briggs (1561 - 1630) was a friend of Napier's and Professor of Mathematics at Oxford in 1617 when he modified Napier's "natural" logarithms to produce the better known and more useful logarithms to the base 10. These are also known as "Briggsian" logarithms.

Brown: In 1633 Thomas Brown arranged a Gunter line in a "spiral" consisting of circles gradually increasing in diameter. In 1662 John Brown (son of Thomas) used the term "sliding rule" in print for the first time, and it was in general use by the end of the century. John Brown is the Brown mentioned by Pepys as making him a slide rule.

C

Circular: The earliest form of slide rule which consisted of one disc with logarithmic scales and required a form of compasses as a cursor or opening index. Alternative forms have two or more discs which rotate relative to each other, with or without a cursor or cursors.

Clairault: In 1727 J.B. Clairault described a 21 inch cardboard circular slide rule.

Closed Frame: This is a form of construction for rectilinear slide rules where the stock is such that the reverse of the slide cannot be seen from the reverse of the rule, other than through apertures or cut-outs to do so. It is the standard form for most wood/celluloid slide rules.

Coggeshall: Henry Coggeshall (1623 - 1690), an English Mathematician, added a slide into a two foot folding rule in 1677. This was specially adapted for timber measure and use by carpenters. The Coggeshall slide rule is made of boxwood with a brass hinge and end caps and a brass slider in one of the arms. The slider has a Gunter type logarithmic scale, on one side a conventional double 1 - 10 scale, on the other a broken scale from 4 - 40 called the girt line for measuring volume of timber. Also on the slider is a foot rule enabling a yard to be measured, and on the edges is a scale of a foot in 100 parts. The Coggeshall slide rule became one of the earliest standards, with large sales over about 200 years to 1870.

Cursor: A plastic, glass or mica sliding element covering stock and slide and capable of lateral movement, with one or more hairlines engraved on it to enable more accurate translation of figures between scales. On duplex slide rules the cursor must also translate information between the two sides of the rule.

Cylindrical: A form of slide rule consisting of a number of concentric cylinders capable of sliding, rotating, or sliding and rotating against each other. Some cylindrical rules have helical scales.

D

Darmstadt: A standard arrangement of scales designed in the Institute for Applied Mathematics at the Technical High School in the German city of Darmstadt whose name the standard took rather than its proposer, Professor Dr. Alwin Walther. This standard included log-log scales and is the last of the standards for rectilinear slide rules proposed in 1934, but subject to some degree of interpretation amongst makers.

de Prony: A French nobleman, Baron Riche de Prony (1755 - 1839), carried out a survey in 1833 of United Kingdom technology. He noted that the United Kingdom were " ... *commonly using 'sliding rules' at all levels, including shop keepers and artisans.*"

Delamain: Richard Delamain (1590 - 1645), a teacher of practical mathematics, published the first description of a circular slide rule in about 1631. There was considerable ill feeling between Delamain and Oughtred, it being mooted that at the very least Delamain was responsible for some plagiarism, at the worst for stealing Oughtred's designs. He went on to become mathematical instrument maker to King Charles I and became well known with his Mathematical Ring being covered by Royal Letters Patent.

Dennert: Johann Christian Dennert (1829 - 1920) is the founder of the firm Dennert & Pape set up with Martin Pape in 1862. Dennert and Pape in Altona, Germany introduced white celluloid (Ivorine) in 1886 as a material for the inscription of scales, instead of the traditional marking on wood, brass or ivory which had been in use until then. This vital invention allowed much-improved accuracy and legibility to be achieved as a result of the lack of grain and colour of the material.

Displaced Scale: The "A" and "B" scale positions on a slide rule are sometimes occupied by a displaced scale where the indices are not at the ends of the scale, but are nearer the middle of the rule. This can ease calculation by reducing the number of movements required by the slide. These scales are displaced by different factors by various manufacturers, the most common being $\sqrt{10}$ and π. The earliest displaced scale is by *Beghin*.

Duplex: A form of slide rule which has scales on both sides of the slide and stock. The scales have coincident indices, and a double-sided cursor allows translation of the figures between the two sides. Construction must be open frame, and allows a large number of different scales without excess width.

E

Evans: Dr. J.D. Evans described a grid-iron slide rule in 1866. This is the earliest form of long scale rule, whereby the scale length is increased by having multiple portions of the scale in a grid.

Everard: Thomas Everard, a gauger (customs officer) working in the excise department in Southampton, in 1683 adapted a slide rule with slides on opposite sides of a square rule, for gauging. According to Everard many thousands of his rules were sold between 1683 and 1705. The Everard slide rule is a foot long and an inch square with inset Gunter's scales and is marked with standard points important for excise use, including WG - wine gauge, AG - Ale gauge and MB - Malt Bushel. The Everard slide rule became a standard slide rule, with many additions by Shirtcliffe, Leadbetter, Vero and others who added scales as well as slides so that the final design had four slides, one on each face of the square section.

F

Fuller: Professor George Fuller of Dublin University is credited with inventing a cylindrical instrument in about 1860. In 1878 Fuller patented a tubular slide rule which was produce with papier-mâché cylinders. This large and very accurate instrument, which took his name, the Fuller Calculator, was available for almost 100 years.

G

Galileo: Galileo Galilei (1564 - 1642) was the Italian inventor originally credited with the invention of the sector in 1598. It has since been accepted, however, that it was probably a friend of his, Guidobaldo del Monte, who designed the original Sector, and that Galileo simply added to the design and popularised it.

Gauge mark: An additional mark on a scale, indicating a constant value which is of use in a particular type of calculation. (See Appendix 5).

Grid-iron: A form of construction of long scale slide rules which takes its name from the grid of scales which make up the total scale length. The splitting up of scales is done in various ways; see Evans, Cherry, Thacher, Proell etc.

Gunter: Edmund Gunter (1581 - 1626), Professor of Astronomy at Gresham College in London, was a prolific inventor who produced the first mechanical implementation of the logarithmic scale when he drew up Henry Briggs's table of logarithms as the Line of Numbers, or Line of Proportion, also known as Gunter's rule or Gunter's line, which ultimately became the slide rule. Oughtred credited Gunter as the next most important character in the invention of logarithms in his *Circles of Proportion;*

> *The honour of the invention of logarithms, next to the Lord of Merchiston and our Mr Briggs, belongeth to Master Gunter, who exposed their numbers upon a straight line.*

H

Helical scale: A method of extending the scale length used on tubular/cylindrical slide rules. The scale is wrapped as a helix around the tube thus enabling many feet of scale in a very small length of tube/cylinder.

I

Index: A slide rule scale was usually marked '1' at the left-hand end of the scale, '10' at the right-hand end of the single cycle scales, and '100' at the right-hand end of two cycle scales.

Indicator: Another name for a cursor.

L

Leadbetter: In 1750 Charles Leadbetter added a third slide to the Everard rule. He also wrote about the Everard rule and the work of gauging.

Linear: Also rectilinear. A form of slide rule with the scales arranged linearly, used as the generic term for the different forms of construction of this type of slide rule, probably the most common type.

Long scale: The generic term for slide rules with extended scale length, i.e. spiral scale circular, helical scale tubular, grid iron etc.

M

Mannheim: Colonel Victor Mayer Amédée Mannheim (1831 - 1906), a French artillery officer gave his name to a slide rule with an arrangement of scales and a cursor. The first modern arrangement of scales dated from 1850, and in 1851 Mannheim added a cursor. This was the first modern slide rule and was manufactured as a standard from 1859. Note that the designation "Mannheim Rule" was regularly used up to World War II.

Moxon: Joseph Moxon published the *Mechanick Exercises* in 1677. This described the secrets of the instrument makers at that time, thus opening up all the tricks of the trade as common knowledge.

de Morgan: Augustus de Morgan is best known as the publisher of a number of technical volumes, his *Penny Cyclopaedia* of 1842 contained an article on the Slide (or Sliding) Rule. This gives some of the history of this device at that time.

N

Napier: John Napier, Baron of Murchiston (1550 - 1617) was the inventor of logarithms and Napier's bones. The logarithm as finally modified by Briggs was the basis of Gunter's rule and the slide rule scales.

Nicholson: Between 1787-97 William Nicholson (1753 - 1815), editor of *Nicholson's Journal*, designed many types of slide rule, including linear with 1 slide and cursor, as well as circular and spiral types. He is credited with many designs.

O

Open Frame: A form of construction for rectilinear slide rules where the stock is in two separate pieces joined by some form of bridge with the slide in between. Both sides of slide and stock are visible and thus useable for the engraving of scales which are adjusted so that the indices coincide on the two sides. Duplex slide rules use this form of construction.

Oughtred: Sometime between 1620 and 1625 the Reverend William Oughtred (1575 - 1660) arranged two scales to slide alongside each other. In 1632 he produced his Circles of Proportion, also a circular slide rule with concentric logarithmic scales, and in 1633 invention of the slide rule by Oughtred was announced in London. He stated that as early as 1621 he had put two lines together in a circle with dividers from the centre.

P

Pape: Martin Pape (1834 - 1884) was Johann Dennert's partner in Dennert & Pape.

Partridge: In 1657 Seth Partridge (1603 - 1686), Surveyor and Mathematician, describes the development of the moving slide/fixed stock principle. In 1658 Partridge describes a rule made in 1654 with a small scale sliding between two fixed rules, he " ... *joined two lines of Proportion together*", which gives some doubt as to whether he preceded Bissaker. He recommended all the well known instrument makers including Allen, the Thomsons and the Browns.

Perry: Professor John Perry of the Royal College of Science, London, patented another version of log-log scale in 1901. (See also Roget and Thompson)

de Prony: Baron Riche de Prony (1755 - 1839) a French nobleman who investigated technology for their government.

R

Rietz: A standard arrangement of scales designed by Max Rietz (1872 - 1956), a steam engineer at the Trenck Steam Boiler factory at Erfurt in Germany, which became the best known and most common arrangement of scales for rectilinear slide rules. This arrangement is subject to some interpretation amongst manufacturers, and later Rietz arrangements bear little resemblance to the original proposed in 1902. Also spelt "Reitz" and in one Japanese case, "Ritz".

Robertson: The development of the slide rule cursor is credited to John Robertson (1712 - 1776) Headmaster of the Royal Academy, Portsmouth England, and then Librarian of the Royal Society in London. In 1775 he describes an ... *improved sliding Gunter* with Runner or Cursor.

Roget: Dr. Peter Mark Roget (1779 - 1869), an English physician, presented his invention of the log-log scale principle in the Philosophical Transactions of the Royal Society, London, in 1815.

Runner: Another name for the cursor.

S

Sector: The sector, invented in approximately 1598, was a hinged and graduated rule which used the theory of similar triangles, with the aid of a pair of dividers, for the calculation of natural numbers, squares, cubes, reciprocals, chords, tangents etc. It was in use until the late 18th century. See also *Galileo.*

Scott: Benjamin Scott (fl1714 - 1751) was a Mathematical Instrument Maker who was attracted to Russia with a number of other makers to start an Instrument making trade in the city of St. Petersberg where he died. In 1733 just prior to his departure for Russia, Scott produced an 18" diameter circular slide rule with 20 scales.

Scale: The logarithmic interpretation of numbers engraved in various formats onto a slide rule to enable different calculations to be performed.

Shirtcliffe: In 1740 Robert Shirtcliffe, a writer, added the SR scale (for ullaging) to the Everard slide rule.

Simplex: A slide rule with scales on one face only. Construction may be open or closed frame.

Slide: That part of the slide rule which moves in a linear way within the stock of the rule. In some cases the slide has scales on both sides and has to be withdrawn and reinserted upside down into the stock to use these scales; in other cases the scales are used with apertures in the back of the rule to work with the front face scales. The slide can be used all four ways, upside down and back to front to perform calculations.

Slide Rule: The term first used by John Brown in about 1662 to describe the mathematical calculating instrument that is now known world-wide by that name.

Soho: A standard engineers slide rule designed by Boulton and Watt in about 1779. Used by many steam engineers.

Spiral Scale: A method of extending the scale of a circular slide rule by presenting the scale in a spiral. In some cases there are in excess of ten turns to the spiral, allowing a scale many feet long to be put on a small circular slide rule.

Stock: That part of the slide rule which carries the slide in between its sides. It can be in two pieces (open frame) or effectively a single piece (closed frame).

T

Thacher: Edwin Thacher, a bridge engineer, patented a very large cylindrical slide rule in America in 1881. This was made in New York by the Keuffel and Esser Company.

Thompson: Captain J.H. Thompson re-invented the log-log scale in 1881 (see also Roget and Perry).

Tubular: A form of slide rule consisting of concentric tubes which can slide and/or rotate relative to each other. The scales are helical.

W

Watt: James Watt (1736 - 1819), better known as a designer of steam engines, worked with Mathew Boulton to produce a standard engineers slide rule in 1779, the Soho, named after the Soho Foundry in Birmingham where they worked.

Well: The gap in between the two sides of the stock in a closed frame slide rule, above which the slide moves, where additional scales are sometimes placed.

Wingate: In 1626 Edmund Wingate (1596 - 1657) was a JP and MP who attended lectures given by Edmund Gunter in London. He claimed to have taken the Gunter's line to France in 1624 when he went there to tutor Princess Henrietta. He added some modifications to Gunter's line, these being made in France by Melchior Tavernier (later Tavernier-Gravet). He is also credited with adding a slide to a fixed body and with arranging two scales against each other. Wingate was credited by some authorities as producing some of the earliest slide rules in France, though this is now known to be untrue. His work is now known as being an extension of the Gunter's line, the Echelle Angloise (as he called it), or Line of Proportion.

Appendix 2.
Key Dates in the History of Slide Rules.

This chronology has been taken from a variety of sources, and includes key dates as well as dates of general interest. As can be seen, there is some disagreement about some dates, and also some error in what was previously reported or thought. As this book is not intended as a definitive treatise on the invention of the slide rule, the anomalies are left in to highlight the differences. Equally, the list of dates is not intended to be absolutely comprehensive, but to provide the reader with a tabular summary showing the gathering pace of development, and the sudden termination of the slide rule with the invention of the cheap electronic calculator.

The dates start pre-slide rule to enable some of the parallel paths of contemporary calculating equipment to be compared with the development of the slide rule.

1598	Invention of the sector. This is the date generally attributed to its invention, with Galileo the inventor. However, it is now believed that Guidobaldi del Monte, who later made sectors for Galileo, actually invented the device. Hood and Gunter in the UK are also supposed to have invented a sector at about the same time as Galileo.
1614	Invention of logarithms by John Napier.
1617	Development of logarithms to "base 10" by Henry Briggs. They are also known as Briggsian logarithms after the inventor.
1617	Napier invented his "bones". These were used for multiplying and dividing numbers, and used small rods with numbers set out in a special pattern.
1620	Interpretation of logarithms into a linear scale form, Gunter's lines or Gunter's rule, by Edmund Gunter.
1624	Gunter puts a logarithmic scale on a copper disc or rule. This required one or two dividers to perform calculations. This device was used by the British Navy for many years.
1624	The Reverend William Oughtred arranged two scales to slide alongside each other. It is now generally agreed that this was the forerunner of the slide rule. The date could be any year from 1620 to 1625 as he did not announce his invention at the time.
1626	Edmund Wingate added a slide to a fixed body. It is now generally agreed that all references to Wingate's work apply to improvements in Gunter's scales.
1626	Benoit in France said that Wingate arranged two scales against each other.

1627   In Hutton's *Mathematical Tables* published in 1811, he wrote that Wingate had arranged linear scales to slide against each other, and Oughtred proposed a circular slide rule in 1627. See also Cajori, 1909.

1630   Delamain produced his book *The Mathematical Ring*, the first published description of a circular slide rule in flat and cylindrical forms with either a flat compass, or with an index/cursor to assist in calculation.

1632   Oughtred produced his Circles of Proportion, which was also a circular slide rule with concentric logarithmic scales. This may have been the first slide rule with sliding parts, said Cajori, and Dunlop 1911.

1633   The announcement of the invention of the slide rule by Oughtred in London. He made a statement that *as early as 1621* he had put two lines together in a circle with dividers from the centre. A pupil of his, Foster, published a translation from Oughtred's Latin manuscript, and started the acrimonious debate with Delamain as to who invented the first slide rule.

1633   Thomas Brown arranged a Gunter's line in a "spiral" consisting of gradually increasing diameter circles.

1654   Robert Bissaker introduced the double faced slide. Some authorities also credit him with the invention of the slide/fixed stock principle.

1657   Development of the moving slide/fixed stock principle by Seth Partridge.

1662   Partridge described a rule made in 1654 with a small scale sliding between two fixed rules.

1662   John Brown, possibly the son of Thomas Brown, used the term "sliding rule" for the first time. It was in general use by the end of the 17th century.

1663   Samuel Pepys mentioned a "slide rule" in his diary. This reference to a slide rule is made within one year of the original reference to such a device by Brown.

1673   Flamsteed, the Astronomer Royal, purchased for 10 shillings a two foot double "Rulers of Proportion".

1675   Newton suggested three parallel logarithmic scales for solving cubic equations.

1677   Henry Coggeshall put a logarithmic scaled slide in a two foot folding rule, adapted for timber measure. This rule was very popular in England and became the first standard which was made by a number of mathematical instrument makers. It was improved and updated a number of times over the next 200 years.

1682	An updated and revised pamphlet for Coggeshall's rule was produced.
1683	Thomas Everard adapted a slide rule to have slides on opposite sides of a square rule, for gauging. This slide rule was also the first to have an inverted scale, and in parallel with Coggeshall's rule, became a standard which was updated and improved by various people during the next 150 years.
late 1600s	Newton used logarithmic scales for cubic and higher order equation solution. He also suggested a straight line across several scales, was this the first runner or indicator - the hair-line?
1700-1800	The Coggeshall and Everard rules continued as standard designs made by many instrument makers. Considerable numbers of slide rules were in common use, according to Everard many thousands of his rules were sold between 1683 and 1705. Note that these rules were relatively crude and inaccurate.
1722	There was yet a further update to the pamphlet for the Coggeshall rule.
1723	The instrument maker Scott produced an 18" diameter circular slide rule with 20 circles. See also 1733.
1727	J.B. Clairault described a 21" cardboard circular slide rule.
1733	Benjamin Scott produced an 18" diameter circular slide rule with 20 scales. See also 1723. This may have been the same device as described for 1723, though one describes circles the other scales.
1740	Shirtcliffe added the SR scale (for ullaging) to the Everard slide rule.
1748	George Adams produced a 12" brass circular slide rule with 10 windings.
1750	Leadbetter added a third slide to the Everard rule.
1767	The seventh revision to the book of the Coggeshall rule, this time by another author.
1775	Development of the slide rule cursor by John Robertson. He described an "...*improved sliding Gunter*" with runner or cursor. See also 1851. This was a rule for navigation purposes.
1775	Matthew Boulton & James Watt designed an engineering slide rule, the Soho slide rule, which was used during the design of their steam engine. See also 1779.

1779   Boulton and Watt increased the accuracy of the slide rule by improving the graduations. They produced a special engineers rule, the Soho rule, which became a standard for the engineering profession.

1787-

1797   William Nicholson designed many types of slide rule and produced a number of improvements, many of which only found favour much later when "invented" by someone else. These included linear slide rules with one slide and cursor, as well as circular and spiral types. These slide rules had increased accuracy, one had a logarithmic scale broken into ten parallel parts, equivalent to 20 foot scale length (this must have been a two foot rule). This example required a beam compass with a movable centre as a "cursor".

1700-

1800   Slide rules were used more in England than France and Germany. They were probably not known in any other European country. This was postulated as being due to the British procedure of teaching decimal fractions in schools.

1800s   The slide rule was little known in Germany, but the circular slide rule was popular in France. It was later displaced by the straight Mannheim rule. There was a government order made in France, that knowledge of slide rule calculation was a prerequisite for entry to the administration grades of their public service.

1807   Farey used "logarithmic" logarithms for musical scales

1815   Invention of the log-log scale principles by Roget. It was presented in the *Philosophical Transactions of the Royal Society*. This was probably the most important development of the 19th century.

1833   A French survey by Baron Riche de Prony of their technology, noted that the United Kingdom was ... *commonly using 'sliding rules' at all levels, including shop keepers and artisans*.

1842   LeMorgan publishes his *Penny Cyclopaedia* which includes an article on the early history of the slide rule.

<1850   The slide rule was little known in America.

1850   Colonel Amédée Mannheim produced the modern arrangement of scales.

1851   Mannheim added a cursor to his slide rule. This rule was made by Tavernier-Gravet in Paris, and was very good for general calculation. It was adopted by the French Army.

1851   Someone added a magnifying glass to the runner. Note that J.F. Arthur attempted to use a vernier on a runner at a later date.

1859   Mannheim is credited with the invention of the first modern slide rule. Note that the designation "Mannheim Rule" was regularly used in literature up to World War II.

c1860   A cylindrical instrument is invented by Professor George Fuller in Dublin.

1866   Dr. J.D. Evans describes a grid iron slide rule.

1874   Coggeshall rules were still in use.

1876   A circular watch type slide rule was produced by Stanley, the Boucher calculator. This is the design made by numerous manufacturers.

1878   Fuller produced his cylindrical calculator with papier-mâché cylinders.

1880   Henry Cherry described two fixed index points for a long scale slide rule to avoid the duplication of portions of the scales.

1880 on   General development of the cursor/indicator.

1881   Edwin Thacher's cylindrical slide rule was patented in the United States.

1881   Captain J.H. Thompson re-invented the log-log scale.

1886   Dennert and Pape in Altona (Hamburg), Germany, introduced white celluloid (Ivorine) as a material for the inscription of scales, instead of the traditional marking on wood, brass and ivory in use until then.

c1890   A survey showed that England, France and Germany were the only manufacturers of slide rules.

1890   The Mannheim rule was adopted as a standard design in the United States.

1890   William Cox in the United States patented the duplex slide rule.

1894   The start of slide rule production in Japan followed the visit of two Japanese engineers to Europe. This resulted in the formation of the firm Sun/Hemmi who became one of the largest slide rule manufacturers world-wide.

1898   Beghin's Règle à Calcul was the first design with displaced scales.

>1900   Great increases in the accuracy of slide rules became possible, particularly as a result of using engine divided scales.

1901	Re-introduction of log-log scales by Professor John Perry. Early log-log scales were found to have little practical use, and so were re-invented as the requirement to perform complex calculations for thermodynamics, physics, electrical engineering, and so on became necessary.
1902	Max Rietz proposed the Rietz arrangement of scales. This was the most common arrangement of scales used on rectilinear slide rules.
1909	Florian Cajori published his *History of the Logartithmic Slide Rule* in which he described 256 designs of slide rule since 1800, 90 of which were produced in the first nine years of the 20th century.
1933	Differential trigonometrical and log-log scales were invented by Hubert Boardman of Radcliffe, Lancashire.
1935	Dr. Alwin Walther proposed the Darmstadt arrangement of scales. System Darmstadt slide rules were probably the last standard arrangement of slide rule scales to become accepted.
1940-1950	Development continued of slide rules with scales on both faces, plus trigonometrical and log-log scales.
1976	The final slide rule made by K&E donated to the Smithsonian Institute in Washington, United States.
1985	Final United Kingdom slide rule patent granted for a disc calculator.
1992	Hemmi in Japan were still making a few specialist medical slide rules.
1998	Blundell Harling in England are still making slide rules for specialist applications in small quantities.
1998	IWA in Germany are still making slide charts, as are a number of firms in the United States.

Appendix 3.

Bibliography.

Books on slide rules are as old as slide rules themselves. This bibliography covers a wide range of the available literature including instruction books, but has specifically ignored instruction <u>leaflets</u> even though some of these do contain relevant historical and other non- instructional information. The bibliography also includes books on Napier's bones, the sector and Gunter's rule to illustrate just how long after the invention of the slide rule these instruments were still used by sections of the calculating community.

The first author on slide rules was Delamain with his *Grammelogia* published in 1630, closely followed by Oughtred with a Latin manuscript describing his *Circles of Proportion*, translated into English by William Forster in 1632. The heated debate on priority which sprang up between Delamain and Oughtred has been covered very thoroughly by both Cajori (1920) and Bryden (1974). There followed books by Brown (pere & fils), Coggeshall, Everard, Leybourn and many others.

Authors in Europe writing about Gunter's rule followed surprisingly quickly, with French manuscripts from Henrion in 1626 and Boisseau in 1637 as well as the many manuscripts from Wingate. French authors do not appear to have written about the slide rule for almost 100 years after its invention in England. Biler in Germany covers a wide variety of mathematical instruments in the manuscript of 1697.

The numerous books written about using slide rules all have something of interest for the student of the slide rule, if only for a small item that gives a clue to the age of a particular example. The large majority of these books are now out of print, but can be obtained from specialist or secondhand bookshops at very reasonable prices. Some have been reprinted by specialist publishers. Most ran to a number of editions, and many had several revisions. Each edition and some revisions cover a different selection of slide rules. Some of the older books also included an advertising section at the back which is very useful in providing dates as to when models were introduced. Many of these are available to read in the major science museums or to study by arrangement.

Wherever possible the first names of the authors have been given. The titles are capable of considerable variation within bibliographies, and many of the earlier ones have been considerably abbreviated in other bibliographies. Both title and sub-title are cited as fully as possible, and these give a fascinating insight into what were considered the problem areas of the day.

Many articles from the *Journal of the Oughtred Society* are included as they provide considerable detail of the particular slide rule type or manufacturer and are to be recommended for anyone who is interested in the study of slide rules. The Society can be contacted at the following address: *The Oughtred Society, P.O. Box 99077, Emeryville, California 94662, USA.*

The date quoted is that of the first edition where known. The dates of subsequent editions and revisions or reprints which are known or believed to contain different information are given.

SLIDE RULE BIBLIOGRAPHY - ALPHABETICAL.

Abel, Thomas.: *Subtensial plain Trigonometry, wrought with a sliding-rule, with Gunter's lines: and also* ... Philadelphia, 1761.

Allen, C.J.A.: *A Century of Scientific Instrument Making.* (History of W.F. Stanley), Harley, 1953. 50pp, commissioned by Stanley for their Centenary in 1953. There is very little on slide rules in the book, but it provides a fascinating insight into one of the UK's premier instrument manufacturers.

Allen, R.K.: *Systematic slide rule technique, an analytical survey of modern slide rule theory and practise for technical students and all who make calculations.* Pitman, 1962.

Aldinger, Henry.: *The Gilson Slide Rule Company.* Journal of the Oughtred Society Vol, 1, No, 1. February 1992. pp 14-16.

Aldinger, Henry.: *The Gilson Slide Rule Revisited.* Journal of the Oughtred Society Vol, 1, No, 2. August 1992. pp 13-15.

Aldinger, Henry.: *The Dietzgen Company.* Journal of the Oughtred Society Vol, 5, No, 2. October 1996. pp 40-41.

Alvarez, ??.: *Using the slide rule in Electronics Technology.* 1962.

Anderson, F.J.: *The Anderson Slide Rule.* 1904.

Anon.: *Use of Coggeshall's Sliding Rule.* Mount & Page, London, 1787.

Anon.: *Mensuration made perfectly easy, by the assistance of a new improved sliding rule; which* ... Birmingham, 1819.

Anon.: *Colby's Slide Rule, Stadia.* Mahn & Co,. 1890.

Anon.: *Slide Rule Trigonometry Simplified.* 1943. 24 page soft cover booklet.

Anon.: *Use and Working of the Slide Rule and Watch Calculator.* Scientific Pub. Co. 1900.

Anon.: *The Jack Burton Collection.* Journal of the Oughtred Society Vol, 2, No, 1. March 1993. pp 15-18.

Anthony, Gordon.: *The K&E Braille Slide Rule.* Journal of the Oughtred Society Vol, 2, No, 1. March 1993. pp 19-21.

Appleton, ??: *Dictionary of ... Engineering.* New York, 1868.

Arnold, J.N.: *Special Slide Rule Course.* Purdue University, 1932.

Arnold, J.N.: *The slide rule, principles and applications.* Prentice-Hall, 1954.

Arnold, J.N.: *Complete Slide Rule Handbook.* 1954.

Artur, J.F.: *(Treatise on Slide-Rule - no title given).* London, 1827, 1845.

Ashworth, John. (Carder).: *The Slide Rule Instructor for the Improved Slide Rule.* "... containing over 200 questions and solutions for the Improved slide rule adapted for engineers. Hyde, Manchester, 1869. Deals with calculation problems for the Textile trade.

Asimov, Isaac.: *An Easy Introduction to the Slide Rule.* Houghton-Miflin, 1965, Ronald Whiting & Wheaton, London, 1966. A "How to" book which probably only got its popularity from the author's name.

Ayoub. Prof. Raymond.: *Napier and the Invention of Logarithms.* Journal of the Oughtred Society Vol, 3, No, 2. September 1994. pp 7-13.

Babcock. Ph.D. Bruce E.: *George Washington Richardson's Direct Reading Slide Rules.* Journal of the Oughtred Society Vol, 1, No, 1. February 1992. pp 9-14.

Babcock. Ph.D. Bruce E.: *An Error on a Slide Rule for 50 Years?* Journal of the Oughtred Society Vol, 2, No, 2. October 1993. pp 15-18.

Babcock. Ph.D. Bruce E.: *A Guided Tour of an 18th Century Carpenter's Rule.* Journal of the Oughtred Society Vol, 3, No, 1. March 1994. pp 26-35.

Babcock. Ph.D. Bruce E.: *Some notes on the History and Use of Gunter's Scale.* Journal of the Oughtred Society Vol, 3, No, 2. September 1994. pp 14-20.

Babcock. Ph.D. Bruce E.: *George Ward and "The Gauger's Practice ...".* Journal of the Oughtred Society Vol, 4, No, 1. March 1995. pp 21-25.

Babcock. Ph.D. Bruce E.: *Two Noble Attempts to Improve the Slide Rule.* (Laurie's Precision Slide Rule & The Richardson Pyramid Slide Rule). Journal of the Oughtred Society Vol, 4, No, 1. March 1995. pp 41-45.

Babcock. Ph.D. Bruce E.: *K&E Student's and Beginner's Slide Rules - 1897 to 1954.* Journal of the Oughtred Society Vol, 4, No, 2. October 1995. pp 41-49.

Babcock. Ph.D. Bruce E.: *Lawrence Engineering Service - A Tale from an American Small Town.* Journal of the Oughtred Society Vol, 5, No, 2. October 1996. pp 55-61.

Babcock. Ph.D. Bruce E.: *A K&E Slide Rule for Planer Work.* Journal of the Oughtred Society Vol, 6, No, 1. March 1997. pp 25-26.

Baldermann, Herbert.: *Wir rechnen mit dem Rechenstab.* Germany, c1960. Recommended by Aristo as a suitable self-instruction book, also Stender.

Balogh, Arthur.: *A logarlec.* Budapest, 1970.

Barlow, Peter.: *Dictionary of Pure and Mixed Mathematics.* London, 1814.

Barnes, Colin.: *The Customs and Excise Gauging Slide Rule.* Journal of the Oughtred Society Vol, 4, No, 2. October 1995. pp 53-57.

Barnes, Colin.: *T.S. Emblem Series - Oral Histories.* Journal of the Oughtred Society Vol, 6, No, 1. March 1997. p 15.

Barnes, Colin.: *John Davis & Son (Derby) Ltd - A chronology of Slide Rule Production.* Private Publication, 1997.

Bateman, Joseph.: *The construction and application of the Sliding Rule used by officers of Her Majesty's Revenue.* Declares that "... navigators still used a Gunter's rule with a pair of Compasses". c1842.

Baxendall, D. & Pugh, J.: *Catalogue of the collections of the Science Museum. Calculating Machines and Instruments.* Science Museum, 1975. Lists over 200 slide rules in the collection and provides details as well as dates of acquisition which can be a help in dating.

Bayley, W.H.: *Hand-Book of the Slide Rule.* London, 1861.

Bedini, Silvio A.: *Early American Scientific Instruments and their makers.* Smithsonian Inst., 1964.

Beghin, Auguste.: *Règle à calcul, modèle spècial,* Paris, 1898, 1902.

Beghin, Auguste.: *Règle à calculs. Instruction - application numériqués. 100 problèmes pratiques et industriels.* Paris, 5th edn, 1912.

Bennett, Al.: *Small Wonder. (The SMALL Calculator).* Journal of the Oughtred Society Vol, 1, No, 1. February 1992. pp 16-18.

Bennett, Al.: *The "Cooper" 100-inch Slide Rule.* Journal of the Oughtred Society Vol, 3, No, 1. March 1994. pp 18-20.

Bennett, Al.: *Schact & Westerich Calculator.* Journal of the Oughtred Society Vol, 4, No, 1. March 1994. pp 26.

Benoit, M.P.-M.-N.: *La Régle à Calcul expliquée.* Paris, 1853.

Benson, R.: *Use of Slide Rule for Forge calculations.* c1895.

Berle, A.: *Systematic Collection of Exercises and problems on the Slide Rule, System Cherepashinskii.* St. Petersberg, 1890.

Bernoulli III, Johann.: *Encyclopédie méthodique (mathématiques), Article "Logarithmiques,".* 17--.

Bevan, Benjamin.: *Practical Treatise on the Sliding Rule.* London, 1822; Carey, London, 1838.

Bevan, Benjamin.: *Guide to the Carpenter's Rule.* London, 1832.

Bevan, Silvanus.: *Slide Rule.* Philosophical Magazine Vol. 49, London, 1817.

Biler, Johann Mathes.: *Descriptio instrumenti Mathematici universalis, quo mediante omnes proportiones sine circino atque calculo methodo facillima inveniuntur.* Germany, 1696.

Bion, Nicolas.: *Traité de la construction et des principaux usages des instruments de mathematique. Avec les ...* Paris, 1709; Paris, 2nd edn, 1716; The Hague, 3rd edn, 1723; Paris, 4th edn, 1752. See also **Edmund Stone.**

[Bion, Nicolas]. DOPPELMAYR: *Neu eröffnete mathematische werkschule, oder gründliche anweisung, wie die ...* Frankfurt, (German) 1712.

Bishop, Calvin Collier.: *Practical Use of the Slide Rule.* Barnes & Noble. College Outline Series, New York, 1944; 1947. An earlier version of the next book with virtually identical text and illustrations. 147 pages.

Bishop, Calvin C.: *Slide rules, a practical guide to its use.* Barnes & Noble. 1955. 152 pages profusely illustrated with K&E, Post, and Dietzgen slide rules.

Black, R.D. & M'Cord, M.: *Introducing the Slide Rule.* Nixon Enterprises Inc., Wabash, 1938.

Blaine, R.G.: *Some Quick and Easy Methods of Calculating.* London. 2nd edn. 1903.

Blaine, R.G.: *The slide rule as an aid to calculating.* Spon. 1952. This 152 page hardback book covers a wide field of calculation including commercial, mechanics and engineering, as well as a wide variety of slide rule types including Fullers, Fowlers and the normal linear rule scales.

Blandi, T.: *The slide rule simplified.* 1942.

Boardman, D.: *Book of Instructions (T) for the Use of P.I.C. Differential Scales for Trigonometrical Computations. Supplement to the P.I.C. Book of General Instructions/Introducing the Use of P.I.C. Differential Scales for Trigonometrical Computations.* P.I.C. Slide Rules, Manchester, Booklet No 81.

Boardman, D.: *Appendix B/Instructions for ∫(S) scale of √1-S².* For use with slide rules No 121, 131, P123, P133, 122, 132, VP124, VP134. P.I.C. Slide Rules, Manchester, Booklet No 81.

Boardman, D.: *Book of Instructions (T) for the Use of P.I.C. Direct and Inverse Log-Log Differential Scales for (1+x)ⁿ Expansions and Finance Calculations/ Supplement to the P.I.C. Book of Instructions / Instructing the Use of P.I.C. Direct & Inverse Log-Log Differential Scales for Finance Calculations of the order of Compounding, Amortization of Assets, Sinking Fund Accumulations, Loan Charges and Deferred Payments Assurance Values etc.* P.I.C. Slide Rules, Manchester, Booklet No 80.

Boisseau, Jean.: *Methode tres facile pour se servie de sinopse ou tableau circulaire tres Utile et necessaire.* Paris, Broadside, 1637.

Bonneycastle, John.: *Introduction to Mensuration & Practical Geometry.* Kimber & Sharpless, Phil. 1807, 1833.

Bowditch. ?? : *Navigator.* (describes Gunter's Line usage). 1802

Boys, C.V.: *Slide Rule.* Nature 64, 265-8. 1885.

Breckenridge, William E.: *The Polyphase duplex slide rule: a self teaching manual with tables of settings, equivalents & gauge points.* Keuffel & Esser. 1924.

Breckenridge, William E.: *The Polyphase slide rule No. 4053.* Keuffel & Esser. 1925, 1938.

Breckenridge, William E.: *The Log Log duplex slide rule.* Keuffel & Esser. 1926.

Breckenridge, William E.: *The Mannheim slide rule, Nos. 4031S, N4035S, N4041 and 4051.* Keuffel & Esser. 1938.

Breckenridge, William E.: *The Polyphase Duplex slide rule No. 4088.* Keuffel & Esser. 1938.

Brown, John.: *The Description and Use of a Joynt-Rule: ... also the use of Mr. White's Rule for measuring of Board and Timber, round and square; With the manner of Using the Serpentine-line of Numbers, sines, Tangents and Versed Sines.* John Brown & Henry Sutton. 1661.

Brown, John.: *A Collection of Centers and Useful Proportions on the Line of Numbers.* London, 1662(?).

Brown, John.: *Description and Use of the Carpenter's-rule: Together with the use of the line of numbers.* J. Brown(e) & T. Brown(e). 1684; London, 4th edn 1704.

Brown, John.: *The Use of the Line of Numbers on a Sliding or Glasier's Rule.* and, with a separate title-page *The Description and Use of a Carpenter's Rule.* W. Fisher & R. Mount. 1688.

Brown, John.: *The Description and Use of the Carpenter's Rule, together with the Use of the Line of Numbers commonly called Gunter's Line.* (probably Posthumous). 1704.

Brown, Joyce.: *Mathematical Instrument Makers in The Grocers Company 1688-1800.* London Science Museum. 1979. An intriguing insight into the generations of mathematical instrument makers from the Grocer's Company who founded many world famous manufacturers making all instruments including slide rules. Detailed lists of addresses and apprentices allows the full history of a maker to be traced.

Brown(e), Thomas.: *Description of the Serpentine Slide-Rule, with an engraving of the Instrument.* London. 1631.

Bryden, D.J.: *Scottish Scientific Instrument makers, 1600-1900.* Edinburgh, 1972.

Bryden, D.J.: *A Patchery and confusion of disjointed stuffe : Richard Delamain's Grammelogia of 1631/32.* Transactions of the Cambridge Bibliographical Society 6, pt 3 1974. pp 158-66. Highly detailed comparison of various editions of Delamain's Grammelogia with carefully drawn conclusions showing the plagiarism.

Burns, J.A.: *Useful Engineers' Constants for the Slide Rule and How They are Obtained.* Percival Marshall & Co. London, 1921.

Cajori, Florian.: *A History of the Logarithmic Slide Rule and allied Instruments.* Constable & Co. 1909. Astragal Press reprint 1994. The "Bible" of slide rule history with a comprehensive Bibliography and list of slide rules invented up to publication date in 1909. The Appendix to the book is an essential part as here Cajori finally agrees that Oughtred was the inventor of the slide rule, and not Wingate as the rest of the book attempts to prove. There are noticeable gaps in his information, and while it is very comprehensive it is by no means complete. The reprint at least is an essential part of any slide rule collector's library.

Cajori, Florian.: *Aaron Palmer's Computing Scale.* Colorado College Publications (Engineering Series). 1909.

Cajori, Florian.: *John E. Fuller's Circular Slide Rules.* Colorado College Publications (Engineering Series). 1909.

Cajori, Florian.: *On the invention of the slide rule.* Colorado College Publications (Engineering Series) I. 1910.

Cajori, Florian.: *William Oughtred, a Great Seventeenth-Century Teacher of Mathematics.* Open Court Publishing Co. Chicago. 1916.

Cajori, Florian.: *On the History of Gunter's Scale and the slide rule during the 17th century.* Univ. of California Pubs. in Maths Vol.1, No9. 1920.

Camm, Frederick J.: *Newnes' Slide Rule Manual.* Newnes. 1944, 1948, 1957, 1960. Contains the standard "how to use" information for linear rules, also examples of how to use Fowlers watch calculators, Halden Calculex and a brief introduction to other non-linear slide rules.

Camus, Charles E-L.: *Instrument propre à jauger les tonneaux et les autres vaisseaux qui servent à contenir des liqueurs.* Paris, 1741.

Campbell, H.: *Using a slide rule. Part 1 etc.* Stillit. 1966.

Campbell, Lt. Col. J.R.: *The Theory and Practise of the slide rule.* 1886.

Carper, Don.: *Slide rule in electronics.* Foulsham. 1967. 150 pages of How-to illustrated, as the title suggests, with examples from electronics. Unusually for an English publication it uses 110 Volts and 60 Hz American standards in the examples.

Carver, Isaac.: *The Description and Use of a new Sliding Rule, projected from the Tables of the Gauger's Magazine.* printed for William Hunt. 1687.

Carwitham, T.: *Description and Use of an Architectonic Sector.* 1723.

Chadwick, John.: *Slide rule instructor for Instruction on Chadwick's improved slide rule.* Manchester, 1883.

Chamberlain, Edwin J.: *The Voith Slide Rule and Mechanical Pencil Combination.* Journal of the Oughtred Society Vol, 6, No, 1. March 1997. pp 27-28.

Chatfield, John.: *The Trigonal Sector.* R. Leybourne. 1650.

Cherepashinskii, Prof M.M.: *Instruction in the Use of the Slide Rule, New System.* (In Russian). Moscow, 1883.

Chew, R.A. Capt. A.C.: *The Slide Rule, Logarithms & Co.* Published by the Author. After 1907. Aimed particularly at the military use, as a private publication it suffers from numerous errors, some extremely basic, e.g a^b equals a times b!

Clairaut, Jean Baptiste.: *Histoire de l'académie royale des sciences.* 1727.

Clapham, C.B.: *Arithmetic for engineers, including simple algebra, mensuration, logarithms, graphs, trigonometry and the slide rule etc.* Chapman & Hall. 1955.

Clark, J.J.: *The Slide Rule, An Elementary Treatise.* Tech. Supply Co. Scranton PA & New York, 1909, 1929.

Clark, J.J.: *The Slide Rule and Logarithmic Tables including a ten-place table of logarithms. A concise ...* Fredk. J. Drake, Chicago. 1929, 1941; 3rd edn, 1943, 1945; Wilmette 3rd edn, 1954. 219 pages covering linear and watch type rules, and logarithm systems. Rules look to be K&E.

Clark, Samuel.: *The British gauger: or, trader and officer's instructor, in the Royal revenue of the excise and ...* London, 1765.

Clason, Clyde.: *Delights of the Slide Rule.* Crowell. New York, 1963.

Clifton, Gloria.: *Directory of British Scientific Instrument Makers. 1550 - 1850.* Zwemmer in association with the National Maritime Museum, 1995. A huge compendium of information on all Instrument makers with details of their work and instruments produced. Difficult to find specific slide rule makers, but if the maker is known there is information available, not always with mention of slide rules which may have been an adjunct to their main business.

Clouth, F.M.: *Anleitung zum Gebrauch der Rechenschiebe.* Hamburg, 1872.

Coggeshall, Henry.: *Timber Measure by a Line of more Ease, Dispatch and Exactness than any other way now in use, by a Double Scale ...* London, 1677.

Coggeshall, Henry.: *A Treatise of Measuring by a Two-foot Rule, which slides to a Foot.* London. (A second edn. of 1677), 1682.

Coggeshall, Henry.: *The Art of Practical Measuring easily performed by a Two-foot Rule which slides to a Foot.* London. (A third edn. of 1677), 1722. 4th edn. revised by John Ham, London, 1729.

Coggeshall, Henry.: *The Art of Practical Measuring by the Sliding Rule.* 5th edn. 1732. 7th edn. 1767.

Coggeshall, Henry; & Ham, John.: *The Art of Practical Measuring by the Sliding Rule: shewing how to measure round, square ...* 6th edn. 1745.

Cohen, I Bernard.: *Some early tools of American Science. An account of early scientific instruments in Harvard University.* Harvard Univ Press. 1950.

Coles, P.: *How to use the slide rule.* Coles notes service. 1968.

Coley, Henry.: *The Description and Use of a Portable Instrument vulgarly known by the Name of Gunter's Quadrant. To which is added the Use of Napier's Bones ... by a true lover of the Mathematicks.* 1685. (Reprinted 50 years later as by William Leybourne).

Collardeau, du Héaulme, Felix.: *Instruction sur la Règle à Calculs.* Paris, 1820.

Collardeau, du Héaulme, Felix.: *Instruction sur L'Usage de la Règle à Calculs Portative, a une Seule Coulisse.* Paris, 1833.

Cooper, Henry O.: *Instructions for the use of A.W. Faber "Castell" Precision Calculating Rules.* Berwick Bros. 1905?

Cooper, Henry O.: *Slide Rule Calculations.* 1931.

Cordingley, J.L. & others.: *Instructions for the Sliding Rule.* A. Derrough. Cincinatti, 1841.

Coulson, S.: *Coulson's treatise on his newly invented Engineers and Mechanics slide rule.* Stokesly. 1842.

Cox, William M.: *The Slide Rule.* Keuffel & Esser. 3rd edn 1891.

Cox, William M.: *The Mannheim slide rule. Complete manual with table of settings, equivalents and gauge points.* Keuffel & Esser. New York, 1891, 1917.

Cox, William M.: *The Mannheim slide rule (Complete Manual) and the Duplex Slide Rules.* Keuffel & Esser. New York, 1891; 1909.

Cox, William M.: *The Mannheim (Polyphase) and the Duplex (Polyphase-Duplex) Slide Rules.* Keuffel & Esser. 1917.

Craig, John D.: *Hawney's COMPLETE MEASURER ..., Corrected and improved by T. Keith, Third Edition ...* Lucas & Neal, Wills & Cole. Baltimore, 1813.

Croarken, Mary.: *Early Scientific Computing in Britain.* Oxford Science Pubs. 1990. Short description of the history of slide rules illustrated with excellent photographs of a Mannheim slide rule from the London Science Museum.

Crockett, C.W.: *Explanation of the Principle and Operation of the Mannheim Slide Rule.* Troy, New York, 1891.

Crutis, G,A.: *Treatise on Gunter's Rule, together with a description & use of the sector, protractor, plain scale and line of chords.* E. Adams. Whitehall, N.Y. 1824.

Cullimore, Alan Reginald.: *The Use of the Slide Rule.* Keuffel & Esser. 1915. A 40 page book including adverts for K&E slide rules of the period to go with the dense print and poor examples of this How-to book.

Cullimore, Alan Reginald.: *The Mannheim Slide Rule.* Dietzgen, Chicago, 1925.

Cullimore, Alan Reginald.: *The Philips Slide Rule.* Dietzgen, 1925.

Culmann, C.: *Die Graphische Static.* Zurich, 1875.

Cunn, Samuel.: *A new treatise of the Construction and Use of the Sector ...* Revised by Edmund Stone. London, 1729.

Curtiz, George.: *Treatise on Gunter's Scale and the Sliding Rule.* 1824.

D'Ocagne, Maurice.: *Le Calcul Simplifié.* Gauthier-Villars et Fils, Paris, 1894. 2nd edn. 1905.

D'Ocagne, Maurice.: *Calcul Graphique et Nomographie.* Octave Doin et Fils, Paris, 1914.

D'Ocagne, Maurice.: *Vue d'Ensemble sur les Machines à Calculer.* Gauthier-Villars et Cie, 1922.

Darling, John.: *The Carpenter's Rule made Easie.* 1658, reprinted regularly to 1711.

Daumas, Maurice.: *Scientific Instruments of the 17th & 18th Centuries and their makers.* Batsford, London, 1972.

Davies, Charles.: *Practical Mathematics, with Drawing and Mensuration Applied to the Mechanic Arts.* Discusses the Carpenters rule. A.S. Barnes & Co NY. 1852.

Davies, Charles.: *Elements of Surveying and Navigation with Descriptions of the Instruments and the Necessary Tables.* A.S. Barnes & Co NY. 1853, 1868.

Day, Jeremiah.: *The Principles of Plane Trigonometry, Mensuration, Navigation and Surveying. Adapted to the method of Instruction in American Colleges.* Hezekiah Howe, New Haven, 3rd edn 1831.

Delamain, Richard.: *Grammelogia: or the Mathematical Ring.* 1630.

Delahain, Richard.: *The Making, Description and Use of a small portable Instrument ... called a Horizontal Quadrant.* London, 1631.

Delehar, Peter.: *Notes on Slide Rules.* Bulletin of Scientific Instrument Society, No 3, 1984, pp 3-10.

De Morgan, Arthur.: *Penny Cyclopaedia. Article on "Slide (or sliding) rule".* 1842.

De Morgan, Arthur.: *Arithmetical Books from the Invention of Printing to the present time.* London, 1847.

Dennert, Hans.: *Dennert & Pape and Keuffel & Esser.* Journal of the Oughtred Society Vol, 3, No, 1. March 1994. pp 3-7.

Dennert, Hans.: *An update on D & P and K & E Slide Rules.* Journal of the Oughtred Society Vol, 3, No, 2. September 1994. pp 27-28.

Dennert, Hans.: *DENNERT & PAPE and Aristo Slide Rules 1872 - 1978.* Journal of the Oughtred Society Vol, 6, No, 1. March 1997. pp 4-14.

Dennert & Pape.: *100 Jahre Dennert & Pape - Aristo Werke.* Dennert & Pape, Hamburg - Altona, 1962.

Diestelkamp, F.: *Das Jubiläum "Addiator" und Der Lebenslauf: Karl Kübler, Erfinder des "Addiator".* Historische Bürowelt No 32. S, Köln Feb 1992.

Dietzgen Company.: *How to use a slide rule.* Chicago, 1925.

Dietzgen, E.: *The Maniphase Slide Rule.* Eugene Dietzgen Co. 1928.

Dietzgen, E.: *How to Use a Mannheim Slide Rule.* Eugene Dietzgen Co. 1933.

Dilworth, Thos.: *Schoolmasters Assistant.* London, 1793. Robert Patterson, Philadelphia, 1805.

Dingler, J.G.: *"Polytechnic Journal".* Various volumes. Berlin, 1820.

Dixon, ??.: *Treatise on the combined circular, spiral, multi-index slide rule.* 1882.

Dobson, Alan B.: *The Slide Rule in every day use.* Blundell Harling. 1971. 85 pages of How-To instructions illustrated with B-H slide rules. Alternative editions are available published by W.H Smith to accompany their slide rules which were made by B-H.

Dodrum, P. & Itard, J.: *Maths and Mathematicians.* Open Univ Press. 1973.

Dreyssé, A.: *Règle à calcul Mannheim.* Paris, 1903.

Dreyssé, A.: *Instruction Détaillèe sur la Règle à Calcul Mannheim et Méthode Pratique.* Librarie Vuibert, Paris, 1925.

Drooyan, Irving & Wooton, William. *Manual for the Slide Rule.* Wadsworth Publishing. 1969.

Dunlop, Lt.Col. H.C. & Jackson, C.S.: *Slide rule notes.* Longman. 1911, 1913. Deals exclusively and comprehensively with the Mannheim rule. The original 1911 publication is described as a pamphlet and presumably shorter than the 127 page hardback book of 1913.

Dyck, Walther.: *"Katalog mathematischer u. mathematisch-physicalischer Modelle, Apparate und Instrumente.* Munchen 1892 und Nachtrag 1893.

EA.: *Working with a slide rule.* Engineer Apprentice. 1954.

ETL.: *Understanding and using the slide rule.* Electronic Teaching Labs. Foulsham. 1963, 1967.

Elliott, Rev. William.: *A Treatise on the slide rule with description of Lalanne's Glass Slide Rule.* Elliott & Sons. London, 1851, 1890.

Elliott, William.: *A description of the slide rule. With particular directions for the use of the glass slide rule.* London, 1852.

Ellis, John P.: *The theory and operation of the slide rule.* Constable. 1961. Dover Pubs. Inc. New York, 1961. 280 pp of hugely detailed instructions on how to use every scale imaginable, and including amongst a number of appendices, one on the history of the slide rule taken from Cajori including the erroneous claim to Wingate as predating Oughtred as the inventor of the slide rule, and no mention of Delamain. Emphasis on K&E slide rules.

Everard, Tho.: *An appendix Containing the description and use of another New Rule, very useful in the Gauging of Worts.* 1683.

Everard, Tho.: *Stereometry made easie: or the description and use of a new Gauging Rod or Sliding Rule.* 1684.

Everard, Tho.: *Stereometry, or, The Art of Gauging made easie, by the Help of a Sliding Rule.* London, 1705.

Everitt, M.J.: *A Vernier method of using the slide rule.* 1920?

Ewing, Alexander.: *A Synopsis of Practical Mathematics.* London, 1799 4th edn.

Eyre, John.: *The Exact Surveyor.* London, 1654

Faber, A.W.: *Anleitung zum Gebrauch des Rechenstabes.* Nuremburg, 1906.

Faber-Castell, A.W.: *Anleitung zum Gebrauch des A.W. Faber-Castell-Präzisions-Rechenstabes für Kaufleute "Columbus" System Rohrburg, 5 Auflage.* Nuremburg, 1937.

Faber-Castell.: *Directions for use of A.W.Faber-Castell precision Slide Rules.* Nuremburg, 1958.

Farey, John.: *A Treatise on the Steam Engine.* Contains about 50 pages of information on the use of slide rules "for calculating the dimensions for the parts of Steam Engines" In two volumes. London, 1827. David & Charles reprint 1971.

Favaro, Antonio.: *Veneto Istituto Atti (5).5. Article on the History of the slide rule.* Venice, 1878.

Favaro, Antonio.: *Leçons de Statique graphique 2ième partie, calcul graphique.* Translated into French by P. Terrier. Paris, 1885.

Favaro, Antonio.: *Sulla elica calcolatoria di Fuller con cenni storica sopra gli strumenti calcolatori a divisione ...* undated extract.

Feazel, Bobby. ed.: *Richard A. Gilson.* Journal of the Oughtred Society Vol, 2, No, 2. October 1993. pp 8-13.

Feazel, Bobby.: *Palmer's Computing Scale.* Journal of the Oughtred Society Vol, 3, No, 1. March 1994. pp 9-18.

Feazel, Bobby.: *Palmer's Computing Scale Revisited.* Journal of the Oughtred Society Vol, 4, No, 1. March 1995. pp 5-8.

Feazel, Bobby.: *Special Purpose Slide Rules.* Journal of the Oughtred Society Vol, 3, No, 2. September 1994. pp 43-44.

Feazel, Bobby.: *Colby's Slide Rule.* Journal of the Oughtred Society Vol, 4, No, 2. October 1995. pp 27-29.

Feazel, Bobby.: *A Burton "Special".* Journal of the Oughtred Society Vol, 5, No, 1. March 1996. p 20.

Feazel, Bobby.: *The Pilot Balloon Slide Rule.* Journal of the Oughtred Society Vol, 5, No, 2. October 1996. p 51.

Feely, Wayne & Schure, Conrad.: *Thacher Slide Rule Production.* Journal of the Oughtred Society Vol, 3, No, 2. September 1994. pp 38-42.

Feely, Wayne & Schure, Conrad.: *The Fuller Calculating Instrument* Journal of the Oughtred Society Vol, 4, No, 1. March 1995. pp 33-40.

Feely, Wayne.: *Update of Known Fuller and Thacher Rules.* Journal of the Oughtred Society Vol, 5, No, 1. March 1996. pp 25-29.

Fiesenheiser, E.I et al: *The VERSALOG Slide Rule.* The Frederick Post Co. 1951.

"Fisher, George". (Mrs. Slack): *Arithmetic.* London, 1794.

Flower, Rev. William.: *A key to the Sliding rule, and also Two Improved Sliding Rules together with a new Instrument of sliding Sines and Tangents.* G. Kearsley. London, 1768.

Forster, William.: *The Circles of Proportion and the Horizontal Instrument.* London, 1632.

Forster, William.: *The Circles of Proportion and the Horizontal Instrument with an Addition etc., and an Appendix; The Declaration of the Two Rulers for Calculation.* London, 1633.

Fortin, ??: (translator) *Atlas céleste de Flamsteed.* 1776.

Fowler, W.H.F.: *Fowlers Pocket Books.* (Mechanical, Electrical, etc.) Scientific Publishing Co. 1900 - annually

French, Robert, W.: *The Engineers Slide Rule.* Wm. C. Brown (USA), 1941. 100 pages of highly detailed "How To" instructions on all scales of a slide rule. Spiral bound, with high quality "typescript", his short section on the history of the slide rule has one or two unique statements, including one ascribed to Hutton that Oughtred had proposed the circular slide rule in 1627, some 5 years earlier than most authorities would give.

Fricke, H.W.: *Der Rechenschieber.* Leipzig, 1958.

Fritzen, J.: *Der praktische rechenhelfer. Tabellen, logarithmen und rechenschieber in ihrem wesen und ...* Stuttgart, 3rd edn, 1922.

Frost, W. & Withnold, T.: *Mensuration made perfectly Easy, by the Assistance of a new improved Sliding Rule.* 1768.

Fuller, George.: *Spiral slide Rule, Equivalent to a straight rule 83 feet 4 inches long, or a circular rule 13 feet 3 inches in diameter.* London, 1878.

Fuller, George.: *Professor Fuller's calculators having a logarithmic scale of numbers 41 feet 8 inches in length.* London. 1st edn undated.

Fuller, John E.: *Key to Fuller's Computing Telegraph. (Improvement to Palmer's Endless Self-Computing SCALE AND KEY Adapting it to the Different Professions, ...).* Fuller. New York, 1846.

Fuller, John E.: *Key to Palmer's Computing Scale. (Improvement to Palmer's Endless Self-Computing Scale and Key.).* Fuller. New York, 1846.

Fuller, John E.: *Fuller's Computing Telegraph.* New York, 1845? 1847.

Fuller, John E.: *Time Telegraph (Palmer's Improved by Fuller) Computing Scale.* Boston, 1852.

Fürle, Herman.: *Zur Theorie des Rechenschiebers.* Berlin, 1899.

Fürle, Herman.: *Rechenblätter.* Berlin, 1902.

Galbraith, Joseph A.: *A description of Major Hannyngton's Slide rule.* Dublin Univ. Press. 1890. 20 pages describing the various sizes of Hannyngton's washboard slide rule and how to use it. See also **Hannyngton.**

Galileo, Galilei,: *Operazioni del Compasso Geometrico e Militare.* Description and use of Gallileo's Sector. 1606.

Garcelon, David.: *Solving the Keuffel & Esser Catalog Problem.* Journal of the Oughtred Society Vol, 5, No, 2. October 1996. p 52-53.

Garrood, H.A.: *The Slide Rule for Calculation in the Timber & Building Industries.* Classic. 19--. 18 pages describing the use of Classic Series III slide rules for calculations involving Timber in the UK building industry, would that houses were still so cheap to build!

Garvan, Anthony N.B.: *Slide rule and sector: a study in Science, Technology and Society.* Actes du dixieme Congrè Internationale d'Histoire des Sciences. Ithaca 26 VIII 1962-2 IX, Paris. 1964.

Gattey, François.: *Instructions sur l'Usage des Cadrans logarithmiques.* Paris, 1799.

Gattey, François.: *Explication et usage de l'arithmographe.* Paris, 1810.

Gattey, François.: *Usage du Calculateur, instrument portatif au moyen duquel on peut en un instant, et sans etre obligé d'ecrire aucun chiffre, se procurer les résultats de toutes sortes de calcul.* Paris, 1819.

Gauge, Viscount.: *The Universal Ready Reckoner, or Royal Road to Arithmetic being instructions in the use of a sliding rule, by an Idle Gentleman.* John Lee. 1839.

Geerts, W.: *Working with a slide rule.* (Contained a special "Unique" plastic slide rule the J180, which was an 18 cm. version of the "Unique STUDY 900". Book plus Slide Rule cost 28/-.) 1965, translated from Dutch in 1969.

Gilbert, William.: *Description and Use of a New Sliding Rule.* (by Dr P.M. Roget). 1828.

Gill, ??: *Technological Repository.* Vol IV. London, 1829.

Goering, A.: *Der Rechenstab aus dem Mechanisch-mathematischen Institut von Debnnert & Pape, Altona.* Altona, 1873.

Goldstine, H.H.: *The Computer from Pascal to von Neuman.* Princeton University Press. 1972.

Goldstine, H.H.: *A History of Numerical Analysis from the 16th through the 19th Century.* Springer-Verlag. NY. 1977.

Good, John.: *Measuring made Easy, or the Description and Use of Coggeshall's Sliding Rule containing ...* Edited and corrected by James Atkinson. 1717; 2nd edn, 1726; 4th edn, 1751; 5th edn 1760.

Göschen, ?.: *Das Rechnen in der Technik und seine Hilfsmittel (Rechenschieber, Rechentafeln, Rechenmaschinen usw.).* Leipzig, 1908.

Guedon, E.C.; Canden, N.J.: *Miniature Vest Pocket Slide Rule.* 1930.

Guedon, E.C.: *Miniature Slide Rule.* 1925.

Gunter, Edmund.: *De Sectore. Descriptio et Usus.* London, 1607.

Gunter, Edmund.: *Canon Triangulorum or a Table of artificial Sines and Tangents.* London, 1620.

Gunter, Edmund.: *De Sectore et Radio.* London, 1623.

Gunter, Edmund.: *The description and use of the Sector, Cross-Staff and Other Instruments.* London, 1623.

Gunter, Edmund.: *The works of Edmund Gunter ... amended by Henry Bond.* London, 3rd edn. 1653.

Gunn, G.A.: *The Slide Rule.* 1928. 40 pages. Lists some unusual Aston & Mander slide rules.

Hadéry, Aug.: *(Treatise on slide-Rule - no title given).* Paris, 1845.

Halden & Co. J.: *Rules, tables and Formulae for Patent Circular Slide Rule (Halden Calculex).* Manchester, 1906.

Hall, B.J.: *"P.I.C." Calculating Slide Rules.* B.J. Hall, London, c1940.

Hall, John L.: *Slide Rule manual.* New York, 1907.

Halsey, F.A.: *The use of the Slide Rule.* van Nostrand. 1899, 1907, 1915.

Hammer. E.: *Der logarithmische Rechenschieber und sein Gebrauch.* Stuttgart, 4th edn 1908.

Hammond, John; Warner ??.: *The practical Surveyor: shewing, ready and certain methods for measuring, mapping and ...* London, 2nd edn. 1725 & 1731.

Hannyngton, Maj. Gen.: *The Slide Rule Extended.* Spon. 1884. 31 pages describing of the use of his "washboard" slide rule, descriptions of the three sizes available. Very similar to **Galbraith.**

Harling, W.H.: *Drawing Instruments.* London, 1906.

Harling, W.H.: *Mathematical Drawing and Surveying Instruments.* London, 13th edn 1906.

Harris, John.: *Lexicon Technicum.* 1716.

Harris, Charles. O.: *Slide rule simplified.* American Technical Society. 1943.

Hartley-Smith, S.: *Mechanical Methods of Calculation.* 1946.

Hartmuth, Max.: *Vom Abakus zum Rechenscieber* Hamburg, 1942.

Hartung, M.L.: *How to Use Basic Rules.* Pickett, Inc. c1950.

Hartung, M.L.: *How to Use Log Log Slide Rules.* Various numbers of pages. Pickett, Inc. 1953.

Hartung, M.L.: *How to Use Trig Slide Rules.* Pickett, Inc. 1960.

Haslett, Charles & Hackley, Chas. W.: *Mechanics' and Engineers' Book of Reference and Engineers' Field Book.* New York, 1856.

Hatton, Edward.: *The Gauger's Guide.* 1729.

H(aughton), A(rthur). (editor).: *Description and Use of the Circles of Proportion etc.* (by William Oughtred). A revised & corrected version, also the *Use ... for Nautical Questions.* 1660.

Hawney, William: *The Complete Measurer.* London, also Baltimore, 1717; 6th edn. 1748. 1813.

Hawney, William: *The Complete Measurer; or the whole Art of Measuring with an Appendix on Gauging & Land Measuring.* London, 1729.

Hawthorn, R.: *Instructions for the Use of the Mechanics Improved Sliding Rule.* Newcastle-on-Tyne, 1832.

Heather, J.F.: *Mathematical Instruments, Volume 1. Drawing & Measuring Instruments.* Crosby, Lockwood & Co. London, 1849.

Heather, J.F.: *A treatise on Mathematical Instruments Their Construction, Adjustment and Use Concisely Explained.* 2nd Edition with corrections. George Woodfall & Son. London, 1851. Virtue, 1864.

Heather, J.F.: *Drawing and measuring instruments.* London, 1871.

Hellman, C.D.: *John Bird (1709-76) Mathematical instrument maker in the Strand.* Isis XVII, 127-53 1932.

Henrion, Denis.: *Logocanon ou régle proportionnelle.* Paris, 1626.

Herold, J.F. (Don?): *How to Choose a Slide Rule.* Keuffel & Esser. 1940, 1952.

Herold, J.F.: *A slide rule with intellect and horse sense, K&E Log-Log Duplex Decitrig.* Keuffel & Esser. 1956.

Herold, J.F.: *The consistent Log-Log Duplex Decitrig.* Keuffel & Esser. 1956.

Hills, E. Justin.: *A Course in the Slide Rule and Logarithms.* Ginn and Company, Boston & N.Y. 1928.

Hill, Thomas.: *Practical Gauging Improved, By the Help of a New Portable Sliding-Rule.* 1734.

Hinckley, F.C. & Ramsey, W.F.: *The Slide Rule.* Wright & Potter Printing, Boston, 1910.

Hoare, C.E. Charles.: *The Slide Rule and How to use it: Containing full, easy and simple instructions to perform all calculations.* (with a slide rule in the tuck of the front flap). Weales Scientific & Tech Series. 1867. 2nd edn, 1875; The Technical Press. 15th Printing. 1937. Versions of this book exist in hard and soft cover, some without and probably never had the slide rule included. Hoare covers a large variety of subjects suitable for slide rule calculation using the slightly unusual scale arrangement advocated on his slide rule. In print over 60 years.

Hodgson, F.T.: *Complete Mechanical Slide Rule and How to Use it.* New York, 1907.

Hogg, ??.: *The Hogg Slide Rule.* 1880.

Hood, Thomas.: *The making and use of the Geometrical Instrument called a Sector ...* London, 1598.

Hopp, Peter, M.: *Otis-King Update.* Journal of the Oughtred Society Vol, 4, No, 2. October 1995. pp 33-40.

Hopp, Peter, M.: *Otis-King - Conclusions?* Journal of the Oughtred Society Vol, 5, No, 2. October 1996. pp 62-65.

Hopp, Peter, M; Barnes, Colin; & Knott, John V.: *UNIQUE Slide Rules.* Journal of the Oughtred Society Vol, 6, No, 1. March 1997. pp 32-44.

Horsburgh, E.M. ed.: *Handbook of the Napier tercentenary celebrations, or modern instruments and methods of calculation.* Bell London, 1914. Reprinted 1982. The 1982 Preface notes the inaccuracy of the dates in the slide rule section by Stokes, however the sections on Napier, and the 25 pages on slide rules with additional adverts and examples is well worth having for the variety of examples illustrated.

Howell, John.: *A sure guide to the practical Surveyor.* London, 1678.

Hunt, W.: *A Mathematical Companion, or the Description and Use of a New Sliding-Rule ...* 1697.

Hutton, Charles.: *Mathematical Tables.* 1811.

Hutton, Charles.: *Philosophical and Mathematical Dictionary.* 1815.

ICS: *The Slide Rule.* International Textbook Co. 1879; 1939.

ICS Reference Library: *Mathematics, Slide Rule, etc.* International Correspondence Schools. 1900.

ICS Reference Library: *Trigonometry, Slide Rule, Mechanics, etc.* International Correspondence Schools. 1930.

Interbook International, II.: *Dictionary of British Scientific Instruments. 1921.* Interbook International. Reprint 1976.

Iley, Matthew.: *The New Practical Gauger.* Includes the use of the sliding rule in practical geometry and mensuration. 1819.

Imeson, Charles V.: *The Maya calendar adapted to the slide rule.* Offprint. New York, 1935.

Jacob, L.: *Le Calcul Mécanique.* Paris, 1911.

Jackson, Thos.: *Slide Rules & How to Use them.* Chapman & Hall Ltd. undated, but post 1902 as Rietz is mentioned.

Jerrmann, L.: *Die Gunterscale. Vollständige Erklärung der Gunterlinien und Nachweis ihrer Entstehung nebst zahlreichen Beispielen fü praktischen Gebrauch.* Hamburg, 1888.

von Jezierski, Dieter.: *The Rietz-System and Max Rietz.* Journal of the Oughtred Society Vol, 2, No, 2. October 1993. pp 3-5.

von Jezierski, Dieter.: *The A.W. Faber-Castell Columbus System.* Journal of the Oughtred Society Vol, 4, No, 1. March 1995. pp 12-14.

von Jezierski, Dieter.: *Special Slide Rules of Faber-Castell.* Journal of the Oughtred Society Vol, 4, No, 2. October 1995. pp 50-52.

von Jezierski, Dieter.: *Faber-Castell Combination Rules.* Journal of the Oughtred Society Vol, 5, No, 1. March 1996. p 24.

von Jezierski, Dieter.: *Rechenschieber - eine Dokumentation.* Private Publication, Stein, Germany. 1997. 120 pages of Text with 55 pictures in German describing various manufacturers and their rules.

Jillson, Arnold.: *Utility of the Slide Rule.* Case, Lockwood & Brainard, Hartford. 1859.

Johnson, Lee H.: *The Slide Rule.* Van Nostrand & Co. New York, 1949, 1961, 1966.

Johnson, Lee H.: *Centre Drift Method for Slide Rule Computation.* Article, Kansas Engineer. University Of Kansas. Lawrence, Oct 1947.

Johnston, C.L.: (Professor of Mathematics, East Los Angeles College. *Slide Rule, A textbook for Classroom and Self Instruction.* Wm. C. Brown Company. Dubuque, Iowa. 5th edn, 1953, 1956, 1960, 1967 & 1971.

Jordon, W.: *Handbuch der Vermessungskunde.* Stuttgart, 1897.

Joss, Heinz.: *A Circular Slide Rule in the Shape of a Ring.* Journal of the Oughtred Society Vol, 6, No, 1. March 1997. p 22.

Kadokura, Katsunori.: *Chronology of Japanese Slide Rules.* Journal of the Oughtred Society Vol, 1, No, 2. August 1992. pp 34-39.

Kebabian, P.B.: *The English Carpenter's Rule, Notes on It's Origin.* Chronicle of Early American Industries, June 1988, vol. 41, no. 2.

Kebabian, P.B.: *Further Notes on the Early English Three Fold Ship Carpenter's Rule.* Chronicle of Early American Industries, March 1989, vol. 42, no. 1.

Keith, I.R.: *The ABCD of Slide Rules*. Brighton College of Technology. 196? This is a "scrambled" book which means that you do not read the pages in the usual order, but go through in an order dictated by the way the questions are answered. A simple how to use book with examples illustrated with Unique all-plastic slide rules.

Kells, Lyman M., Kern, Willis F., & Bland, J.R.: *K&E Polyphase Duplex Decitrig Slide Rule No. 4071A. A manual*. Keuffel & Esser. Hoboken, 1939.

Kells, L.M., Kern, W.F., & Bland, J.R.: *Log-log Duplex Trig Slide Rule. No. 4080. A manual*. Keuffel & Esser. Hoboken, 1939, 1948.

Kells, L.M., Kern, W.F., & Bland, J.R.: *K&E Slide Rule Manual Log Log Duplex Decitrig Trade Mark No. 4081*. Keuffel & Esser. 1943, 1955.

Kells, L.M., Kern, W.F., & Bland, J.R.: *K&E Slide Rules, A Self Instruction Manual*. Keuffel & Esser. 1955.

Kells, L.M., Kern, W.F., & Bland, J.R.: *K&E Slide Rule Manual Log-log Duplex Decitrig No 4081. For Mod 68-2047 (Old Mod 4187S)*. Keuffel & Esser. 1955.

Kentish, Thomas.: *Treatise on a Box of Instruments, and a Slide Rule for the use of Excisemen, Engineers, Seamen & Schools*. London, 1839; Philadelphia, 1847.

Kentish, Thomas.: *Treatise on a Box of Instruments, and the Slide Rule for the use of Gaugers, Engineers, Seamen & Schools*. Philadelphia, 1852; 2nd edn, 1877.

Keuffel, A.W.: *Slide rule - Encyclopedia Britannica*. Vol 20, 1971.

Keuffel & Esser: *Elementary Instructions for Operating the Slide Rule*. Keuffel & Esser. 1909, 1928.

Keuffel & Esser: *Instructions for Operating the Ever-There Slide Rule No. 4097C*. Keuffel & Esser. 1936.

Keuffel & Esser: *The log log trig slide rule No. 4090. A self teaching manual with tables of settings, ...* Keuffel & Esser. Hoboken, 1936.

Keuffel & Esser: *K&E Slide Rules, DECI-LON, an Instruction Manual*. Similar to Kells et al. Keuffel & Esser. 1962.

Keuffel, W.L.E.: - *extracts from "The First Quarter Century of my Life"*. Journal of the Oughtred Society Vol, 2, No, 1. March 1993. pp 5-14.

Kinahan, S. & Lynch, M.: *Notes on the slide rule*. Methuen. 1980.

Kirby, Joshua.: *The Architectronic Sector*. 1761. Possibly a reprint of Carwitham, 1723.

Kitchenmun, T.: *On the Use of Kitchenmun's new authentic sliding rule*. 1808.

Klügel, G.S.: *Mathematisches Wörterbuch*. Leipzig, 1808.

Knapen, A.M.: *The MECHANIC'S ASSISTANT: A Thorough Practical Treatise on MENSURATION and THE SLIDING RULE ...* Appleton & Co. New York, 1850.

Knight, Richard.: *Branan's Rule - An Undervalued Slide Rule*. Journal of The Oughtred Society, Vol 4, No 1, March 1994, pp 16-20.

Knott, John, V.: *Fowler & Company. 1898-1988*. Journal of the Oughtred Society Vol, 4, No, 2. October 1995. pp 16-17.

Knott, John, V.: *Joshua Routledge. 1775-1829*. Journal of the Oughtred Society Vol, 4, No, 2. October 1995. pp 25.

Knott, John, V.: *British Thornton - A Slide Rule Manufacturer of Manchester England*. Journal of the Oughtred Society Vol, 6, No, 1. March 1997. pp 29-31.

Knowles, W.: *Calculating scale; a substitute for the Slide Rule*. Spon & Chamberlain. New York, 1903.

Kolesh & Co.: *Slide Rule Instructions*. New York, 1908.

Krause, Rudolf.: *Rechnen mit dem Rechenschieber nach dem Dreiskalsystem der Firmen Dennert & Pape, A.W. Faber, Nestler u. A.* Mittwerda. 19--

Krebs, Erich.: *Die Rechenstäbe und Rechenmaschinen, einst und jetst. Aus "Beiträge zur Geschichte der Technik und Industrie"*. Jahrbuch des VDI, Berlin, 1911.

La Croix, ?, & Ragot, ?: *A Graphic Table Combining Logarithms and Antilogarithms*. Contains two graphic scales in book form, one on 40 pages is equivalent to a 360 foot slide rule, another on 6 pages is equivalent to 44 foot. Macmillan, New York, 1924.

Labosne, A.: *Instruction sur la règle a calcul, contenant les applications de cet instrument au calcul des ...* Paris, 1872.

Lagrange, Joseph Louis.: *De la résolution des équations numériques de tous les degrés*. Paris, 1798.

Lalande, Joseph Jérôme Lefrancais de: *Encyclopédie Méthodique (marine), Article "Echelle anglaise"*. 17--

Lalanne, M Léon.: *Instruction sur les règles à calcul et particulièrement sur la nouvelle règle à enveloppe de verre.* (see also Elliot). Paris, 1851, 1854, 1863.

Lamb, Joseph.: *Description of a Concentric Circular Proportioner, an Instrument for Facilitating Calculation*. 1827.

Lambert, Johann H.: *Beschreibung und Gebrauch der Logarithmischen Rechenstäbe*. Augsburg, 1761, 1772.

Lambert, Johann H.: *Briefwechsel, Vol IV*. 1777.

Lanchester, Frederick William.: *The radial cursor: a new addition to the slide rule. From the Philosophical Magazine*. London, 1896.

Lanchester, F.W.: *Lanchester's potted logs – a concise tabulation (Slide Rule Annex)*. 1900?

Larsen, H.D.: *Introduction to the Slide Rule*. 1940.

Larson, Mel.: *The Runner*. Journal of the Oughtred Society Vol, 2, No, 1. March 1993. pp 40-47.

Larson, Mel.: *Runner, Indicator or Cursor*. Journal of the Oughtred Society Vol, 2, No, 1. March 1993. pp 47-49.

Laurie, A.N.: *A Precision Slide Rule*. Using extended graduations to extend scale length. Scientific American Supplement. 1910.

Leach, H.W.: *The slide rule and technical problem solving*. Macmillan. 1963.

Leadbetter, Charles.: *Royal Gauger; or, gauging made easy, as it is actually practised by the officers of his majesty's excise*. 1739, 2nd edn, 1743; 4th edn, 1755.

Leadbetter, Charles.: *Everard's Slide Rule Improved*. 1760.

Leblond, A.S.: *Cadrans logarithmiques adaptés aux poids et mésures*. Paris, 1799.

Lemoine, Jean.: *Instruction pratique sur la règle à calculs universelle Barbotheu "J. D."* Vincennes, 1925.

Lemonnier, ??: *Abrégé du Pilotage*. Paris, 1766.

Le Patourel, Charles.: *The slide rule manual for the use of seamen*. London, 1928; 2nd edn, 1937.

Leupold, Jacob.: *Theatrum Arithmetico-Geometricum, Das ist: Schau-Platz der Rechen - und Mess Kunst*. Leipzig, 1727.

Leybourn, William.: *The Line of Proportion or Numbers, commonly called Gunter's Line, made easie: By the which ...* 1667; 5th edn 1678; 6th edn 1684.

Leybourn, William.: *The Art of Numbering by Speaking Rods, vulgarly termed Napier's Bones*. 1667.

Leybourn, William.: *Panorganon: or, a Universal Instrument ...* London, 1672.

Leybourn, William.: *The Compleat Surveyor (with Supplement)*. London, 3rd edn 1674.

Leybourn, William.: *A Second Part of the Line of Proportion, commonly called Gunter's Line ...* With a Supplement by John Brown. 1677.

Leybourne. ??: *Cursus Mathematicus.* 1690.

Lipka, J.: *Graphical and Mechanical Computation.* John Wiley & Sons. NY. 1918.

Loftus, W.R.: *The Brewer: A description of the new & Improved brewing saccharometer & slide rule.* London, 1857.

Long, Joseph.: *Description & Use of a Sliding Rule for Gauging, Ullaging, Valuing and Reducing as practised at the Port of London.* 1829.

Long, Joseph.: *Description and Use of the Sliding Rule for Gauging Valuing and Reductions.* 1895.

Lovat-Higgins, A.: *The slide rule : Its operation & Digit Rules.* Sir Isaac Pitman & Sons. 1942. Probably physically the smallest book on how to use the basic scales on a slide rule, it measures 3.75" by 2.5" and contains 14 pages of simple instructions.

Lucas, ??: *The Slide Ruler's Guide.* Maynard's Catalogue, 1828.

MacGregor, John.: *A Complete Treatise on Practical Mathematics.* Edinburgh, 1792.

Machinova, P.E.: *Slide Rule Manual for Engineering Students.* Ohio State University. 1947.

MacKay, Andrew.: *The description & use of the sliding rule, in the mensuration of wood, stone, bales &c.* London, 1799.

MacKay, Andrew.: *The description & use of the sliding rule also description of ships carpenter sliding rule, gauging rule, gauging rod and ullage rule.* London, 1806.

MacKay, Andrew.: *The description & use of the sliding Gunter in navigation.* London, 2nd edn, 1812.

MacKay, Andrew.: *The description & use of the sliding rule, in arithmetic, and in the mensuration of surfaces.* London, 2nd edn, 1806, Edinburgh, 2nd Issue, 1811.

Mackintosh, Thomas.: *Description and Use of the Sliding Rule.* 1799.

Mackintosh, Thomas.: *Use of the Sliding Gunter in Navigation.* 1802.

Mackintosh, Thomas.: *The Complete Navigator.* 1804.

McCully, H.M.: *Slide Rule Photography.* Journal of the Oughtred Society Vol, 6, No, 1. March 1997. p 48.

McLeod, T.H.: *Instrumental Calculation: or a Treatise on the Sliding Rule.* George Smith, Publisher. 1846.

Mair, John.: *Arithmetic.* 1794.

Marks, Robert W.: *Simplifying the Slide Rule.* Bantam Books, 1938, 1964, 1971. Strictly a "how to use" book with questions and answers.

Markquart, John.: *Mechanical Slide Rule. The Carpenters Rule.* 1850.

Martin, Benj.: *The description and uses of an universal sliding rule.* London, 1770.

Mannheim, Amédée.: *Règle à calculs modifiée. Instruction.* Metz, 1851.

Mehmke, R.: *Encyklopädie der Mathematischen Wissenschaften,* Vol 1. Leipzig, 1898.

Mehmke, R.: *"Der Rechenschieber in Deutschland" und "Soho-Rules".* aus Zeitschr. f. Mathematik und Physik, 1901, 49.Bd.

Merrill, Arthur. A.: *How do you use a slide rule. For non-mathematicians and casual calculators, Ten Easy fifteen-minute lessons in multiplication and division.* Dover Publications. 1961. 36 pages with graded examples with illustrations on a K&E slide rule.

Meyer zur Capellen, W.: *Mathematische Instrumente.* Leipzig, 1949.

M.I.M.E.: *The Slide Rule & How to use it.* J. Halden & Co. 1920? 4 pages of actual instructions with another 4 pages of lists of rules available from J. Halden.

Middlemiss, Ross Raymond.: *Instructions for Post Trig & Mannheim Trig Slide Rules.* Fredk. Post Co. 1945. Middlemiss is described as Technical Consultant to the Frederick Post Company, the 81 pages are specific to Post rules of this type.

Middlemiss, Ross Raymond.: *Instructions for using Log Log slide rules.* 1946.

Miller, Robert, C.: *Nystrom's Calculator.* Journal of the Oughtred Society Vol, 4, No, 2. October 1995. pp 7-13.

Miller, Robert, C.: *Gurley Slide Rule.* Journal of the Oughtred Society Vol, 5, No, 1. March 1996. pp 14-16.

Mitchell, Donald.: *Computing with Slide Rule and Mathematical tables.* ?? ??

Mitchell, James.: *Dictionary of Mathematical and Physical Sciences.* London, 1823.

Mittelstadt, W.S.: *Basic slide rule operation, a program of self-instruction.* McGraw-Hill. 1964. Lecture notes from a series of slide rule lecture he gave when he worked for Eastman Kodak. Illustrated with K&E Polyphase Duplex slide rules.

Mittelstadt, W.S.: *Instructions for Trig and Log-Log slide rules.* McGraw-Hill. 1964. Lecture notes from lectures.

Miyazaki, Jisuke.: *Developing History of the Slide Rule,* (In Japanese). 1956.

Montferrier, ??: *Mathematical dictionary.* 1841.

Moore, Jonas.: *The many uses of the spirall or serpentine line etc., usefull for all mathematicians in generall and for ingenious architects, measurers of land, carpenters and other mechanics in particular. First applied to these several practices by William Milburne M.A., and reduced into this form with many additions by Jonas Moore, Professor of the Mathematicks.* London, 1659.

Moore, Sir Jonas.: *Modern Fortification ... with the Use of a Joynt-Rule or Sector, for the speedy Description of any Fortification.* London, 1673.

Moore, Sir J.: *A Mathematical Compendium; or, Useful Practices, etc.* London, 1705.

Morellon, P.: *Méthode de Vulgarisation Simple et Détaillée Pour L'Emploi de la Règle a Calcul.* Paris, c1940.

Morris, Alan.: *Model Designations of Modern Era K&E Slide Rules.* Journal of the Oughtred Society Vol, 4, No, 2. October 1995. pp 18-24.

Moss, Thomas.: *A Treatise on gauging, Containing not only what is common on the subject, but likewise a ...* London, 1765.

Mountaine, William.: *Gunter's Scale : A Description of the Lines drawn On Gunter's Scale, as Improved by Mr John Robertson.* Nairne & Blunt. 1778.

Mouton, William.: *A treatise on Gunter's Sliding Rule - Designed as a Pocket Companion for the Mechanic, Manager and Artisan.* Benjamin B. Musey, Boston. 1843.

Mouzin, Ph.: *Instruction sur la manière de se servir de la Règle à calcul, dite Règle Anglaise ou Sliding Rule.* Paris. 2nd edn. 1824. Paris, 1837.

[Mouzin, Ph.]: *Instruction sur la manière de se servir de la règle à calcul, instrument à l'aide duquel on peut etre ...* Paris & Dijon, 2nd edn, 1825.

Moxon, Joseph.: *A Book of Napier's Bones.* London, 1674.

Moxon, Joseph.: *Mechaniks Exercises.* London, 1677.

Naish, Edmund.: *The Logarithmicon. A mechanical contrivance for facilitating calculations.* Dublin, 1898.

Napier, John, (Baron of Murchiston): *Rabdology. An Instruction Book for Napier's Bones.* 1617. MIT Press reprint, 1991.

Nasmith, John W.: *The slide rule: Its principles and application.* Manchester, 1902.

Nelting, Robert.: *Der-Nautisch-Astronomische und Universal-Rechenstab und seine Verwendung.* Altona-Hamburg, 1912.

Nesbit, Anthony.: *A Treatise on Practical Mensuration. New Edition thoroughly revised and greatly improved by the Rev. John Hunter M.A.* Longmans Green & Co. London, 1875.

Nestler, Albert.: *Instructions for the calculating slide rule.* Lahr, 1906.

Nestler, Albert.: *Der logarithmische rechenschieber und sein gebrauch.* Lahr, 1907.

Newnes.: *Newnes' Slide Rule Manual.* George Newnes Limited, London, 1963. These are later editions of F.J, Camm's book of the same name which continued publication until at least 1966.

Newson, ?? Larsen ??: *Basic maths for pilots and flight crews.* With a circular slide rule. 1943.

Nicholson, William.: *Philosophical Transactions, Pt II.* London, 1787.

Nicholson, William.: *Nicholson's Journal Vol. I* 1797, 1802.

Nickel, F.F.: *The Runnerless Slide Rule.* New Jersey. 19--

Noble, J.: *Noble's and Routledge's Slide Rule.* 1922.

Norwood, Matthew.: *System of Navigation: Teaching the whole Art ... Spherical Triangles and their Use in Great Circle sailing. The Use of the Globes, Azimuth Compass for Variation ... The Description and the Use of a New Instrument, called Encyclogium, or the Sliding Circles.* T.Wall. Bristol, 1685.

Ohkura Denki Electric Co.: *50 Years of Company History.* (In Japanese).

Otnes, Robert K.: *The Otis King Slide Rule.* Journal of the Oughtred Society Vol, 0, No, 0. August 1991. pp 7-9.

Otnes, Robert K.: *The Charpentier Calculator.* Journal of the Oughtred Society Vol, 0, No, 0. August 1991. pp 9-12.

Otnes, Robert K.: *Log-Log Scales.* Journal of the Oughtred Society Vol, 1, No, 1. February 1992. pp 19-24.

Otnes, Robert K.: *Slide Rule Pencils.* Journal of the Oughtred Society Vol, 1, No, 2. August 1992. pp 11-13.

Otnes, Robert K.: *Thacher Notes.* Journal of the Oughtred Society Vol, 2, No, 1. March 1993. pp 21-26.

Otnes, Robert K.: *Bonnycastle on Gauging.* Journal of the Oughtred Society Vol, 3, No, 2. September 1994. pp 21-26.

Otnes, Robert K.: *Log-log scales - revisited.* Journal of the Oughtred Society Vol, 4, No, 1. March 1995. pp 9-11.

Otnes, Robert K.: *An Example of Naperian Logarithms* Journal of the Oughtred Society Vol, 4, No, 1. March 1995. pp 46-47.

Otnes, Robert K.: *How Briggs Computed Logarithms* Journal of the Oughtred Society Vol, 4, No, 2. October 1995. p 26.

Otnes, Robert K.: *The Thomson Log-Log Slide Rule.* Journal of the Oughtred Society Vol, 5, No, 1. March 1996. p 11.

Otnes, Robert K.: *Dietzgen Patents, Runners and Log-Log Scales.* Journal of the Oughtred Society Vol, 5, No, 2. October 1996. pp 45-48.

Otnes, Robert K.: *Dietzgen Calculation Devices.* Journal of the Oughtred Society Vol, 5, No, 2. October 1996. pp 49-50.

Otnes, Robert K.: *K&E Instruction Manuals.* Journal of the Oughtred Society Vol, 6, No, 1. March 1997. pp 18-19.

Otnes, Robert K. & Schure, Conrad.: *The Blundell Vector Slide Rule.* Journal of the Oughtred Society Vol, 5, No, 1. March 1996. p 19.

von Ott, Karl.: *Der logarithmische Recheschieber: Theorie und Gebrauch desselben.* Prag, 1891.

Oughtred, Rev. William.: *The Circles of Proportion and the Horizontal Instrument &c.,* 1631 Latin. Forster, William. 1632 English.

Oughtred, Rev. William.: *To the English Gentrie ... the just Apologie of W. Oughtred against the ... insinuations of R. Delamain in a pamphlet called Grammelogia or the Mathematical Ring.* Published with Forsters translation. 1632.

Oughtred, Rev. William.: *The New Artificial Gauging Line or Rod: Together with Rules concerning the use thereof : Invented and written by William Oughtred.* London, 1633.

Oughtred, Rev. William.: *An Addition unto the Use of the Circles of Proportion for the working of Nautical Questions.* 1633.

P.I.C.: *Instructions for PIC Differential Scales, Booklet No 81.* William Morris. 194? (Walker, M.?). One of a series of booklets specifically aimed at PICK slide rules, see also **Walker, M.**

Palmer, Aaron.: *Palmer's computing scale.* Boston, 1843.

Palmer, Aaron.: *A Key to the Endless Self-Computing Scale, showing it's application to the Rules of Arithmetic, etc.* Smith & Palmer, Boston, 1844.

Palmer, Aaron.: *Palmer's Pocket Scale with Rules for its use in solving Arithmetical and Geometrical problems.* Rochester, 3rd edn. 1845.

Partridge, Seth.: *The Description and Use of an Instrument called the Double Scale of Proportion: by which ... all questions in ... Astronomy, Geography, Navigation ... Dialling may be performed.* Describes a slide rule. (written in 1657). London, 1661; 3rd edn, 1671; 1672, 1685, 1692.

Patten, William.: *The Complete Slide Rule Instructor.* 1890.

Peacock, Thomas.: *The Practical Measurer.* with "... the Uses of Logarithms, Gunter's Scale, the Carpenter's Rule, & the Sliding Rule". 1798.

Pedie, Alexander.: *The Practical Measurer, New Edition Improved & Enlarged.* Blackie & Son. Glasgow, 1843.

Peraux, E.: *Instruction sur la règle a calcul a deux règlettes, de E. Peraux.* Paris, 1893.

Perrine, Dr. J.O.: *Slide Rule Handbook.* Gordon & Barry, Baywood Publishing Co. N.Y. 1965. Rather a good, clear hand-book, hardback with typeface which looks to be photocopy from typeface.

Petrie Palmedo, D.: *How to Use Slide Rules.* New York, 1908.

Pezenas, R.P.: *Elémens du Pilotage.* 17--

Pezenas, R.P.: *Pratique du Pilotage.* 17--

Pezenas, R.P.: *Nouveaux essais pour déterminer les longitudes en mer par les mouvements de la lune et par une seule observation.* Paris, 1768.

Pickworth, C.N.: *Instructions for Nestlers calculating slide rule.* Newark, 1883.

Pickworth, Charles N.: *The Slide Rule; A Practical Manual.* Emmott & Co. 1897. (4th edn); 1920 (17th edn); 1938 (21st edn); 1945. (24th edn). Probably one of the best books describing a wide variety of slide rules available on the market and how to use them. The various editions are updated to include more or less information on the rules as they went out of production, and new rules and scales are then described as they are introduced. Later editions continue with less information on older slide rules. Early editions also contain useful adverts for slide rule manufacturers and retailers. 122 pages with illustrations.

Pickworth, C.N.: *Revised instructions for the use of A.W. Faber's calculating rule.* A.W. Faber. London, 1897.

Pickworth, C.N.: *Instructions for the Use of A.W. Faber's Improved Calculating Rule.* London, c1903 (2nd edn); c1906 (3rd edn).

Pike, Nicolas.: *A new & complete System of Arithmetic.* Isiah Thomas, USA. 1788. 1797, 2nd edn.

Podmore, J.C.: *The slide Rule for Sea and Air Navigation.* Brown Son & Ferguson. Glasgow, 1974.

Podmore, J.C.: *The Slide Rule for Ship's Officers.* Brown Son & Ferguson. Glasgow, reprinted 1943.

Poland, John.: *The Quick and Easy Slide Rule Instruction Book. How to get answers instantly without pencil and paper.* Peru, c1930.

Poland, John.: *The Quick and Easy "Lawrence" Slide Rule Instruction Book.* Lawrence Engineering Service. 1940.

Porro, J.: *La Tachéométrie*. Turin, 1850.

Post, Frederick & Co.: *Slide Rule Instructions*. Hemmi, Japan. ??

Post, Frederick & Co.: *Post Versalog Slide Rule instructions (For Mod 4450)*. Fredk. Post Co. 1945

Pouchet, ??: *Arithmétique linéaire*. 17--

Powers, B.D.: *Strategy without slide rule*. A book about air power in Europe, 1914-39, not slide rules, despite the title. London, 1976.

Price, Osborne I.: *The K&E 4110 Slide Rule*. Journal of the Oughtred Society Vol, 2, No, 1. March 1993. pp 26-28.

Price, Osborne I.: *Keuffel & Esser Patents*. Journal of the Oughtred Society Vol, 2, No, 1. March 1993. pp 34-38.

Puscariu, Johan Ritter von: *Das Stereometer, Köper-Messinstrument*. Budapest, 1877.

Rabone & Sons, John.: *The Improved Slide Rule, arranged by W.E. CARRETT, Esq, Engineers, Leeds*. J. Rabone & Sons. 1865.

Rabone & Sons, John.: *Instructions for the Use of the Practical Engineers' and Mechanics' Improved Slide Rule, as Arranged by J. Routledge, Engineer*. J. Rabone & Sons. 1867. Ken Roberts Reprint. 1983. See also **Routledge**.

Rabone & Sons, John.: *The Carpenters Slide Rule, It's History & Use*. J. Rabone & Sons. 1867. Ken Roberts Reprint. 1982. 32 pages of instructions for Sir William Armstrong's Carpenter's rule.

Rabone & Sons, John.: *Hawthorns Loco Engineers Slide Rule*. J. Rabone & Sons. c1880.

Randier, Jean.: *Nautical Antiques for the Collector*. (English Translation). Barrie & Jenkins. London, 1976.

Rappolt, F.A.: *Simplified Mathematics and how to use the Slide Rule*. 1943.

Rawlings, G.P.: *The slide rule in theory and practise*. Marshall. 1950. 127 pages of theory, practise and examples illustrated by a PIC wood/celluloid slide rule. Extremely comprehensive book.

Richardson, George W.: *Direct reading Slide Rule*. George Washington Richardson. 1911.

Richardson, George W.: *The Slide Rule Simplified*. George Washington Richardson, Chicago, 1912, 5th Ed. 1917.

Richardson, George W. & Clark, J.J.: *The Slide Rule Simplified*. A 100 page manual. The 7th edn., published sometime after 1922 indicates the Gilson Slide Rule Co., of Niles, Michigan; and mentions the Gilson Midget slide rule patented in 1922.

Riddell, Robert.: *The Slide Rule Simplified*. Philadelphia, 1881.

Riddell, Robert.: *The Slide Rule. Lessons in Carpentry*. 1881.

Riley, J.: *Instrumental Arithmetic or ... the Use of the Sliding Rule*. 1836.

Ritow, H.: *Elementary Slide Rule Manual*. Fredk. Post Co. Chicago, 1925.

Ritow, H.: *Complete Book of Slide Rule Use*. Doubleday. 1964.

Roberts, E.: *A Programmed Sequence on the Slide Rule*. W.H. Freeman & Co. San Francisco, 1962.

Roberts, Kenneth D.: *An Introduction to Rule Collecting*. Ken Roberts Publishing, 1982.

Roberts, Kenneth D.: *Carpenters and Engineers Slide Rules (Part 1, History)*. The Chronicle of the early American Industries Association, 36: 1, 1983.

Robertson, John.: *Treatise on Mathematical Instruments*. 1775.

Robichon, André.: *La règle à calculs (Instructions). Méthode d'utilasation rationnelle des règles modernes (y compris les échelles Log-log et leurs inverses)*. Intended primarily for Tavernier-Gravet slide rules, it includes instructions for Mannheim, electricien, and Rietz rules. Les éditions Foucher, Paris, undated.

Roget, M.D. Peter, M.: *Description of a New Instrument for performing mechanically the Involution and Evolution of numbers*. Philosophical Transactions, Part 1. London, 1815.

Rohrburg, Albert.: *Theorie und Praxis des Logarithmischen Rechenschiebers*. Leipzig, 1916. 11th edn 1953.

Rohrburg, Albert.: *Der Rechenstab im Unterict aller Schularten. Eine methodische Anleitung*. Berlin u. Munchen, 1928.

Root, Oren.: *New Treatise on Surveying and Navigation, Theoretical and Practical*. 1½ pages on Gunter's. Blackman & Co, 1867.

Rosenthal, L.W.: *Mannheim and Multiplex Slide Rules*. Dietzgen, Chicago. 1905, 1908.

Ross, ??: *Ross Precision Computer*. 1910.

Routledge, J.: *Instructions for the Use of the Practical Engineers' & Mechanics' Improved Sliding Rule*. Leeds, 1805, 4th edn 1813, 6th edn 1818, reprint 1823, 7th edn 1826. Ken Roberts reprint, c 1995. A fascinating 32 page booklet on how to use this rule, the additional short history of the rule type in the reprint is a valuable addition.

Roylance, L. St D.: *K&E Roylance Electrical Slide Rule*. 1913.

Rozé, P.: *Règle à calculs*. Paris, 1907.

Rozé, P.: *Théorie et usage de la règle à calculs*. Paris, 1907.

Rudowski, Werner H.: *German Patents and the Columbia*. Journal of the Oughtred Society Vol, 2, No, 2. October 1993. pp 13-15.

Russell, ??: *Rapid Calculations*. 1956.

Ruth, Franz.: *Theorie der Logarithmischen Rechenschieber. Als anleitung für die benützung der fünf ...* Graz, 1878.

S., E.: *The Description, Nature and General Use of the Sector and Plain-Scale, briefly and plainly laid down. As also a Short Account of the Uses of the Lines of Numbers, Artificial Sines and Tangents*. 1721.

Saffod, R. & Smalley, A.: *The Slide Rule*. Doubleday & Co. NY. 1962.

Saffold, R.: *Slide Rule*. Consultant Editor : R.Goodman. E.U.P. Tutor Text Series. 1964.

Sauer, Robert. J.: *Two Scarce K&E Rules*. Journal of the Oughtred Society Vol, 6, No, 1. March 1997. pp 20-21.

Saverien, ??: *Tome I. Dictionnaire universel de mathematique et de physique. Article "Echèlle Angloise"*. Paris, 1753.

Scheffelt, Michael.: *Pes mechanicus artificialis d. i. neu erfundener Maassstab, mi welchem alle Proportionen der ganzen Mathematik ohne mühsames Rechnen und so weiter können gefunden werden*. Ulm, 1699, 1718.

Schuitema, Ijzebrand.: *Slide Rule Cross Sections*. Journal of the Oughtred Society Vol, 1, No, 2. August 1992. pp 19-34.

Schuitema, Ijzebrand.: *The ALRO Circular Slide Rule*. Journal of the Oughtred Society Vol, 2, No, 2. October 1993. pp 24-38.

Schuitema, Ijzebrand.: *Spirit Rules*. Journal of the Oughtred Society Vol, 3, No, 2. September 1994. pp 31-37.

Schuitema, Ijzebrand.: *Logomat Werbeartikel, Logomat Advertising Articles*. Journal of the Oughtred Society Vol, 4, No, 1. March 1995. pp 29-32.

Schuitema, Ijzebrand.: *The Appoulet Circular Slide Rule*. Journal of the Oughtred Society Vol, 4, No, 1. March 1995. pp 48-52.

Schuitema, Ijzebrand.: *A Beer Slide Rule*. Journal of the Oughtred Society Vol, 5, No, 1. March 1996. p 17.

Schuitema, Ijzebrand.: *A Geological Slide Rule*. Journal of the Oughtred Society Vol, 5, No, 1. March 1996. p 21.

Schuitema, Ijzebrand.: *Slide Rule Hunting in Ireland*. Journal of the Oughtred Society Vol, 5, No, 2. October 1996. p 66.

Schuitema, Ijzebrand.: *IWAMATIC 1650 and VÖTSCH-variant*. Journal of the Oughtred Society Vol, 5, No, 2. October 1996. p 66-67.

Schure, Conrad.: *The Hart Equationer*. Journal of the Oughtred Society Vol, 1, No, 2. August 1992. pp 15-17.

Schure, Conrad.: *The SMALL Calculator - Part II*. (see also Bennett. A.) Journal of the Oughtred Society Vol, 1, No, 2. August 1992. pp 17-19.

Schure, Conrad.: *The Conrad Schure Collection.* Journal of the Oughtred Society Vol, 2, No, 1. March 1993. pp 28-33.

Schure, Conrad.: *The Schofield-Thacher Slide Rule.* Journal of the Oughtred Society Vol, 3, No, 1. March 1994. pp 20-26.

Schure, Conrad.: *The Wollaston Chemical Slide Rule.* Journal of the Oughtred Society Vol, 5, No, 1. March 1996. pp 22-23.

Schure, Conrad.: *Slide Rule Watches.* Journal of the Oughtred Society Vol, 6, No, 1. March 1997. p 47.

Schwartz. Ph.D. Benjamin, L.: *Some Speciality Slide Rules.* Journal of the Oughtred Society Vol, 6, No, 1. March 1997. pp 23-24.

Scofield, C.W.: *Basic Maths for Technicians.* Edward Arnold Books for Technicians, 1978. A short section on how to use slide rules in general, illustrated with British Thornton Mk II slide rules.

Scofield, E.M.: *The Slide Rule.* Dietzgen. Chicago, 1902.

Scofield, E.M.: *Slide Rule.* New York, 1904.

Scott, Benjamin.: *The Description and Use of an Universal and Perpetual Mathematical Instrument, shewing the most Expeditious and Exact method of solving all Questions in Arithmetic, Trigonometry, in Surveying, Theodolites, Semicircles, Circumferenters, Plain Tables, Drawing-Tables, Water Levels, Measuring Wheels etc. In Navigation, Quadrants, Forestaffs, Gunter's Sea-Compasses, both Meridian and Azimuth, etc.* London, 1733.

Scherer, ??: *Logarithmischgraphische Rechenstafel.* Kassel, 1893.

Schulz von Strassnitzki, L.C.: *Anweisung zum Gebrauche des englischen Rechenschiebers.* Wien, 1843.

Schumacher, Dr. Soh.: *Ein Rechenschieber mit Teilung in gleiche Intervalle auf Grundlaged er zahlentheoretischen Indizes. Fur den Unterricht konstruiert.* Munich, 1909.

Sedlaczek, Ernest.: *Anleitung z. Gebrauche einiger logarith. getheilter Rechenschieber.* Wien, 1851.

Sedlaczek. Ernest.: *Neber Visir- und Recheninstrumente.* Wien, 1856.

Selden, Joseph.: *The Trades-man's Help. An Introduction to Arithmetic with the Use of a New Instrument called: The Joynted Sliding Rule; and also a Perpetual Almanack.* Tunbridge Wells, 1694.

Sella, Quintino.: *Teorica e pratica del regolo calcolatore.* Torino, 1859, 1886.

Seller, John.: *A Circular Table of Proportion, resolving many questions of Multiplication, Division ... only by moving an Index upon the Centre of the said Instrument.* 1680.

Sellins, A.: *Stanley Folding Rules, A History and Descriptive Inventory.* Privately published. 1984.

Sexton, ??: *Sexton's Omnimeter.* Engineering News. 1897.

Sharp, K.J.: *Computations - a flow chart method for success in basic numerical operations, percentage ratio, proportion, use of mathematical tables, logarithms, slide rules etc.* Heineman. 1976.

Shepherd, Rodger.: *Pickett Metal slide Rules.* Journal of the Oughtred Society Vol, 1, No, 1. February 1992. pp 5-9.

Shepherd, Rodger.: *Pickett's "Eye-Saver Yellow".* Journal of the Oughtred Society Vol, 1, No, 1. February 1992. pp 18-19.

Shepherd, Rodger.: *The End of the Aristo Slide Rules: a Case Study.* Journal of the Oughtred Society Vol, 2, No, 2. October 1993. pp 5-8.

Shepherd, Rodger.: *Translation of an Article on F.Blanc's Slide Rule.* Journal of the Oughtred Society Vol, 4, No, 2. October 1995. pp 31-32.

Shepherd, Rodger.: *Some Distinctive Features of Dietzgen Slide Rules.* Journal of the Oughtred Society Vol, 5, No, 2. October 1996. pp 42-45.

Shireby, R.M.: *The slide rule applied to commercial calculations.* Pitman. 1922, 1930. Aimed at filling the gap in "the many excellent books" on slide rules, which "fail to give adequate attention to such calculations as continually present themselves in the daily routine of the counting-house of every commercial firm".

Shirtcliffe, Robert.: *The Theory and Practice of Gauging.* London, 1740.

Shuster, Carl N.: *A Study of the Problems in Teaching the Slide Rule. New York.* Article, Bureau of Publications, Columbia University, 1940.

Shute (Norway), Neville.: *Slide Rule.* (his autobiography). Heinemann. 1954. Apart from the title there is only really one small paragraph which describes the use of Fullers tubular slide rules by his team of "calculators" to calculate the stresses within the geodetic sections of the R100 airship, a task taking one week per arm, one month for a complete section and checked instantly by adding together all stresses with the expected result being a nett zero. Failure to achieve this meant a complete recalculation!

Simm, B.A.: *Slide Rule Manual, A Text Book for use with Standard Types of Slide Rules.* Melbourne, Australia. Undated. Uses W&G slide rules to illustrate examples. A number of versions with slightly different information showing different rules.

Simms, F.W.: *A treatise on the Principal Mathematical Drawing Instruments.* (20 pages on the Sector). Weale. 1845.

Skelton, John.: *Slide Rule Simplified.* Ponsford, Newman & Benson Ltd., Melbourne, Australia. 1969. Based on Hemmi slide rules.

Slater, Derek.: *The Astronomical Scales of Oughtred's Calculator.* Journal of the Oughtred Society Vol, 6, No, 1. March 1997. pp 16-17.

Slater, Alfred L.: *The slide rule, a complete manual.* Rinehart & Winston. 1967.

Sloane, T., O'Connor, ?. & Thompson, J.E.: *Speed and Fun with Figures.* Includes a complete copy of Thompson's "A Manual of the Slide Rule". Van Nostrand. 1939.

Smith, ??. Capt. R.E.: *Description of the Gunner's Sliding Rule.* 1835.

Smith, George E.: *Smith's Slide Rule of Weights.* 1925.

Snodgrass, Burns.: *Teach yourself the Slide Rule.* English University Press. 1955, 1958, 1965, 1971. Snodgrass was the founder of Unique in 1921, this book is a "how to use" book written round Unique slide rules which are illustrated in some detail. The last editions were updated by his son Donald, though this is not credited.

Sommers, Hobart, H., Drell, Harry, & Wallschlaeger, T.W.: *The Slide Rule and how to use it.* Bailey & Swinfen. 1976. (No slide Rule).

Sommers, Hobart, H., Drell, Harry, & Wallschlaeger, T.W.: *The Slide Rule and How to Use It.* With a 10" Lawrence 10-B (sometimes Lietz?) slide rule in a box built into the cover. Grosset & Dumlap, NY. 1941.

Spang, David A.: *All about those "Gunters".* The Chronicle of the Early American Industries Association 35, 1982, 10-14.

Stanley, Philip.E.: *Boxwood & Ivory - Stanley Traditional Rules, 1855 - 1975.* American Stanley Company. Stanley Publishing Co, Mass. 1975?

Stanley, Philip E.: *Carpenters and Engineers Slide Rules (Part II, Routledge's Rule).* See also Roberts. The Chronicle of the Early American Industries Association, 37: 2, 1984.

Stanley, Philip E.: *Carpenters and Engineers Slide Rules (Part III, Errors in the Data Tables).* The Chronicle of the Early American Industries Association, 40: 1, 1987.

Stanley, W.F. (revised by Tallack): *Drawing and Mathematical Instruments.* Spon. 1925.

Stanley, W.: *Stanley's Slide Rule Practice.* W. Stanley & Co. USA. 1938.

Stender, Dr. Richard.: *Der moderne Rechenstab*. Recommended by Aristo, together with **Baldermann**, as suitable for home learning. Frankfurt am Main u. Hamburg, 1950.

Stender, Dr. Richard & McKelvey, K. K.: *The Modern Slide Rule, a manual of self-instruction for schools, colleges and the professional user*. Uses ARISTO slide rules in Illustrations. Cleaver Hume. 1960.

Stephens & Co.: *Improved slide rule*. 1881.

S(tirrup), T(homas).: *The Carpenter's New Rule shewing how ... to take Heights and Distances and to draw the Plat of a Town or City*. 1659.

Stokes, G.D.C.: *The slide rule*. (In Horsburgh). 1914.

Stone, Edmund.: *New Mathematical Dictionary. Article "Sliding Rules"*. London, 1726, 1743.

Stone Edmund, translation of Bion, Nicolas.: *The construction and principal uses of Mathematical Instruments*. London 1723, 1726, 1758. Reprinted Holland Press, 1972. Contains good instructions for use of the Sector. Extremely comprehensive descriptions of the manufacture and usage of all mathematical instruments available at that date. The 18th century spelling and printing make it a bit hard reading to start with, but this is soon overcome in the wealth of fascinating detail.

Strohm. R.T. & De-Groot, A.: *The Slide Rule*. International Text Book Co. Scranton, 1939, 1945.

Strohm. R.T. & De-Groot, A.: *The Slide Rule, How to Use it*. Laurel Publishers. 1948, 1951.

Strubecker, Karl: *Einführung in die Höhere Mathematik, Bd.I, Grundlagen, 2 verb. Auflage*. Munich and Vienna, 1966.

Suxspeach, John.: *Instrument called The Catholic Organon or Universal Sliding Foot rule*. London, 1856.

Suzuki, Hisao: *Developing History of Calculating Instruments*. (In Japanese). 1967.

Symons, W.: *The Practical Gauger*. London, 1754. 8th edn. 1811.

Taton, R.: *Le Calcul Mécanique*. Paris, 1939.

Tavernier & Gravet.: *Instructions concernant l'usage des Règles a Calculs - Mannheim, Rietz, Electricien*. No 22.   ????

Taylor, Eva G.R.: *The Mathematical Practitioners of Tudor and Stuart England*. Cambridge University Press. 1954. See below.

Taylor, E.G.R.: *The Mathematical Practitioners of Hanoverian England. 1714-1840*. Cambridge University Press. 1966. Eva Taylor's two books are the most comprehensive tabulation of the works of a huge number of mathematical practitioners including most of the early workers with slide rules. The historical background is an essential part of understanding why various enhancements to mathematical instruments happened when they did.

Thacher, Edwin.: *Thacher's Patent Calculating Instrument, or Cylindrical Slide Rule*. Van Nostrand, New York, 1884. K&E, New York, 1903.

Thacher, Edwin.: *Directions for using Thacher's calculating instrument*. K&E. New York, 5th edn 1914.

Thomson, Prof. J.E.: *A manual of the slide rule; its history, principle and operation*. Chapman & Hall, 1930. New York, 5th edn, 1942, 1953.

Thomson, Prof. J.E.: *The Standard Manual of the Slide Rule*. 1930, 1952.

Thompson, H. Loren, & Eshbach, ?.: *Dietzgen Decimal Trig Log-log Slide Rule Manual*. No 1786 *(Models 1733, 1734, 1738)* 1960.

Thompson, H.L., & Eshbach, ?.: *The Dietzgen Maniphase, Multiplex Decimal trig type Log-log Slide Rule Manual. No 1786-32. (For Model No 1732 slide rule)*. 1945.

Thrivikraman, R.: *The slide rule guide (for students of engineering and technology)*. Macmillan. 1955. 1961.

Tonkes, W.: *Engineers' slide Rule*. New York, 19--

Turner, Anthony. J.: *Early Scientific Instruments. Europe 1400 - 1800*. Sotheby's Pubs. 1987. Several pages on slide rules with some excellent illustrations of early rules including Oughtred, Thomas Brown spiral and others.

Turner, A.J.: *Mathematical instruments and the education of Gentlemen*. Anals of Science 30, 51-88. 1973.

Turner, A.J.: *William Oughtred, Richard Delamain, and the Horizontal Instrument in 17th C. England*. Analli dell Istituto e Museo di Storia della Scienza di Firenz VI 2 1981, 99-125.

Turner, A.J.: *From Pleasure and Profit to Science and Security - Etienne Lenoir and the transformation of precision instrument-making in France 1760-1830*. Whipple Museum of the History of Science, Cambridge, 1989.

Turner, A.J.: *Of Time and Measurement*. Ashgate. 1993.

Turner, Gerard L'Estrange.: *Nineteenth Century Scientific Instruments*. Sotheby Publications. 1983.

Turner, G.L'E.: *Antique Scientific Instruments*. Blandford. 1980.

Tuttle, Thomas.: *Practical Application of the Slide Rule*. 1883.

Vero, J.: *Excise and Malt-Examiners' Assistant, Useful also for Supervisors and Officers*. London, 17--

Wade, ??: *College Joe Slide Rule*. 1942.

Wakely, Andrew.: *The Mariners Compass Rectified*. London, 1664. James Atkinson added an appendix: *An Appendix containing Use of Instruments most useful in Navigation*. London, 1694. This describes Gunters and Sliding Gunters.

Walker, Miles.: *Book of Instructions for PIC AC & Electrical Rule, Mod E & F*. William Morris. 1940. Part of a series of booklets describing the use of PIC slide rules, see also the following titles.

Walker, Miles.: *Book of Instructions for PIC slide rules, Standard, Reitz & Electrical. No 3660*. William Morris. 1942, 1946.

Walker, Miles.: *Book of instructions for P.I.C. Slide Rules, Standard & Reitz. No 3659*. William Morris. 1946.

Wallington, C.E.: *An atmospheric Diffusion Slide Rule*. Meteorological Office Scientific paper No 24. HMSO. 1966. 32pp. The paper describes a slide rule that can be used to calculate concentrations and dosages in clouds of aerosols being transported and diffused by atmospheric wind and turbulence. Many different scales are shown for assessing depths and widths of the clouds.

Wallis, P.J.: *William Oughtred's "Circles of Proportion" and "Trigonometries"*. Transactions of the Cambridge Bibliographical Society IV, 1968, 372-82.

Ward, George.: *The Gauger's Practice*. London, 1693.

Ward, John.: *The young Mathematician's Guide ... with an Appendix on Practical Gauging*. London, 1707.

Ward, John.: *Lives of the Professors of Gresham College*. 1740.

Ward, T.G.C.: *The slide rule for students of science and engineering*. English Univ. Press. 1961.

Weber, R.: *Anleitung zum Gebrauche des Rechenkreises*. Aschaffenburg, 1872.

Weimbach, M.P.: *K&E Log Log Duplex Vector Slide Rule No. 4083*. Keuffel & Esser. 1939.

Wells, ??: *Slide Rule Gauge Points*. 1916.

Whiblin, John.: *Description of the Line of Numbers*. Gunters Rule. 1669.

Willers, Adolf.: *Mathematische Instrumente*. Munich & Berlin, 1943.

Williams, Michael R.: *A History of Computing Technology*. Prentice Hall. 1985.

Wingate, Edmund.: *L'usage de la règle de proportion en l'arithmétique et géométrie*. A modification of a Gunter's rule. Paris, 1624.

Wingate, E.: *Construction, description et usage de la règle de proportion.* Paris, 1624.

Wingate, E.: *The use of the Rule of Proportion.* London, 1626, 1628, 1645, 1658, 1683.

Wingate, E.: *Arithmétique logarithmique.* Paris, 1626.

Wingate, E.: *Construction and Use of the Line of Proportion.* London, 1628.

Wingate, E.: *Of Natural and Artificial Arithmetic.* London, 1630.

Wingate, E. (editor): *Posthuma Fosteri: the Description of a Ruler, Upon which is described divers Scales, and the Uses thereof by Mr. Samuel Foster ... by which the most usual Propositions in Astronomie, Navigation, and Dialling, are facily performed.* R. & W. Leybourne. London, 1652.

Wolaston ??.: *Chemical slide Rule. Part 1.* 1814.

Wolf, R.: *Geschichte der Astronomie.* Munich, 1877.

Woollgar, J.W.: *The Mathematical Lines usually inserted in the Plain Scale, Sector, & Gunter's Scale.* Mechanics Magazine. 1827.

Woollgar, J.W.: *Description of an improved six-inch plotting and projecting scale, designed by J.W. Woollgar Esq.* 1828.

Worgan, John.: *A short Treatise of the Description of the Sector.* London, 1699.

Worgan, John.: *A new set of Tables calculated after a Plain, easy & Correct method ... being of great use to Surveyors, Carpenters, Masons.* London, 1740.

Wyman, Tom.: *The Thomas Dixon Engineer's Slide Rule.* Journal of the Oughtred Society Vol, 5, No, 2. October 1996. p 68.

Wyman, Tom.: *A Five-Piece Combination Rod and Slide Rule.* Journal of the Oughtred Society Vol, 6, No, 1. March 1997. pp 45-46.

Wynter, Harriet & Turner, Anthony.: *Scientific Instruments.* Studio Vista. London, 1975.

Yeldham, E.A.: *The Story of Reckoning in the Middle Ages.* Harrap & Co. 1926.

Young, Neville W.: *A complete slide rule manual.* West Publishing Corp Pty Ltd, Sydney, Australia. 1971. David & Charles. 1973.

Zühlke, M.: *Wirtschaftlich Rechnen, Nomographie, Rechentafeln und Sonderrechenstäbe.* Braunschweig, 1952.

Appendix 4.

Patents.

The study of slide rule patents allows one to see who was involved with rule development, identifies trends in the designs, and gives clear indications of the relative importance of the various elements at various times.

Patents and other forms of protection for inventions have been used for many hundreds of years. The earliest Letter of Protection in England was granted to John Kempe of Flanders in 1331 for the production of woollen cloth. The first patent grant for a new invention was granted by Henry VI, in 1449, to John of Utynam for the production of coloured glass as used in the Eton College chapel. The second was granted by Edward VI, in 1552, to Henry Smyth for the production of Normandy glass. In 1617 the Clerk of Letters Patent in the Court of Chancery started his docket book. This was used to number and record all patent applications until 1852, when the Patent Law Amendment Act was passed which changed the procedure. Bennet Woodcraft, who was appointed Superintendent of the Specifications at this time, compiled a series of indexes of the 14,359 patent applications granted between 1617 and 1852. These are an invaluable aid to identifying the dates of the earliest inventions.

One might hope that reviewing the Letters Patent granted in the United Kingdom by the Lord Chancellor's Office, starting in the year 1617, would offer an opportunity to trace the development of the slide rule from its earliest dates. However Woodcraft's, *Alphabetical Index of Patentees of Inventions, 1617 - 1852*, originally published in 1854 and reprinted in 1969, lists very few patents pertaining to slide rules, and even these are often difficult to identify as belonging specifically to slide rules. A number of Royal Letters Patent, which were apparently granted by various kings of England to slide rule inventors, are not listed by Woodcraft, e.g., the Letters Patent granted to Delamain by King Charles I. This is surprising since, later on, other slide rule designs claimed Royal Letter Patent protection, examples being Fearnley in 1881, and Watson in the early part of the 20th century.

One can only speculate as to why so few patents were granted in the first 100 years of slide rule development. A possible reason may be that one of the basic tenets of patent law is that "mathematical methods" are not patentable, and the slide rule may have been considered a "mathematical method". We do not know whether the first inventors even tried to patent the earliest slide rules and were stopped from doing so by this principle. On the other hand, Delamain was granted Letters Patent. The majority of patents appear to have been granted after 1852; the last, relating to a specialist circular slide rule, was in 1985.

A patent number on a slide rule, its case, or its instructions is a useful indicator of the date on which it could have been produced. It must be noted that the patent date serves only to indicate a date before which the particular patented item could not have been made. However, similar slide rules may have been produced earlier than the patent date and not marked with a patent number while waiting for the grant of the patent, and it was possible for the patent to have been issued for a number of years before the item was produced with the patent number marked on the rule. Also, as a patent lasted variously for 14, 16 or 20 years in Great Britain, this is not a particularly accurate method of dating, but does serve to indicate the "era" of the design.

It is worth noting some of the idiosyncrasies of the British patent numbering system, and the effects of changes thereto at the various times:

— From **1617** to **1852** one continuous sequence of numbers was used, the application number being the same as the specification (patent) number. The number thus gives some idea of age, and only a hundred or so patents were granted each year.

— From **1852** to **1916** a new sequence of numbers, starting at zero, was begun each year. This can cause some confusion as it is impossible to know what year a number was issued. The application and patent number stayed the same.

— From **1916** to **1979** a continuous series of numbers, starting at 100,001 was begun. The start number for each year is given in **Table 15.** Until the end of 1978, each patent application (provisional patent) consisted of a serial number followed by the final two digits of the year, e.g., 5123/77. Note that even though provisional patent numbers are shown on some slide rules, this is no guarantee that a full patent was issued, and in some cases the patent application may even have been withdrawn.

— From January 1st **1979** applications were numbered according to an internationally agreed system of seven digit numbers, the first two being the year e.g., 7900001.

There is some difficulty in correlating the dates shown on patent applications and the issued numbers as listed in **Table 15.** This probably arises from the delay between application and issue, the result of awaiting the completion of all searches, which in some cases may be as long as three years. The number may *initially* be issued in the year listed, but may *finally* be issued years later. In some cases the date is from the actual specification; in others it is deduced from the number and the table.

United States patent numbers work in a slightly different way. The first patents were issued in 1836, starting at number 1, and have followed on continuously since then. United States patents last for only 17 years. See **Table 12.** A list of the first patent number for each year up to 1988 is given in **Table 15.** Since 1927 American law has required the patent number to be placed on the item. A number of manufacturers did put the number on before this date, but it was by no means universal.

German patents - DR/DRP, Deutsches Reichs Patent - should not be confused with the German equivalent of registered trademark - DRGM, Deutsches Reichs Gebrauchsmuster, which were necessary, as patents could only be used to cover manufacturing techniques. **Table 15** shows DRP numbers to 1942; from 1942 until 1970 there is no obvious correlation between patent number and date. After 1970 "50" must be added to the first two digits of the patent number to get the year, e.g. patent number 33 14 234 is from 1983 (33 + 50 = 83). For the DRGM, after 1970 the first two digits of the DRGM number give the year of notification. There are over 100 German slide rule patents. See also **Table 13.**

French patents are marked Brevette S.G.D.G. 'Brévété' is the French for patented, and the S.G.D.G. stands for "Sans Guarantee de Gouvernement" - literally, without the support of the Government, though why this should be is not known. See **Table 14.**

See **Table 14** also for Swiss, Japanese, Dutch, Belgian and Canadian patent numbers and dates.

The patent numbers usually relate to design features on the rule or cursor, or an arrangement of the scales. In the following list of patents, those annotated with an asterisk (*) are not necessarily related particularly to slide rules, but to interesting mathematical instruments, methods of manufacture of similar types of instrument requiring similar accuracy, or to give an idea of the areas of science which were being explored in parallel with the development of slide rules. The hash sign (#) is used to annotate the first patent granted to a particular company in Great Britain.

TABLE 11 : BRITISH PATENT NUMBERS AND DATES.

Date	Name	Number	Subject
163?	Richard Delamain	??	By Royal Letters Patent. The Grammelogia or Mathematical Ring.
1731	William Bucknall	528*	Mathematical machines for improvement in astronomy and navigation.
1753	John Suxspeach	676	Instrument called "The Catholic Organon or Universal Sliding Foot Rule". This is the first slide rule patent for a combined sliding foot rule, telescope and quadrant.
1753	Jonathan Hulls	686	Machine for weighing gold coins etc, also a sliding rule for taking the contents of solids and superficials.
1780	John Dicas	1,259	Constructing hydrometers with sliding-rules to ascertain spirit strength. This is actually a line of numbers, 0 - 360, on a rod opposite to which is placed the strength of spirits. This shows the strength of spirits at any degree of heat.
1790	Estienne Leguin	1,753*	Instrument for calculating longitude.
1812	Henry Ewington	3,566*	Navigators sector.
1812	C.A. Schmalcalder	3,545*	Mathematical instruments.
1824	James Rogers	4,924	Instrument for ascertaining the cubic contents of standing timber.
1833	David Rowland	6,528*	Manufacture of sextants etc.
1842	Francis Marston	9,235*	Apparatus for making calculations.
1842	James Chesterman	9,214	Improved cattle gauge and tapes for measuring & boxes for same.
1843	David Isaac Wertheimber	9,616*	Calculating machine.
1845	William Piggott	10,625*	Mathematical, nautical and astronomical instruments.
1847	Jean Jayet	11,928*	Calculating machine.
1851	Charles DeColmer	13,504*	Calculating machine. The Arithmometer.
1857	A.V. Newton	1,121	Box tapes with different marks on the reverse to give cubes, diameter from circumference etc. for timber calculations. See also J.C. Watkins, 1887.
1858	L. Cunq	1,647	Provisional patent for a circular slide rule with four discs to calculate the volume of liquid flowing through an orifice.
1858	H. Bevan	2,770	Numeric discs in any order on a rod for calculation - not specifically a slide rule with logarithmic scales.
1860	J. Ingham & G. Hinchliffe	200	Circular slide rule specifically adapted to calculate the weight of different warps. Six circular scales, A to F, B&C move on one disc, the others are fixed.
1867	W.H. Barlow	3,506	Circular slide rule for railway calculations - not logarithmic scales.
1869	J.P. Nolan	1,566	Range finder and attached slide rule showing the distance of a target from the angle, and the time of flight to set the fuzes.
1870	J.W.S. Watkin (H.S. Watkin)	1,877	From the height of the observer and the angle of dip, range can be calculated from a special slide rule.
1871	H. Lefevre	1,665	Unusual special slide rule.
1876	H.E. Newton	4,310	A.E.M. Boucher watch calculator. The seminal patent.
1878	Prof. George Fuller	1,044	Fuller's invention of the helical scale tubular slide rule.
1878	Frederick H. Sheppard	3,059	Stanley rule (of Great Turnstile, Holborn), boxwood slide rule with 2 adjoining slides, reversible or otherwise with decimal or duodecimal scales for timber calculation. Improvements in stock design.
1881	J.B. Fearnley	863	Fearnley's Universal circular calculator. Quoted as "By Royal Letters Patent".
1881	E.L. Walford	2,875	Slide rule for money, goods, stocks and shares. Provisional patent not allowed.
1881	Capt. J. H. Thomson	5,540	Re-invention of log-log scales. "Calculating apparatus for raising a number to any power, fractional or integral ...". Patent shows to scales, a log scale against a log-of-the-log scale.
1882	A.J. Boult	2.119	G. Charpentier-Page slide rule and computing scale (Charpentier circular calculator). Provisional patent showing outer and inner disc between which is a moveable scale.
1883	A.J. Boult	278	Unusual calculator, not specifically slide rule, provisional patent not allowed.
1883	C.E. Kelway	3,539	Unusual calculator which gives the distance of an object from the angles subtended, for ship's use.
1884	Ganga Ram	2,149	Special slide rule for stability of retaining walls.

1885	W.R. Lake	831	Along one side of, or upon the back of the tape measures, or scale A, is printed a correspondng logarithmic scale B, so that calculations of cubical contents, weights of cattle and so on may be raedily made.
1885	A.P. Trotter	1,210	Calculator for wire area calculation from wire gauge.
1885	C. Perks	10,465	Calculating apparatus for problems in navigation.
1885	W.C. Carter	13,041	Boiler Joint Calculator. Two adjacent slides for calculating of steam boiler and other rivet work.
1885	E. Halsey	13,298	Multiplying, dividing, adding and so on for calculating taxes. A seties of slides depending on the application.
1886	J.C. Watkins	12,189	A measuring tape marked on one side with inches, and on the other side with figures showing the cubic content of a linear foot of timber having a "quarter girt" shown by the tape.
1886	G. Bousfield	12,806	Linear slide rule for the timber trade.
1887	N. Harrison	1,884	A navigational calculator using three bearing points to calculate.
1887	E.A.P. Burt	4,359	Burt's patent for a measuring and cubing slide rule, (timber, stone etc.).
1888	Sir. D.L. Solomons	3,615	Sir David Solomons' Photographic Ratio Rule. A slide rule specifically graduated to calculate the "rapidity" of photographic lenses.
1888	F. Hurter & V.C. Driffield	5,545	Actinometer, Hurter & Driffield Actinograph. Sole agents Marion & Co. Slide rule for calculating photographic exposures.
1888	G.H. Meire	6,227	Slide rule for calculating the ratio of live weight of an animal to the weight of dressed carcase.
1888	H. Cheesman	6,747	Rotating disc "nomogram" for different calculations.
1888	T. Mudd	10,311	Watch type calculator with a third "recording" hand.
1889	T.G. Baron & J.C. Harding	6,050	Steam engine indicator calculations with an extra slide to calculate horsepower.
1889	C.F. Martin	7,905	Measuring tape with special (logarithmic?) markings, which in a special telemeter (container?) allows the tape to be doubled over so as to be used as a slide rule.
1889	S.W. Cuttriss	8,285	Construction of computer scales on discs and rotary rings rather than on straight rules - Cuttriss Computing Disc?
1889	F. Drha	11,194	Circular slide rule for calculating interest and capital - not logarithmic.
1889	J.H. Young	17,153	Linear slide rule with two adjacent slides for calculation of timber, masonary, stonework etc.
1890	A.P. Trotter	1,302	Slide Rule for electrical calculations with 2 slides - "Wiring Slide Rule".
1890	A. Watkins	1,388	Tubular exposure time slide rule around a densinometer with an eyepiece.
1890	A. Honeysett & W.F. Stanley	4,803	Log scales which give the flow of liquid in pipes. Provided witha transparent courser (sic). Honeysett is well known for such calculations.
1890	F.G. Stone	9,333	Unusual military calcultor with non-logarithmic scales. Also 15,978 and 19,824.
1890	J.J.W Crufurd	10,314	Circular slide rule for railway cuttings and curve calculation.
1890	G.H. Meire	11,493	Two sided (non duplex?) metal slide rule indexs with chisel edges are formed from a pair of blanks by bending and interlocking. The whole is held together by an elastic band which does not press the unit against the stock.
1891	A.J. Watson	1,676	Military "cover" calculator using a two foot, two fold rule with a plotting table to calculate the cover provided by a berm.
1891	C.L. Hett	6,680	Linear A/B,C/D scales for horsepower and transmission belting size calculation.
1891	J. Smith	12,101	Slide rule for navigation: A = logcosec and height/angular distance; B = miles.
1891	H. Allison	16,939	Keuffel & Esser Co. Two sided slide rule.
1891	S. Milne	17,794	Slide rule with 2 adjacent slides. Patent never issued.
1892	W.P. Thompson (G. Meyer)	3,062	Ribbons on rollers allowing calculation of days and interest accrued.
1892	D.G. Prinsep	13,915	Circular slide rule for rangefinding and triangulation calculations.
1892	E. Frampton	14,358	Sophisticated circular slide rule for calculating wages. Uses a spiral scale.
1892	W.T. Goold	16,807	Design for a slide rule with a second slide fitted with pointers for use to mark intermediate values and constants.
1892	C. Manners	16,846	Logarithmic numbers on the reverse of a measuring tape.
1892	J.A. Scott	17,642	With the Britannia Works Co. Photographic exposure circular slide rule.
1892	L. Lord	22,210	Circular slide rule for calculation in the textile industry. Lord's Calculator.
1893	E.A.P. Burt	3,551	Slide rule No 101 for calculating timber and transport costs (patent abandoned).
1893	Thomas Wynne	10,617	Photographic slide rule with separate actinometer.

??	McPherson	24,616	A.G. Thornton, covers gauging rule in inches and millions of gallons.
1895	??	2,166	To determine the cost of woven fabric.
1895	C.N. Pickworth	#	
1896	??	???	Sexton's Omnimeter. The amount of water flowing over a gauge weir.
1896	R. Belshaw	3,281	Slide rule for the quantity in partly filled casks.
1896	M.J. Sheridan	14,623	Slide rule for the amount of water to reduce spirits.
1899	Robert H. Smith	17,357	Professor R.H. Smith's tubular calculator, sold by J.H. Steward.
1899	F.C. Farmar	19,288	An improved slide rule for computing and calculating purposes.
1899	F.C. Farmar	19,289	An improved dip rod.
1901	J.T. Ashworth	3,873	Imperial Exposure Meter (actinometer) with slide rule.
1901	F.C. Farmar	8,309	A slide rule for all calculation purposes in connection with wines and spirits.
1901	Prof. John Perry	23,236	Re-invention of log-log scales as on A.G. Thornton's Perry slide rule.
1902	R.W. Paul	??	Callender Cable Calculator.
1903	A.W. Faber	#10,230	Faber's earliest UK patent.
1903	Lt-Col. Anderson	9,095	Splitting of log-log scales into 4 portions.
1906	W.H. Harling	??	Calculex watch type slide rule.
1907	Seinen Yokota	18,218	Use of log e = 2.718281823 as the basis of log-log scales. John Davis and Sons, Derby, slide rules.
1908	F.C. Farmar	7,364	Improvements in, or connected with, slide rules.
1910	William Henry Fowler	#5,528	Fowler's first patent for a circular calculator with a centre knob.
1910	F. Pivora	13,852/10	"Stelfox" slide rule manufactured for Davis of Derby by Dennert & Pape.
1911	Dr. Rudolf Taussig	25,313	"Presto" calculator for discounts and interest.
1911	F. Pivora	25,314	Micrometer attachment to slide rules.
1911	Richardson	??	Engineer's slide rule.
1912	W.H. Lilly	28,603	Lilly's Improved Spiral Rule - spiral log-log scales.
1912	W.H. & Harold Fowler	20,416	Fowler's patent to improve accuracy of the calculators.
1913	R.F. Agnew	#22,165	For cloth manufacturers, also calculating the weight of paper and other materials.
1913	W.P. Thomson	10,753	Faber patent for frame design.
??	??	2,610	John Davis & Son, Smith-Davis Piecework and Premium calculators. These are probably the largest tubular slide rules made.
1914	W.H. & H. Fowler	3,638	First Fowler calculator with two knobs on the rim of the circular calculator.
1914	Harold Fowler	15,990	Single sided circular calculator with a single knob on the rear face, and a rotatable rim.
1917	Jiro Hemmi	107,562	Sun-Hemmi patent.
1920	E. Jones	164,442	Slide rule for proofing, blending and reducing wines and spirits.
1921	Otis Carter Formby King	22,119/21	Provisional patent for the Otis-King Calculator & monetary scales.
1921	Otis Carter Formby King	#183,723	Patent from provisional specification 22,119 to Otis-King calculator by Carbic Ltd.
1923	O.C.F. King	17,400/23	Provisional patent specification for improvements including several cylinders and 5 cursors.
1923	W.H. & H. Fowler	22,148/23	Provisional patent for 215,648. Improvement to the mechanism of a Fowler circular calculator.
1923	O.C.F. King	207,762	Patent from provisional specification 17,400 to Otis-King/Carbic.
1923	O.C.F. King	207,856	Patent for further improvements including multiple scales with apertures in the cursor for reading, also a desk version with interchangeable cylinders.
1924	W.H. & H. Fowler	215,648	Improvement to the mechanism of a Fowler circular calculator.
1925	L.E. Edwards	232,037	Patent slotted slide, John Davis & Son "Glider" slide rules.
1926	J. Halden & Co.	252,908	Patent on a scale rule.
1926	Addison/Luard	253,758	Addison Luard Course and Wind calculator, 8" circular slide rule made by Hemley Hughes Ltd.
1930	Lewis & Taylor Ltd	335,379	Gripalcalculator slide rule - belt tension.
1933	H. Boardman	#411,090	PIC Patent, differential scales.
1933	A.G. Thornton	413,112	Thornton magnifier fitted to a cursor.
1934	Dargue	#413,308	Dargue patent. Bowed spring on end of stock for constant tension.
1934	G.V. Martin	413,451	Celercertus rotary slide rule.
1936	C.J. Dussel & W. Pasveer	#443,689	ALRO (Holland) circular slide rule.

1937	Dr. A.E. Clayton	468,257	PIC Patent. Arrangement of Sin, Cos & Tan scales & inverse trig scales for electrical engineers.
1939	Unique	#500,237	On Unique Commercial slide rules.
1944	Ionic Labs	1407/44	Provisional patent applied for "Flik-O'-Disk" circular calculators. Not known if awarded.
1944	Unique	576,534	For double length scales (Unique 5-10 and 10-20 rules).
1945	Blundell	#	
1945	Unique	29,187/45	Provisional patent for the Unique Friel Sturdy slide rule.
1946	Unique	583,637	On Log-Log and Universal I slide rules for cursor design changes.
1948	Unique	596,635	Design of scales for income tax purposes.
1948	Blundell	602,286	Blundell (Luton) patent on a Blundell L6 slide rule.
1950	Blundell	644,944	Adjustable stock for slide rules. (also on a Heath "Hezzanith" slide rule).
1951	Blundell/GEC.	15,792/51	Vector Slide Rule by Blundell, probably designed by Dr. Eng E. Friedlander of the GEC Hirst Research Centre originally at Wembley, London.
1952	Blundell/GEC	713,655	Patent from provisional patent 15,792.
1952	Thornton	3470/52	Thornton-Mirless, shaft horsepower slide rule.
1953	Hall & Harding	722,148	Blundell "Magnex" cursor. A half-round lens to clip onto the normal cursor, supplied to fit all standard 10" Omega, Academy and Janus models.
1955	Thornton	754,938	PIC Patent for resilient inserts in stock.
1957	A.G. Thornton	878,056	Scales for quadratic equations - reciprocal scales
1957	A.G. Thornton	878,057	as 878,056, but with cubic equations.
1962	Pickett & Eckel	#912,878	Log scales.
1963	G.W.F. Sanguin	919,063	The Magnameta oil tonnage calcultor. A large linear slide rule.
1963	A.G. Thornton	944,634	Manufacture of injection moulded stock, held by moulded ends.
1967	A&V Kaye Ltd.	2,043/67	Decimal currency calculator.
1967?	Blundell-Harling	6,225/67	Blundell patent. On a 503D slide rule with monetary scales.
1968	Blundell-Harling/W.H. Smith	1,114,917	On a W.H. Smith A505 Master Duplex slide rule.
1969	A.& V. Kaye Ltd.	1,159,805	Issued patent from 2,043/67.
1969	Cosmo Rule	2,563/69	Series of conversion rules.
1978	Blundell-Harling	1,530,487	Portland dead reckoning calculator. A plane angle solver for sea and air navigation, simple disc calculator on the back face.
1985	E.H. Zimmerman	2,157,465	Disc calculator for loading on dowels - the last patent.
????	Thos. W. Watson	??	Watson's Patent Double slide rule (Watson of Newcastle).
????	? Golding	??	Horsepower calculator for Steam Engines (Griffin & Co).

TABLE 12 : AMERICAN PATENT NUMBERS AND DATES.

1851	John W. Nystrom	7,961	Nystrom's Calculating Rule - 8" circular slide rule.
1869	Elizur Wright	??	Arithmeter - a large spiral, cylindrical slide rule.
1878	Prof George Fuller	219,246	Fuller's Calculator - made by K&E?
1881?	??	242,591	Circular slide rule - Publishing Office New York.
1881	Edwin Thacher	249,117	United States Letters Patent for Thacher's Calculating Instrument, made by K&E.
1888	Walter Hart	??	Hart's Equationer or Universal Calculator.
1890	Walter Hart	??	Second patent for Hart's Equationer or Universal Calculator.
1891	William Cox	460,930	Duplex Engineers rule by K&E. Hans Dennert gives this as 460,940. Also seen as 460,430.
1891	William Cox	??	Engineers slide rule.
1894	E.P. Roberts	??	Slide rule for wiring calculations.
1894	J. Billeter	??	Cylindrical slide rule.
1894	T.H. Johnson	??	Slide rule design.
1895	Branch Colby	543,612	K&E slide rule. Civil engineering calculations.
1895	??	??	Sexton's Omnimeter - sold by Dietzgen.
1895	E. Hill	??	Slide rule design.
1898	W.L.E. Keuffel	603,695	K&E slide rule.

1898	John Given Mack	606,388	Prevention of slide rule jamming - split stock design. Dietzgen patent.
1898	Herbert Lutz	611,791	Patent for a combined slide rule & wire gauge for elctrical calculations.
1898	J.E.G. Yalden	??	Spiral slide rule design.
1899	W.L.E. Keuffel	621,348	K&E slide rule - Universal slide rule.
1900	W.L.E. Keuffel	651,142	K&E slide rule.
1900	Small & Co	??	"Small" Pocket Calculator - probably Roger Conant's patent, see following.
1900	Roger W. Conant	657,916	"Calculating Instrument". Small & Co. "Small" Pocket Calculator.
1900	William Cox	-	K&E patented adjustable slide rule (Mannheim).
1901	Edwin Thacher	677,817	Thacher-Schofield engineers slide rule. Patent assigned to Schofield.
1902	J.H.C. Dennert	694,258	Spring between two halves of stock. Dennert & Pape patent.
1902	Dennert & Pape	694,452	Celluloid flat spring bottom plate.
1902	Edwin Thacher	701,103	Thacher-Schofield specialised 24" slide rule - F/EE/F scales.
1903	W.H. Glaser	??	Textile slide rule design.
1904	C.G. Barth et al.	??	Machining slide rule.
1904	L.W. Rosenthal	767,170	Dietzgen's Multiplex slide rule. Design of reciprocal scales.
1904	Lt.Col. F.J. Anderson	??	Patent of 30.08.04 equivalent to Anderson's UK patent 9,095.
1904	? Powell	??	Rotating band slide rule.
1904	Elmer Ambrose Sperry	750,500	Logarithmic calculator, Sperry's watch type calculator.
1904	B.E. Winslow	??	Architects slide rule for steel beams.
1904	J.G. Zollman	??	Calculating machine.
1905	F.S. Beckett	??	Rotating band slide rule.
1905	F.S. Beckett	??	Pocket calculator (slide rule?).
1905	? Cooper	??	Rotating band slide rule design.
1905	F.C. Farmar	??	Wine and spirit slide rule.
1905	J. Wilkes	??	Logarithmic calculating apparatus.
1906	B.D. Coppage	??	Slide rule design.
1906	W.F. Doherty	??	Engineer's slide rule design.
1907	H.T. Hincks	??	Slide rule design.
1907	? Holinger	??	Rotating band slide rule design.
1908	O.L. Baumbach	819,237	Slide rule construction, assigned to Dietzgen.
1908	W.L.E. Keuffel	907,373	K&E slide rule. Adjustable duplex slide rule.
1908	??	996,039	K&E L-L rule - adjusting scales relative to each other.
1908	H. Grondahl	??	Computing slide rule.
1908	J.L. Hall	??	Structural slide rule.
1908	F.F. Nickel	??	Slide rule design (runnerless?).
1911	J.A. Ettler	??	Combined record book and slide rule.
1911	G.H. Gibson	??	Slide rule design.
1911	W.L.E. Keuffel	996,039	Magnifier for K&E slide rule.
1911	George Lange	1,012,660	Assigned to K&E - calculating device, Sperry watch calculator improvements.
1911	G.W. Richardson	-	Slide rule, Richardson's Direct Reading rule.
1911	G.W. Richardson	1-31-11	Provisional patent - not known if related to the two adjacent patents. On a type 1776 rule
1911	B.T. Steber	??	Slide rule design.
1912	G.W. Richardson	-	Slide rule, Richardson's Direct Reading rule.
1912	A.N. Lurie	??	Slide rule with cross-hatched pattern.
1912	D. Love	??	Train speed computer and indicator.
1912	L.A. Quivey	??	Instrument for obtaining averages.
1913	D.W. Brunton	??	Circular slide rule design.
1913	W.C. Colwell	??	Slide rule design.
1913	? Connio	??	Rotating band slide rule design.
1913	R.M. Hamby	??	Slide rule design.
1913	J.B. Moody	??	Interest rate calculator.
1913	J.A. Scheibli	??	Slide disc calculator.
1914	C.D. Allan	??	Design of slide rule.
1914	W.O. Phillips	??	Reciprocal scale for slide rules.
1914	H.A. Schwanger	??	Spring calculator.

1914	I.G. Sundell	??	Navigator's apparatus.
1914	H.W. Tomlinson	??	Slide rule design.
1915	George Lange	1,150,771	Assigned to K&E. Slide rule runner - K&E Frameless glass runner.
1915	C.F. Dieckmann	??	Slide rule design.
1915	C.F. Dieckmann	??	Open face runner for slide rule.
1915	F.C. Farmar	??	Slide rule design.
1915	C.A. Gilson	??	Gilson Slide Rule Co. Slide rule design.
1915?	Keuffel & Esser	8-17-15	Provisional patent on a Duplex 4088-3 slide rule.
1915	E.D.N. Schulte	??	Computing slide rule.
1915	J.M. Spitzglass	??	Calculating device.
1916	G.R.F.H Cuntz	1,168,059	Slide rule calculator design.
1916	W.L.E. Keuffel	1,181,672	Slide rule runner, the formed runner. K&E.
1917	H. Daemen-Schmid	??	Calculating machine (Rechenwalze?).
1918	G.W. Richardson	3-26-12	Cursor patent? Could also be 1911 or 1912. On a type 1776 rule.
1917	F.O.Stillman & H.M.Schleicher	??	Slide rule design.
1918	G. Small	??	Pocket calculator.
1919	W. Eastwood	??	Watch type slide rule design.
1919	Clair A. Gilson	??	Patent pending for the Gilson Midget slide rule, not known if awarded.
1919	S.B. Haskell	??	Automatic formula finder - unusual design.
1919	K. Horine	??	Computer (slide rule?).
1919	V.A. Mayer	??	Calculator design.
1920	C.J. Davidson	??	Slide rule design.
1920	? Nieman	??	Cylindrical slide rule design.
1920	Jiro Hemmi	1,329,902	Sun-HEMMI bamboo slide rule.
1921	R.M. Allen	??	Switching system and indicating device.
1921	E. Boehnke	??	Slide rule design.
1921	? Dicken	??	Rotating band slide rule design.
1921	C.J.R. Holle	??	Slide rule magnifier.
1921	H. O'Neill, S.H. Payne	??	Slide rule for steam boilers.
1921	R.A. Pierce	??	Slide rule design.
1921	F.O.Stillman & H.M.Schleicher	??	Slide rule design.
1922	Clair A. Gilson	??	Possible date for the patent on Gilson's Midget slide rule (see 1919). Information comes from an instruction book by Richardson & Clark.
1922	C.F. Dieckmann	??	Slide rule magnifier.
1922	? Dobson	??	Slide rule with multiple slides.
1922	E.T. Frankel	??	Slide rule design.
1922	G.R. Frickert	??	Belt calculator.
1922	C.A. Gilson	??	Gilson Slide Rule Co. Mechanical calculating device.
1922	W.L. Miller	??	Slide rule for cutting speed of folded sheet metal.
1922	R. Schiske	??	Kilometre per Litre calculator.
1922	H.R. Tidswell	??	Sex calculator.
1922	J.W. Wright	??	Calculating guy slide rule.
1922	J. Crompton & W. Gallagher	??	Calculating device.
1922	E.H. Gibb	??	Computing device.
1922	G.D. Johnson	??	Computing device.
1922	T. Min-Kao	??	Spiral slide rule design.
1922	H. Ritow	??	Duplex slide rule design (Frederick Post).
1922	A. Wolpert	??	Adding and subtracting device.
1923	C.F. Dieckmann	??	Slide rule runner.
1923	Olin D. Parsons	1,459,857	Slide rule - made by K&E? Also on a Sun-Hemmi bamboo/celluloid Duplex slide rule.
1923	F.W. Caldwell	??	Calculator design (slide rule?).
1923	J.F. Keller et al.	??	Circular slide rule design.
1924	A.F. Puchstein	1,487,805	Device for making vector calculations. (Hyperbolic scales).
1924	A.W. Keuffel	1,488,686	K&E LL0 scale, log-log duplex slide rule. See 1939? R.E. 20984. Re-issue of this patent.

277

1925	K. Becker	??	Slide rule design.
1925	C.S. Larkey	??	Design of improved scales.
1925	C.C. Moler	??	Power loss slide rule.
1925	K. Tsuda	??	Weight and stature slide rule.
1925	J.S. Wollks	??	Slide rule design.
1925	F. Wompner	??	Slide rule adjustments. (Faber).
1926	L.C. Bulmer et al.	??	Dairy slide rule design.
1926	G.M. Cook	??	Magnifying indicator design.
1926	J.E. Esbaugh	??	Slide rule design with inverted scales.
1926	S. Lee	??	Interest slide rule.
1926	H.A. McCarter	??	Slide rule indicator.
1926	A.F. Moyer	??	Balance slide rule.
1926	H. Ritow	??	Cylindrical slide rule (Frederick Post).
1926	J. Schauer	??	Slide rule pencil.
1926	R. Schwanda	??	Slide rule for folded sheet metal calculations.
1926	F.A. Sherrer et al.	??	Slide rule for Radium needles.
1926	R. Uhlich	??	Slide rule design.
1927	H. Ritow	??	Exponential slide rule (Frederick Post).
1930	Melville P. Hite	1,780,078	Circular slide rule - improved by Dussel and Pasveer.
1930	? Fuss	??	Rotating band slide rule design.
1931	J. Asper	??	Slide rule design.
1931	N. Shifner	??	Slide rule indicator.
1932	J.R. Dempster	1,849,058	Rotorule design. Dempster were manufacturers.
1932	J.R. Dempster	??	Rotorule design.
1932	A.W. Keuffel	1,875,927	K&E slide rule. Indicator design.
1932	John E. Clemons	1,969,939	Simplified Flight Calculator.
1932	A. Langsner	??	Slide rule indicator.
1932	M.O. Schur	??	Computing slide rule.
1932	H. Seehase	??	Flexible slide rule design.
1933	A.W. Keuffel	1,904,474	K&E slide rule. Indicator with "ears".
1933	A.W. Keuffel	1,930,852	K&E slide rule. On a Cooke radio slide rule and others. Slide rule adjustments.
1933	A.W. Keuffel	1,934,232	K&E for slide rule construction and adjustment.
1933	A. Langsner	??	Industrial slide rule. Dietzgen.
1933	? Summers	??	Tape slide rule.
1934	? Summers	??	Rotating band slide rule.
1934	F. Wompner	??	Slide rule design.
1935	A.W. Keuffel	2,000,337	K&E slide rule case.
1935	J.R. Dempster	1,868,816	Rotorule, J.R. Dempster, manufacturer
1935	C. Norris	??	Slide rule indicator.
1936	H.M. Jensen	2,026,537	Flight Computer made by Goble Aircraft Specialities.
1936	G.D. Shaeffer	??	Slide rule indicator.
1936	A. Svensson	??	Combination drafting and computing slide rule.
1937	E.L. Bowles	??	Photographic slide rule.
1937	R.D. Groch	??	Slide rule design.
1937	Hisaki Okura	2,079,464	Gudermanian scales for Hyperbolic Functions. Hemmi & K&E.
(1937)	(K&E)	(2,086,205)	Wrong patent number shown on K&E 9068 DORIC, see following.
1937	A.W. Keuffel	2,086,502	K&E Improved glass runner. (Has been shown as 2,086,205).
1937	Philip Dalton	2,097,116	Plotting & Computing Device. Dalton's Flight Computer. Weems of Annapolis, Maryland; also Jeppeson and K&E.
1937	F.E. Proal	??	Slide rule for gear calculations.
1937	? Von Forster	??	Rotating band slide rule design.
1938	C.J. Dussel	2,117,155	ALRO Circular Slide Rule. Made by Dussel & Pasveer
1938	A.W. Keuffel	2,136,169	K&E - Indicator for slide rule.
1938	P.B. Poole	2,138,879	Slide rule design.
1938	? Sinitzen-White	??	Cylindrical slide rule.
1938	J.G. Valenzuela	??	Financial slide rule.

1939?	A.W. Keuffel	RE20,984	Reissue of 1,488,686 for LL0 scale.
1939	A.W. Keuffel	2,143,559	K&E slide rule. Laminated manufacture.
1939	C.M. Bernegau	2,168,056	Colouring and slanting of scale numbers for ease of use.
1939	H. Gilmore	??	Slide rule design ("endless").
1939	Kells, Kern & Bland	2,170,144	Pickett patent, assigned to K&E, LL Duplex scales, Trig layout.
1939	C. Kubler	??	Slide rule design with Addiator.
1939	? Ludecke	??	Rotating band slide rule.
1939	G.H. Morse	??	Decimal point slide rule.
1939	P.M. van Doormal	??	Penetration index slide rule.
1940	C.P. Berg	??	Estimating slide rule.
1940	B. Ferrughelli	??	Proportional slide rule design.
1941	C.H. Hansen	??	Span clearance slide rule.
1941	Eimei Hirano	2,235,106	Hemmi Improved Slide Rule.
1941	F.W. Johnson et al.	??	Matrix slide rule.
1941	E.A. Ravenscroft	??	Resettable slide rule.
1942	W. Daemen	??	Circular slide rule.
1942	? Fiala	??	Rotating band slide rule design.
1942	John Tyler et al.	2,283,473	K&E slide rule.
1942	Kells, Kern & Bland	2,285,722	K&E patent, log-log duplex scales.
1942	R. Montgomery	??	Braille slide rule.
1943	F.W. Lang	??	Slide rule for wall patterns and areas.
1944	M.C. Andersen	??	Slide rule indicator.
1944	O.E. Batori	??	Circular slide rule.
1945	J.E. Dietzgen	??	Slide rule design.
1945	T.J. Setera	??	Slide rule indicator.
1945	G.L. Tilbrook	??	Sterility and fertility slide rule.
1946	August H.Krelling	2,407,338	Full vision indicator, assigned to Dietzgen.
1946	S. Lerner	??	Design of slide rule pencil.
1946	A.L. Snedaker	??	Slide rule indicator.
1947	O.E. Batori	??	Circular trig slide rule.
1947	J.R. Bland	2,422,649	K&E slide rule. Disclaimer filed in 1949 allows others to use the log-log principle.
1947	J.J. Smidl	??	Slide rule indicator.
1948	O.E. Batori	??	Circular spherical trig slide rule.
1948	G.W. Sammons	??	Moveable tape slide rule.
1948	? Tellander	??	Rotating band slide rule.
1948	G.F. Wittgenstein	??	Circular slide rule design.
1948	K.R. Wood	??	Statistical slide rule.
1949	J.G. Adiletta	??	Design of slide rule end-lifts.
1949	A.F. Eckel	??	Pickett & Eckel, slide rule design.
1949	H.H. Goldstaub	??	Design of a low cost slide rule.
1949	F.L. Koenig	??	Design of architect's scale and slide rule.
1949	H.S. Lieberman	??	Use of fine adjusting wheels within a slide rule design.
1949	E. Pignone	??	Slide rule indicator design.
1949	J.J. Smidl	??	Slide rule attachment.
1949	O.A. Wiberg	??	Slide rule indicator.
1950	A. Green	??	Sterility slide rule.
1950	Herschel Hunt	2,500,460	K&E slide rule.
1950	B.D. Kahan	??	Cylindrical, telescoping slide rule design.
1950	Ray Lahr	2,528,518	Computer, Air Navigation. J.B. Carrol Co, Chicago; also Jeppeson and Felsenthal.
1950	J.R. Bland	2,534,695	K&E slide rule.
1950	N.F. Dewar	??	Weight slide rule.
1950	Y. Fujisaki	??	Circular slide rule design.
1950	M. Hertig	??	Bowling spot location slide rule.
1950	F.J.A. Huber	??	True air-speed slide rule.

1950	C.A. Posson	??	Slide rule pencil design.
1950	P. Sbernodori	??	Unit conversion slide rule.
1950	J.L. Taaffe	??	Rotating band slide rule.
1951	L.A. Warner	2,545,935	Warner Plotter Computer.
1951	R.G. Bock	??	Productivity and quality slide rule.
1951	D.L. Cole	??	Target range slide rule design.
1951	Clarence P. Davey	2,634,912	Framed full vision indicator, assigned to Dietzgen.
1951	P.E. Gaire	??	Slide rule magnifier.
1951	K.H. Hachmuth	??	Spiral scale slide rule.
1951	C. Solomon	??	Slide rule for fur and pelts.
1951	C.A. Steinkoenig	??	Dividend slide rule.
1951	R. Tarshis	??	Rotating band slide rule.
1951	J. Telasco	??	Musical slide rule.
1952	W.J. Boos	??	ACU Rule. Slide rule design.
1952	M.J. Quillinan	??	Cylindrical slide rule design.
1952	E.S. Russell	??	Circular slide rule design.
1953	J.M.A. Allais	??	Circular slide rule decimal point indicator.
1953	P.E. Gaire	??	Slide rule indicator.
1953	M.J. Quillinan	??	Cylindrical slide rule design.
1953	M.J. Quillinan	??	Cylindrical slide rule design.
1953	M.J. Quillinan	??	Triangular slide rule design.
1953	M.J. Quillinan	??	Cylindrical slide rule design.
1953	H.A. Sawyer Jnr.	??	Circular slide rule.
1953	G.R. Stibitz	??	Matrix slide rule.
1953	P.J. Toien	??	Rotating cylinders slide rule design.
1956	Ray Lahr	2,767,919	Computer, Air Navigation. J.B. Carrol Co, Chicago; also Jeppeson and Felsenthal.
1956	Ray Lahr	2,775,404	Computer, Dalton Air Navigation, Jeppeson.
1956	F.J.A. Huber	??	True air-speed slide rule.
1957	Vari-Vue	2,799,938	Decimal Equivalents Converter.
1957	Vari-Vue	2,815,310	Decimal Equivalents Converter.
1959	T.P. Faulconer	??	The Caliputer, a calliper slide rule.
1963	Telex	3,112,875	Telex Communications Inc. Flight Computer.
1964	Pickett & Eckel	3,120,342	P&E Electronic, Model N515T slide rule.
1964	L.A. Warner	3,131,858	Warner Instruments, Chicago. Flight Computer.
1964	D.L. Geiger	??	Electronics slide rule design.
1965	I. Ezopov	??	Circular slide rule with enlarged scales.
1971	A.W. Learned et al.	??	Slide rule design.
1972	A.W. Learned et al.	??	Slide rule design.
1972	J.H. Jeffries	??	Circular slide rule design.

TABLE 13 : GERMAN PATENT NUMBERS AND DATES.

1883	M. Kloth	26.695	Apparatus for calculating - slide rule.
1886	Dennert & Pape	34,583	Patent for veneering white celluloid to mahogany (wood) stocks.
1892	F. Blanc	??	Use of e = 2.718281823 as base for log-log scales.
1898	John Given Davis Mack	102,599	Equivalent to US patent 606,388. 28.6.1898. Use of spiral springs to draw two halves of stock together and provide tension to slide.
1901	Dennert and Pape	126,499	Equivalent to US patent 694,258. 25.2.1902. Use of a metal or other flexible bottom to join the two halves of stock.
1903	Dennert and Pape	192,052	Patent covering adjustment screws to regulate slide movement.
1905	Albert Nestler	173,660	Use of thin layers of rubber between base and sides of stock. - also F-C?
1907	Robert Nelting	207,234	Two slides, interchangeable and reversible, read via windows within stock, no slide rules manufactured to this patent.

1908	A.W. Faber	206,428	Use of brass strips within stock and slide to counter warping. (also on DIWA rule).
1909	A.W. Faber	215,722	Not known - on Faber 377 slide rule.
1910	A.W. Faber	222,297	Not known - on Faber 363 slide rule.
1912	Friedrich Gerwin	269,175	Slide rule which can be used as a drafting scale with changeable triangular scales.
1920	Rudolf Suchoparek	343,545	Gear-wheel selector data slide. Transferred to IWA.
1921	Joseph Schwanda	385,681	U shaped tin-plate slide rules.
1922	A.W. Faber	365,637	Use of a nut and bolt (screw) mechanism on a leaf spring to adjust spacing between two halves of stock. (also on DIWA slide rule).
1922	Léon Appoullot	418,294	Appoullot circular slide rule.
1923	Dr. Ing. Hans Seehase	394,337	Two rectangular bars and a slide within a transparent case.
1923	J.B. Soellner Nachf	397,777	Company "Reiss" (Reiszeugfabrik A.G. Nurnberg) patent. Use of double tongue and groove between stock and slide.
1924	Albert Nestler	410,565	Incorporation of brass rings between halves of the stock.
1925	Carbic (UK)	418,669	As UK patent 207,856 for Otis-King calculators.
1925	Carbic (UK)	418,814	As UK patent 207,762 for Otis-King calculators.
1926	Georg Welsch	450,305	Double length hinged slide as "Stelfox" slide rule.
1930	IWA	488,264	Tabellenschieber - data slides.
1930	Dr. Ing. Hans Seehase	499,260	Improvement to 394,337 to allow more flexibility in the case.
1932	A.W. Faber-Castell	559,664	Synthetic slide rule manufacture, update of 206,428, use of bolts to brass inserts, may not have manufactured any rules to this patent.
1933	A.W. Faber-Castell	574,594	Synthetic rules using a single adjusting bolt on bottom of stock.
1934	A.W. Faber-Castell	596,286	Improvement to 574,594 to incorporate double bolts. Not used in production.
1937	Dussel & Pasveer	643,571	ALRO circular slide rule (may be 648,571).
1936	Carl Kübler	655,353	Faber-Castell patent covering "Addiator" plus slide rule.
??	A.W. Faber-Castell	821,840	Not known - on 57/62 slide rule (DBP).
??	A.W. Faber-Castell	832,084	(From the book by Ralph Stranger).
??	Harald Bachman	873,455	A.W. Faber-Castell patent. On 2/82 slide rule. (DBP).
??	A.W. Faber-Castell?	881,735	(from slide rule).
1952	Wilhelm Stahl	917,215	Albert Nestler patent, metal rings incorporated in synthetic material. (DBP).
??	A.W. Faber-Castell	1,018,659	Not known - on 3/31 slide rule. (DBP).
1952	W. Bülow	1,062,042	Faber-Castell patent covering "Addiator" in a "window" as a method of fixing onto a plastic slide rule.
1962	Harald Bachman	1,135,686	A.W. Faber-Castell patent, improvement of 873,455 with leaf springs between slide and sides of stock.
1964	Heinrich Tegtmayer	1,170,170	A.W. Faber-Castell patent covering capillary grooves in the tongue of slide on synthetic slide rules.
1968	Karl Remers	1,274,826	Aristo Werke (Dennert & Pape), double diverging tongues into single groove.
1971	IWA	??	"Planetengetriebe" - planetary drive for IWAMATIC 1650 circular slide rule.
1971	Dr-Ing Winfried Lück	??	IWA. Vötch slide rule for climatic scales, Hygromatic data. IWAMATIC circular slide rule.
1971	IWA	2,254,387	Plastic moulded guide rails for data slides.
1981	IWA	3,148,355	Improvements to data slide design.

TABLE 14 : SWISS, JAPANESE, FRENCH, DUTCH & BELGIAN PATENT NUMBERS AND DATES

Swiss Patents and dates.

????	Julius Billeter	43,463	Metal/glass slide rule. Rechentafel.
1916	Heinrich Daemen-Schmid	71,475	Transparent case to slide rule.
1918	Heinrich Daemen-Schmid	77,126	Cardboard bound with tin strips. H D-S appears for LOGA circular calculator.
1931	Raymond Cartier	146,298	U shaped tin plate stock with thickened slide.
1949	Heinz Herbert Goldstaub	262,319	Further transparent case for stocks with slide charts.
1957	Anton Rothenfluh	320,811	Upper transparent slide above normal second slide.

Japanese Patent Numbers and Dates.

1912.11 May	Jiro Hemmi	22,129	Sun Hemmi slide rule - for body design.
??	Jiro Hemmi	58,115	Cursor design.
??	??	68,116	Sun-Hemmi; horseshoe shaped metal frame with glass cursor.
196?	??	381,371	Concise circular slide rule patent (Japanese) (No 28)
196?	??	479,446	Concise circular slide rule (No 32).

French Patent Numbers and Dates.

1876	Boucher	114,520	Boucher watch calculator.
??	Denis	461,479	Patent for circular slide`rule, improved by Dussel & Pasveer.
1921	Léon Appoullot	547,879	Circular slide rule.
1937	Dussel & Pasveer	808,295	ALRO Circular Slide Rule.
????	Faber-Castell	965,466	Faber-Castell 2/66 "Demegraph" slide rule by Schirmer

Dutch Patent Numbers and Dates.

| 1937 | Dussel & Pasveer | 41,324 | ALRO Circular Slide Rule. |
| 1986 | ?? | 0142,892 | MegaPromotions Radio Netherlands calculator. |

Belgian Patent Numbers and Dates.

| 1937 | Dussel & Pasveer | 413,897 | ALRO Circular Slide Rule. |
| 1963? | Telex | 648,698 | Telex Communications Inc, Flight Computer. |

Canadian Patent Numbers and Dates.

| 1963 | Telex | 709,263 | Telex Communications Inc, Flight Computer. |

TABLE 15 : UK, USA & GERMAN PATENT NUMBERS BY YEAR.

The following table lists the first patent number in the UK, USA and Germany for the indicated year:

Year	UK	USA	Germany	Year	UK	USA	Germany
1836	-	1-	-	1913	-	1,049,326	255,201
1837	-	110	-	1914	-	1,083,267	268,801
1838	-	546	-	1915	-	1,123,212	281,50
1839	-	1,061	-	1916	100,001	1,166,419	289,551
1840	-	1,46-	-	1917	102,812	1,210,389	295,851
1841	-	1,923	-	1918	112,131	1,251,458	302,851
1842	-	2,413	-	1919	121,611	1,290,027	310,001
1843	-	2,901	-	1920	136,852	1,326,899	317,801
1844	-	3,395	-	1921	155,801	1,364,063	331,101
1845	-	3,873	-	1922	173,241	1,401,948	346,501
1846	-	4,348	-	1923	190,732	1,440,362	366,201
1847	-	4,914	-	1924	208,731	1,478,996	387,501
1848	-	5,409	-	1925	226,571	1,521,590	407,751
1849	-	5,993	-	1926	244,801	1,568,040	423,351
1850	-	6,891	-	1927	263,501	1,612,790	439,051
1851	-	7,865	-	1928	282,701	1,654,521	454,201
1852	-	8,622	-	1929	302,941	1,696,897	470,001
1853	-	9,512	-	1930	323,171	1,742,181	488,501
1854	-	10,358	-	1931	340,324	1,787,424	516,551
1855	-	12,117	-	1932	364,154	1,941,449	541,001
1856	-	14,009	-	1933	385,638	1,985,878	567,401
1857	-	16,324	-	1934	403,718	1,941,449	590,401

Year	UK	USA	Germany	Year	UK	USA	Germany
1858	-	19,010	-	1935	421,827	1,985,878	607,551
1859	-	22,477	-	1936	440,484	2,026,516	623,701
1860	-	26,642	-	1937	459,084	2,066,309	640,401
1861	-	31,005	-	1938	477,516	2,104,004	654,901
1862	-	34,045	-	1939	498,137	2,142,080	669,651
1863	-	37,266	-	1940	516,338	2,185,170	687,301
1864	-	41,047	-	1941	531,239	2,227,418	700,851
1865	-	45,085	-	1942	542,237	2,268,540	715,351)
							729,850)
1866	-	51,784	-	1943	550,279	2,307,007	Note3
1867	-	60,658	-	1944	558,350	2,338,081	"
1868	-	72,959	-	1945	566,451	2,366,154	-
1869	-	85,503	-	1946	574,319	2,391,856	-
1870	-	98,460	-	1947	583,835	2,413,675	-
1871	-	110,617	-	1948	596,286	2,433,824	-
1872	-	122,304	-	1949	615,104	2,457,797	-
1873	-	134,504	-	1950	634,001	2,492,941	-
1874	-	146,120	-	1951	648,181	2,536,016	-
1875	-	158,350	-	1952	663,941	2,580,379	-
1876	-	171,641	-	1953	685,361	2,624,046	-
1877	-	158,813	1	1954	701,181	2,664,562	-
1878	-	198,733	3,001	1955	721,191	2,698,434	-
1879	-	211,078	5,001	1956	742,701	2,728,913	-
1880	-	223,211	8,501	1957	764,681	2,775,762	-
1881	-	236,137	12,001	1958	788,351	2,818,567	-
1882	-	251,685	16,551	1959	806,871	2,866,973	-
1883	-	269,820	20,451	1960	826,321	2,919,443	-
1884	-	291,016	25,401	1961	857,581	2,966,681	-
1885	Note1.	310,163	30,001	1962	885,891	3,015,103	-
1886	-	333,494	34,101	1963	914,251	3,070,801	-
1887	-	355,291	38,001	1964	945,444	3,116,487	-
1888	-	375,720	41,901	1965	978,901	3,163,865	-
1889	-	395,305	45,651	1966	1,015,491	3,227,729	-
1890	-	418,665	50,101	1967	1,053,401	3,295,143	-
1891	-	443,987	55,051	1968	1,097,211	3,360,800	-
1892	-	466,315	60,451	1969	1,138,301	3,419,907	-
1893	-	488,976	66,401	1970	1,175,851	3,487,470	Note4.
1894	-	511,744	72,651	1971	1,217,901	3,551,909	"
1895	-	531,619	78,901				
1896	-	552,502	84,751	1972	1,258,951	3,631,539	-
1897	-	574,369	90,051	1973	1,301,601	3,707,729	-
1898	-	596,467	95,551	1974	1,342,201	3,781,914	-
1899	-	616,871	100,851	1975	1,378,941	3,858,241	-
1900	-	640,167	107,551	1976	1,419,981	3,930,271	-
1901	-	664,827	116,551	1977	1,460,301	4,000,520	-
1902	-	690,385	126,751	1978	1,496,751	4,065,812	-
1903	-	717,521	137,901	1979	1,537,581)	4,131,952	-
					2,000,001)		
1904	-	748,567	147,201	1980	Note2.	4,180,167	-
1905	-	778,834	157,601	1981	-	4,242,757	-
1906	-	808,618	166,801	1982	-	4,308,622	-
1907	-	839,799	180,001	1983	-	4,366,579	-
1908	-	875,679	193,701	1984	-	4,423,523	-

Year	UK	USA	Germany	Year	UK	USA	Germany
1909	-	908,436	205,451	1985	-	4,490,855	-
1910	-	945,010	217,501	1986	-	4,562,596	-
1911	-	980,178	229,651	1987	-	4,645,548	-
1912	-	1,013,095	242,251	1988	-	4,716,594	-

Notes:

1. After the 1885 Act of Parliament
2. From 1979, when patent numbering changed
3. From 1943 to 1970 there is no obvious correlation between year and German patent number
4. After 1970, it is necessary to add "50" to the first two digits of the patent number to arrive at the year of patent issue, e.g., 3315234 was issued in 1983 (33+50=83)

Appendix 5.

Scales and Gauge Points.

Table 16 : Slide Rule Scales.

Symbol.	Maths. Relationship.	Description.
a	$1/x$	As CI, reciprocal scale.
a_{le}	-	Scale for yarn twisting on textile rule.
a_e	CF/DF	Equivalent to DF/CF on textile rules.
a_m	C/D	Equivalent to C/D, but used on textile rules.
A	x^2	Square of D.
AI	$1/x^2$	Reciprocal of square of D.
b	x	Used on some French rules, as C on most rules.
b^2	x^2	Used on some French rules, as B on most rules.
B	x^2	As A (on slide).
B^2	x^2	Used on some French rules, as A on most rules.
B^3	x^3	Used on some French rules, as K on most rules.
BI	$1/x^2$	As AI (on slide).
c^2	$\cos^2$	Cos squared, used in tacheometric (stadia) calculations.
C	x	Principal scale (on slide).
C2,D2	fx	Shifted scales, different functions used.
Cap	Farads	Capacitance, used in electrical and radio calculations.
CF	πx	Folded principal (on slide).
Ch	$\coth^{-1}x$	Hyperbolic cotangent.
CI or R	$1/x$, or $10/x$	Reciprocal of principal (on slide).
CIF	$1/\pi x$ or $10/\pi x$	Reciprocal of folded principal (on slide).
Cos	Cos x	Cosine x.
dB	decibels	Attenuation for electrical calculations.
D	x	Principal scale.
D2	fx	Shifted scale, different functions used.
DF	πx	Folded principal scale.
DI	$1/x$ or $10/x$	Reciprocal of folded principal scale.
DIF	$1/\pi x$	Reciprocal of folded principal.
Dyn	Efficiency	Efficiency scale for Generators.
E	e^x	Log-log scales, see also LL.
Eff	Efficiency	Efficiency scale for Generators and Motors.
f	Freq	Frequency, electrical and radio calculations.
Fund	n/a	Used in some catalogues for the standard suite of A/B,C/D scales.
GΦ	Hyp°	"Gudermanian" scale, gives SinhΦ on T, and TanhΦ on P.
H	2D	A double scale, i.e. x2.
H1	$\sqrt{(1+x^2)}$	Hyperbolic scale, range 1.005 to 1.5.
H2	$\sqrt{(1+x^2)}$	Hyperbolic scale, range 1.4 to 10.
Ind	Henrys	Inductance, used in electrical and radio calculations.
IL	%	Percentage scale.
IR	%	Percentage scale.
ISd	$X/\sin^{-1}X$	Inverse sine differential scale. (Thornton pat.)
ITd	$X/\tan^{-1}X$	Inverse tangent differential scale. (Thornton pat.)
K	12D	Unusual use of K for monetary scales on Unique financial rules, see also M, U and V.
K	x^3	Cube of D.
K_n	f(t/c)	For distance of zero line in steel/concrete slide rule.
K_d	f(t/c)	For effective depth of section in steel/concrete slide rule.

K_{A_t}	f(t/c)	For cross-section of steel in steel/concrete slide rule.
KZ	-	Fundamental scale on the body, folded at 360.
L	$\log_{10}x$	Mantissa of the common logarithm of D.
LL	ln	"Lower" log-log scale.
LL0	$e^{0.001x}$	Exponent of $x.10^{-3}$
LL1 or ZZ1	$e^{0.01x}$	Exponent of $x.10^{-2}$
LL2 or ZZ2	$e^{0.1x}$	Exponent of $x.10^{-1}$
LL3 or ZZ3	e^x	Exponent of x.
LL00 or LL/0	$e^{-0.001x}$	Exponent of $-x.10^{-3}$
LL01 or LL/1	$e^{-0.01x}$	Exponent of $-x.10^{-2}$
LL02 or LL/2	$e^{-0.1x}$	Exponent of $-x.10^{-1}$
LL03 or LL/3	e^{-x}	Exponent of -x.
LU	ln	"Upper" log-log scale.
mV	milli-volts	For electrical calculations.
M	log x	Mantissa of logarithm to base 10. Normal definition.
M	-	Scale of conversion-factor marks for US/British weights and measures.
M1,M2,M3,M4	£.s.d.	Monetary scales.
M	D	Principal scale.
Mot	Motor	Motor efficiency.
N		Discounts, various.
N_1, N_2	-	Scales for yarn numbers on textile rules.
Ne_1, Ne_2, Ne_3	-	Scales for yarn twisting on textile rules.
Ne_B,Ne_W, Ne_K, Ne_L	-	Scales for yarn numbers on textile rules.
n	x	As C or D.
n^2	x^2	As A or B.
n^3	x^3	As K.
neper	attenuation	For electrical calculations.
P	$\sqrt{(1-(0.1x)^2)}$	Cosine of $\sin^{-1}$D. Pythagorean scale.
P	1/T	Reciprocal of T1.
P_0/P_1	1/T1	Commercial, reciprocal of T1.
P_2	1/T2	Commercial, reciprocal of T2.
P1	$\sqrt{(1-x^2)}$	0.995 to 0.
P2	$\sqrt{(1-x^2)}$	0.99995 to 0.995.
Ps or P	$\sqrt{(1-x^2)}$	Cosine of $\sin^{-1}$D.
Ps	$\sqrt{(1-s^2)}$	Vector analysis scale. (Thornton etc.)
Pt	$\sqrt{(x^2-1)}$	Tangent of $\cos^{-1}$(1/D).
Pt	$\sqrt{(1+t^2)}$	Vector analysis scale. (Thornton etc.)
Q	$\cos\Phi$	Used with Φ or $R\Phi$ and for $Z=R+jX$ with some P scales. 1 - 10.
Q'	$\cos\Phi$	Extension of Q from 10 - 14.13.
Q1,Q2		20" scales.
R	1/x	Reciprocal of principal, see CI.
R1 = W1' = W1	$\sqrt{x}$	1 to 3.2
R2 = W2' = W2	$\sqrt{10x}$	3 to 10
$R\Phi$	rads.	Angle in radians. 0 -1.5+ radians.
Recip	n/a	Used in some catalogues for the standard reciprocal scales.
sc	sin.cos	Sine x cos, used in tacheometric (stadia) calculations.
S	$\sin^{-1}x$	D is the sine of angle S
S1	$\sin^{-1}x$	For sines of angles 5^0 45' to 90^0.
S2	$\sin^{-1}x$	For sines of angles 35' to 5^0 45'.
Sd	$\alpha/\sin\alpha$	Sine differential scale, for sine range 0^0 to 90^0. (Thornton pat.)
Sh1	$\sinh^{-1}x$	Hyperbolic sine of angles S (0.1 - 0.9)
Sh2	$\sinh^{-1}x$	Hyperbolic sine of angles S (0.85 - 1 - 3.0)
S&T	sinx,tanx	Sine & tan for small angles, 0.58^0 to 5.73^0, same as ST. Used on French rules.
ST	sinx,tanx	Sine & tan for small angles, 0.58^0 to 5.73^0, also known as arc x.
ST	$180x/\pi$	D radians in degrees.

Sinh1	sinh⁻¹x	Hyperbolic sine of Sinh1 is D between 0.1 and 1.
Sinh2	sinh⁻¹x	As Sinh1, but D between 1 and 10.
T	tan⁻¹x	D is the tangent of angle T. (5.5⁰ - 45⁰)(also Cot T)
T/1"	-	Scale for yarn twisting on textile rule.
tex	CI	Equivalent to CI on textile rules.
Td	-	Scale for yarn number on textile rule.
Td	α/tanα	Tangent differential scale, for tangent range 0⁰ to 60⁰. (Thornton pat.)
T1	tan⁻¹x	Tangent of angle T. (5⁰ - 45⁰)(also Cot T)
T2	tan⁻¹x	Tangent of angle T. (41⁰ - 85⁰)(also Cot T)
T1	-	Fundamental scale on slide.
T2	-	Fundamental scale on slide, folded at 360.
Th	tanh⁻¹x	D is the hyperbolic tangent of Tanh 1.
Trig	n/a	Used in some catalogues for the standard Suite of Trig scales, S,T,ST.
U	D/12	12U=D=M
U_1	as K	Cube scales in electrical calculations.
U_2	as A	Square scales in electrical calculations.
V	6U or D/2	See U and D.
Vd	Volt drop	On Electro slide rules for volt drop in conductors.
V	Volts	For electrical calculations.
Volts	Volts	Volt drop scales for motor & dynamo calculations.
W_1, W_1'	Root scales	Root scales from 1 to 3.16.
W_2, W_2'	Root scales	Root scales from 3.16 to 10.
Z		Fundamental scale on rule body.
ZZ1	)	
ZZ2	)	Scales for compound interest factors.
ZZ3	)	
%	%x	Percentage calculations.
£.s.d.	money	Calculations in English pre-decimal currency.
£	Pounds sterling	Calculations in English pre-decimal pounds.
s.d.	shillings. pence.	Calculations in English pre-decimal shillings and pence.
Φ	⁰ degrees	Angle in degrees. 0⁰ - 90⁰.
Π	1/f	Wavelength, electrical/radio calculations.

Note that the C2/D2 scales are sometimes "folded" or shifted at √10, others are folded at various values including e, 3.6 (Faber-Castell Business rules) or other values, the attendant CIF scale is thus also folded at the same place. The CF and DF scales are traditionally folded at π.

Log-log scales are based on relatively arbitrary bases to work the terms of $\log_a[\log_a x]$. The value selected for a is a balance between range and resolution. Some examples of a are as follows:

a	Maker	Comment
2	John Davis & Son	poor range, excellent resolution.
3.08	A.W. Faber	
2.9	A.W. Faber	earlier design.
2.7183	Keuffel & Esser	using e has many benefits as well as range and resolution.
10	Roget Log-log	huge range, poor resolution.

287

Table 17 : Slide Rule Gauge Marks

Extra information to assist in certain types of calculation is marked on the scales of all slide rules. This is in the form of the value of certain constants which can speed up a calculation. There are literally dozens of such useful constants, and some slide rules are found which have extra gauge marks added by the owner for his own particular use or specialisation. A fairly typical or normal selection of gauge marks would be made up of a number of the following examples.

There are a number of alternative abbreviations for the constants, some of which are given below.

Gauge. Mark.	Meaning.	Value.
π	Value of π.	3.14159
$\pi/4$	π divided by 4.	0.7854
C	$\sqrt{(4/\pi)}$.	1.128 *(1)*
C'	$\sqrt{(40/\pi)}$.	3.568 *(1)*
M	$100/\pi$, or $1/\pi$.	31.83, or 0.3183
δ' or Q'	minutes in a radian.	$3437.74 = (180 \times 60)/\pi$
δ'' or Q"	seconds in a radian.	$206265 = (180 \times 60 \times 60)/\pi$
$\delta.$ or Q.	seconds in a radian. (French decimal system)	$636620 = (200 \times 100 \times 100)/\pi$ *(2)*
_	Gunners mark. (angles $<20°$ tan = circle, -for artillery calculations).	1146
L	conversion of Log_e to Log_{10}	2.3026
M or ϱ^0	as δ' or Q'	
ϱ''	as $\delta.$ or Q.	
R	resistance of copper conductors.	30.6 *(3)*
S or ϱ'	as δ'' or Q"	
U	$180/\pi$, degrees to radians	57.2958
V or ς	$\pi/180$, rad equiv of small degrees	0.1745
W	weight of copper conductors.	111000 *(3)*
-	watts in one horsepower (A&B).	745.47 (746)
Cu^1	$\sqrt{(4\varrho/\pi)}$ ($\varrho = 0.0175\Omega\text{mm}^2/\text{m}$ at 20^0C)	0.1493 *(4)*
Cu^2	$\sqrt{(4/\pi\gamma)}$ ($\gamma = 8.87\text{g/cm}^3$)	0.379 *(4)*
Al^1	$\sqrt{(4\varrho/\pi)}$ ($\varrho = 0.029\Omega\text{mm}^2/\text{m}$ at 20^0C)	0.1922 *(4)*
Al^2	$\sqrt{(4/\pi\gamma)}$ ($\gamma = 2.69\text{g/cm}^3$)	0.680 *(4)*

Note *(1).* Sometimes C and C' are marked on the "C" and "D" scales respectively at 1.273 which is $4/\pi$.

Note *(2).* Sometimes described as graduated in the French decimal system. 400^g

Note *(3).* These points were regularly used by electrical specialists, and are actually the values of R^2 and W^2 respectively.

Note *(4).* Electrical calculations for copper and aluminium conductors.

In addition to gauge marks, some early rectilinear rules have various less than helpful aides-mémoire engraved at the ends of the scales to help the user identify the position of a decimal point in the calculation. These take the form of a matrix of + and - signs used with movements of the slide.

Some makers give tables of gauge points as a table of regular fractions which are easier to set on the slide rule, e.g 22/7 for π.

Early slide rules have gauge points relating to either excise or proof calulations. In the UK in 1824 the earlier Winchester measure was replaced by the new Imperial measure. Due to the problems of implementing the changes, many rules were not changed to about 1826. The main changes are as follows:

Imperial gallon of 1824 for both wine and ale contained 277.274 cubic inches.

Previously the British wine gallon contained 231 cubic inches - still so in USA.

Note also that the British Ale gallon was different to the Wine Gallon at 282 cu ins

This gives rise to the following Winchester (pre 1824) gauge marks and their values:

  MD = Malt Depth at 2150.4 being cubic inches in a small bushel

  MB = 2150.4 being cubic inches in a Malt Bushel

  A = 282 cubic inches in an ALE gallon

  W = 231 cubic inches in a WINE gallon

  si = 0.707 being the *square inscribed* 0.707 being the side of a square in a circle whose diameter is 1

  se = 0.886 being the side of a *square equal* to the same circle (diameter 1)

  C = circumference at 3.1416 of the same circle (diameter 1) [note this is π]

  oc = 0.0796 for the area of a circle whose circumpherence is 1

  od = 0.7854 for the area of the circle whose diameter is 1

  WG = 17.15 for Wine Guage) being gauge points for round and circular

  AG = 18.95 for Ale Guage) measure found by dividing the square root of 231,

  MR = 52.32 for Malt Round) 282 & 21540.4 by the square root of 0.7854 ($\pi/4$)

  MS = 46.37 for Malt Square, gauge point for square measure being root 2150.4

post Imperial

  MB = 2219 cubic inches in a Malt Bushel

  W   )

  WG   ) become redundant

  IM G = 277.3, replaces AG on scale "C" and A on scale "A"

  IM B = replaces MB

Post 1818 - no Verjuice points, as duty on Verjuice repealed

Appendix 6.

Dating and Valuing Slide Rules.

Dating slide rules.

Part of the pleasure of studying any item with an element of scientific interest stems from being able to date a particular article accurately, and to see how that specific example fits into the historical development of a product, or into a manufacturer's portfolio. It is therefore unfortunate that slide rules are particularly difficult to date. The only reliable way to provide a date is to know when a particular model was actually bought, though even this may not accurately relate to the date of manufacture. This sale information may come from a receipt; in a small number of cases the items themselves may be dated, e.g., Fuller calculators. Unlike the modern electronic calculator, the slide rule was not a device that was exchanged or indeed updated to a later model. Thus when a slide rule is obtained for a collection from an original owner, that owner will usually be able to say when it was obtained. This will then assist in dating other similar rules; however, slide rule designs were in general surprisingly long-lasting, and it is often only the details which may have changed, i.e., cursor, packaging, or indeed the price.

The literature and tool catalogues of a particular period are a good indicator of the age of a rule. On the other hand, the instruction leaflets or instruction books provided with the slide rule are often not even covered by a dated copyright statement. Scanning the books of the period will also give some clues as to the date of a slide rule; for example, the following information has been gleaned from various editions of Pickworth's *The Slide Rule,* and Snodgrass's *Teach Yourself the Slide Rule,* both commonly available. It should be noted that even this information must be treated with skepticism.

— Fuller's Calculating Rule was improved in approximately 1938 by replacing the papier-mâché tubes on which the logarithmic and data scales were mounted with tubes of linenised Bakelite; in addition, the fixed and moveable pointers were changed from brass to Perspex. This last information does not tally with actual models: while some special models do have Perspex cursors, most Bakelite models still have brass pointers.

— Thornton slide rules after 1938 had a differential trig scale or scales in addition to the normal suite of scales.

— Keuffel and Esser log-log trig and log-log decitrig slide rules were introduced in about 1938, and frameless glass cursors were available between the two world wars.

— The Unique Navigational slide rule was introduced in the early part of World War II for the purpose of assisting navigators in the Bomber Command; the Unique Dualistic slide rule was introduced in approximately 1958.

Broad-brush dating is also possible by process of elimination. For example:

— Any celluloid/wood rule must be post-1886 as that is when the patent was granted; however the styles were generally extremely static, so that a particular rule may have remained unchanged for a period of 30 to 40 years. Conversely an all-wood rule is not guaranteed to be pre-1886.

— A Unique rule can be no more than about 75 years old, as the factory only started making the familiar wooden rule with the celluloid-covered paper scales in about 1923.

— The change to imperial gallons from the original wine gallon, in 1824, is an accurate watershed which allows accurate before or after dating to be made, though it should be noted that rules marked in Imperial measure came after 1826 due to difficulties and delays in implementing the system.

— Information as to patent numbers can be used as has been described in **Appendix 4**, though this is not a particularly accurate method. Patent numbers by dates for a number of countries are given in **Appendix 4**.

Dating miscellany.

There are a number of other dates which can provide a clue to the date of a slide rule:

pre-1963:	No ZIP code used in American addresses.
pre-1932:	The description "Made in England" is used.
post-1932:	The description changes to "Made in Great Britain".
post-1858:	London Postal District codes without numbers (e.g., WC) are used in addresses. The plan for postal districts was approved in 1856 and implemented during 1857/8.
post-1916:	London Postal District codes with numbers (e.g., WC1) are used in addresses.
post-1974:	Post Codes (e.g. CM7 5LT) become standard throughout England including London where the Postal District code changes to the longer Post Code (e.g., W1A 4WW). The Postcode system was started in Croydon in 1966, and completed in Norwich in 1974.

Valuing slide rules.

Valuing a slide rule is perhaps easier than dating it. It is worth what you are prepared to pay for it!

As so many slide rules were made and sold, particularly during this century, there are numerous examples available. As a result the price will always be low, as demonstrated by the generally low prices at local sales. Obviously condition and completeness are vital: examples which are in excellent condition and complete with original box and instruction leaflet will be worth at least double the price of a dirty and incomplete example. Unusual boxes, cases or decoration in the form of advertising material may also increase the value.

Slide rules from earlier times are for sale in antique auctions and shops as well as collectors' fairs. The word "antique" will usually increase the price considerably, although this may not be an accurate reflection of the value. The true value of a rule can only be ascertained as between a buyer and a seller at any particular time. An author of books on collecting antique cameras quotes a rule which encapsulates the situation admirably:

McKeown's Law.
The price of an antique camera is entirely dependent on the moods of the buyer and seller at the time of the transaction.

There are several corollaries to McKeown's Law which are equally apposite. The final corollary is attributable to Dan Adams:

1. *If you pass up the chance to buy a camera you really want, you will never have that chance again.*

2. *If you buy a camera because you will never have the chance again, a better example of the same camera will be offered to you a week later for a much lower price.*

3. *The intrinsic value of an antique or classic camera is directly proportional to the owner's certainty that someone else wants it.*

291

The purist may decry this light-hearted look at valuation, as "value" is a deadly serious point with many collectors, but it is another way of describing the law of supply and demand. It must also be remembered that what is rare and valuable in one country may be quite common in another. Many people make money by exploiting this facet of the law of supply and demand and are well rewarded for their efforts.

One final point: *caveat emptor*, let the buyer beware, particularly if buying sight unseen!

Definition of Quality and Condition

When buying a slide rule sight unseen, one quickly becomes aware that perceptions of quality and condition vary considerably and are often only in the eye of the beholder. A number of quality/condition classifications exist, two examples of which are given here.

The classification of condition used within the Oughtred Society is recommended as a good starting point:

1. *Very good, virtually "like new".*
2. *Good.*
3. *Slight traces of wear, but fully functional.*
4. *Strong show of use, scratched and/or slight functional disorder.*
5. *Some defects and/or small parts missing.*
6. *Defective, important parts missing, bad condition.*

This classification has some shortcomings, in that it does not really include anything about the state or existence of such ancillaries as case, instructions, etc. Some valuers will often use additional + and - symbols to show an in-between condition: 3+ is better than 2-, or is it?

The Dutch Circle of Slide Rule Collectors have an alternative system that negates the need for sub-qualifications:

C0 *Mint condition with all extras such as box, case and instruction manual present, also in mint condition, factory clean.*

C1 *Mint condition with all extras such as box, case and instruction manual present but these extras not or not all in mint condition.*

C2 *Mint condition but without one or more of the extras.*

C3 *Very minimal signs of use.*

C4 *Minimal signs of use.*

C5 *Signs of normal use.*

C6 *Signs of heavy use*

C7 *Small damage but for the rest C4 quality or better.*

C8 *Damaged and well worn.*

Either system can be used, but a consistent approach is important to ensure that descriptions are fair and accurate.

Appendix 7.

Care of Slide Rules.

Slide rules and their associated ephemera are frequently of such an age that special efforts are required to ensure that no damage is caused to the instruments while in your possession, and that the paperwork in particular is preserved. Some of the earlier instruction leaflets are on acid-based paper, which is effectively getting frailer by the day.

A second-hand slide rule will normally be found in less than perfect condition and will be soiled at the very least.

For all models of slide rules, whatever the material they are made of, cleaning should only be carried out using a mild detergent or soap and water; any form of proprietary chemical cleaner must be avoided. Application of even soap and water is best employed "a little and often," with the slide rule being carefully and completely dried between applications. In this way the most ingrained grime can be gradually removed and the slide rule restored to its original look. All plastic rules can be cleaned more vigorously with soap and water, as there is no likelihood of the water being absorbed by the plastic. Petrol (gasoline), or other oil-based cleaner, is likely to remove the markings at least, and in all probability to dissolve the surface of the plastic. It must never be used as a cleaner.

Unique slide rules and other rules which have celluloid-covered paper scales are worthy of special mention, due to their particular method of manufacture. A lot of such secondhand rules will be discoloured, and there is no way of getting rid of this as it is a feature of the plastic covering to the paper scales. Only the very gentlest application of soap and water should be used, as excessive water can result in the glue being dissolved and the scales beginning to lift.

If the slide is difficult to move, or has completely stuck, sustained pressure should be applied only following the cleaning process described above. If this fails to move the slide, then the stock should be gently "bent" at right angles to the stock/slide join (at right angles to the direction of travel of the slide) to free the slide in its grooves. Only as a very last resort should the slide be struck, as this will probably result in damage which may be avoided by a more patient approach. Once the slide is out, then careful cleaning of the grooves in the stock and the inside corners of the tongue on the slide with a fine screwdriver or knife blade will remove the build-up of accumulated dirt and previous attempts at lubrication which undoubtedly caused the original problem. In some cases, where the slide rule got wet and was not dried before storage, the glue may have run, consequently gluing the whole together. In this case a more thorough removal of the glue will have to be undertaken, either by scraping or careful dissolving.

Where scales have started to lift, or are broken, they can be carefully glued back in place using a thin glue, such as PVA, which allows the positioning to be carefully controlled. One-shot glues are admirable, but it takes a braver man than I am to use them because of the immediacy and permanence of the bond.

Cursors are also best cleaned with soap and water. If the cursor is scratched, there is little that can be done without affecting the hairline, and it is best left alone. Broken cursor springs can be replaced with thin phosphor bronze such as from watch or small clock springs, suitably trimmed to size. Duplex cursors which are screwed together require particular care with disassembly and re-assembly to ensure that threads are not stripped. The correct size screwdriver is an obvious pre-requisite so that the screw heads are not damaged. It is also well worth noting exactly how the cursor was fitted to the rule, though this is no guarantee that it is correct in that previous removal may have resulted in incorrect re-assembly!

Broken cursor glasses can be glued with modern glass glues. These have the benefit of being able to carefully assemble the broken pieces before showing them to the required ultra-violet light (sunlight!) for bonding. I believe that it is better to glue than to leave broken, as the broken pieces can scratch the scales and get lost. Perspex cursors can be replaced simply (though I will usually put the broken bits into a bag with the rule).

There are some more "brutal" actions which can be taken with a particularly badly marked plastic slide rule, especially where certain inks were used which then leached into the plastic. A metal polish, such as Brasso, will deal with lightly stained items and slightly scratched cursors. A paint cutting agent, such as T-Cut, which is used to clean up motor vehicle paint finishes, is effective in more extreme circumstances. Both these items take off a very fine layer of plastic, and are effectively used to re-polish the whole surface. It must be stressed that this is only worth doing in drastic circumstances!

Restoration.

Pre-20th century slide rules in boxwood, ivory, brass or other material that are susceptible to the ravages of age may require restoration, always a difficult decision.

My personal view is that restoration is perfectly acceptable providing that when it comes to selling the restored article, the new owner is made aware that restoration has been carried out. The extent of the restoration is entirely up to the owner, and no hard and fast rules can be applied. In my view restoration using the tools and materials that would have been available to the craftsman of the day is acceptable and indeed preferable to avoid further damage to the slide rule. Replacing missing woodwork is beyond the scope of restoration; however, ensuring that broken wood is repaired allows the rule to be used as it was intended.

Cleaning should be limited to the "soap and water" level described above, as any more vigorous cleaning will completely ruin any patina and may also remove coloration in the markings. There should be no hesitation in polishing or feeding the wood to extend its life, and a good beeswax-based product will maintain the patina and coloration that only age can bring.

In extreme cases, where the markings have worn with use, I also believe that appropriate application of a suitable coloured wax can restore the markings so that at the very least the scales can be appreciated and their use analysed.

The metal bindings on old rules should only be very gently cleaned to remove rust on steel or iron bindings and verdigris on brass. Shiny, highly polished brass, which results from over-enthusiastic cleaning, is an indicator of a restored slide rule, not the result of long term storage! Hinges on joint rules should be very carefully and sparingly oiled to return the joint to full use and avoid possible breakage of the wood/metal join as a result of a reluctant hinge.

Replacement of pins and other metalwork is unnecessary other than in extreme circumstances. The pins which keep joint rules in alignment often work loose, and can be replaced in the interests of strength, but even these are best left out if they are missing.

Restoration of paper scales is difficult and really too specialised other than in really severe cases of disrepair. New markings would also be extremely obvious, so only if this is the only way of producing a workable device or something for demonstration purposes, should this be attempted.

Long term care.

Long term care is vitally important in preserving slide rules. Cleanliness and careful storage play a large part in this; thereafter, the main effort is to ensure that the slide rules are not subject to extremes of temperature or humidity.

It is worth quoting the **Unique** care instructions from the 1950s verbatim, as they are still entirely relevant not only for Unique slide rules, but indeed for any slide rule — with suitable interpretation obviously:

Care of "Unique" Slide Rules.

Attention to the following points will ensure satisfactory service.

The Rule should not be left in a damp place, nor in direct sunlight, it should be kept in its case, in a cool dry place.

If the slide is too tight, vigorous movement in and out, while the stock is gripped firmly, tending to close it up, will generally ease the movement. The lower surface of the slide may be lightly rubbed on fine sandpaper, laid on a flat surface, and the narrow edges of the wood on its upper surface lightly scraped with a penknife, pressed into the corner to remove any dirt, but these operations should be carried out carefully and not overdone.

If the slide is too loose it should be removed and the stock lightly nipped in a vice, or bound around with tape, so as to diminish the gap between the top and bottom scales. If left under pressure for a time the stock will close up.

DO NOT GREASE THE SLIDE, OR WRENCH OPEN THE STOCK.

If the line of the cursor is not quite square across the rule it will be found possible to force the celluloid round slightly to square it up. A small clearance is left to allow for that adjustment.

Slide rule ephemera.

Instruction leaflets, catalogues, early books and other paperwork associated with slide rules are fast becoming a rare commodity, and thus need to be preserved. They need to be handled even more carefully than the slide rules, as they are less durable than most, apart from early cardboard devices.

Instruction leaflets are often well creased, at the very least, when they come with a slide rule. I believe that they are best stored well away from the rule; they should be carefully ironed with a medium electric iron before being stored in a suitable plastic sleeve. Ironing also seems to put some "body" back into the paper. I am not a great believer in sticky tape and the like for repairs, even though modern versions are far better and do not discolour. Much of the problem with old papers is due to horrible tape of one sort or another.

If the instructions are so badly torn as to be useless, they can be laminated in plastic so that they can at least be read and used. However this is a drastic and permanent last resort solution.

Acid-based paper ephemera from after 1850 will be very frail and may be discoloured. It will ultimately change colour to yellow, to orange, and finally to brown, and inevitably crumble to nothing. Having ascertained that the paper is acid by using a pH testing pen, you can choose from a number of neutralising agents to spray the paper to preserve it. Each stage needs to be approached carefully to ensure that the treatment causes no other problem, such as the ink running. Papers can be stored in suitable non-acid storage slips such as those used for archiving photographs.

Table 18 : Historical Milestones

Year		Monarchs / Events	Period	Milestones
672 - 735		*The Venerable Bede*		
1000				c. 1000 A.D. Arabian mathematics, use of zero in calculation
1010				
1020				
1030				
1040			SAXON	
1050		Harold		
1060	1066	William I		
1070				
1080	1087	William II		
1090			NORMAN	
1100	1100	Henry I		
1110				
1120				
1130				
1135		Stephen		
1140				
1150				
1154		Henry II		
1160				
1170				
1180				
1189		Richard I	ANGEVINS	
1190		*1100 - 1300 The Crusades*		
1199	1199	John		
1200				1202 - Leonard of Pisa (Fibonacci) publishes *"Liber Abaci"*
1210				1214 - 1294 Roger Bacon
1216	1216	Henry III		1220 - Leonardo publishes *"Practica Geometriae"*
1220				1225 - 1275 Thomas Aquinas
1230				c.1230 - earliest English manuscripts using Zero
1240				
1250		Henry III *"The First Awakening"*		
1260				
1270	1272			
1272		Edward I		
1280				
1290		Edward I		1299 - Florentine merchants forbidden the use of Arabic numerals - must use Roman!
1300		Edward I		
1307	1307	Edward II		
1310				
1320				
1327	1327			
1330				1331 - First "Letters of Protection", John Kempe, woollen cloth

296

Timeline (rotated page)

Left decade markers: 1730 · 1740 · 1750 · 1760 · 1770 · 1780 · 1790 · 1800 · 1810 · 1820 · 1830 · 1840 · 1850 · 1860 · 1870 · 1880 · 1890 · 1900 · 1910 · 1920 · 1930 · 1940 · 1950 · 1960 · 1970 · 1980 · 1990 · 2000

Monarchs / general events:

1727 — George II
1750 - 1850. Industrial Revolution
1760 — George III
1789 - 95. French Revolution.
1800 - Maudsley Screw 7', 1/16" out. — HANOVER
1820 — George IV
1830 — William IV
1837 — Victoria
1851 - Great International Exhibition
1879 - 1955 Albert Einstein
1901 — Edward VII · SAXE-COBURG-GOTHA
1910 — George V
1936 — Edward VIII · WINDSOR
— George VI
1952 — Elizabeth II
1964 - SHARP announces transistorised electronic calculators. Slide rule sales decline from now.

Slide rule events:

1751 - Diderot, First type of Dividing Engine
1753 - John SUXPEACH, First Slide Rule Patent. *Catholic Organon and Universal Sliding Rule*
1775 - John Robertson, Runner/Indicator
1777 - Jesse Ramsden, Linear & Circular Dividing Engines
1779 - Boulton & Watt, SOHO Engineers Slide Rule
1815 - Roget, Log-log Scales
1833 - Baron Riche de Prony, French Survey of UK Slide Rule usage
1851 - Mannheim Slide Rule
1852 - Patent Law Amendment Act. 14,357 Pats to date, maybe 6 for sr's
1866 - Dr. J.D.Evans : Grid Iron Slide Rules
1874 - Coggeshall Slide Rules still used
1886 - Dennert & Pape : Laminate celluloid to wood
1901 - Prof J. Perry, re-invention of log-log scales.   1903 - Wright Bros first flight.
1911 - Richardson, Engineers sr.   1912 - Lilley, Spiral log-log scales
1922 - Otis-King calculator.   1927 - UNIQUE slide rules
1932 - P.I.C. Patent Differential Scales
1963 - Sun HEMMI make 1 Million slide rules.
1972 - Sinclair EXECUTIVE calculator - £80. HP-35 "Electronic Slide Rule" - $395
1973 - Sinclair CAMBRIDGE calculator - £24.95, 1976 - £4.95
1980 - Sinclair ZX80 Computer - £99.95
1985 - Last Slide Rule Patent. Specialist Circular Slide Rule

Timeline (chronological chart, read with dates down the left):

Year	Monarch / Dynasty	Event
1336		University of Paris rule that no degree awarded without attending Maths lectures
1340		
		1348 - Black Death.
1350	*Edward III* PLANTAGENET	
1360	*1338 - 1453 "100 Years War"*	
1370		
1377	*Richard II*	
1380		
1390		
1399		
1400	*Henry IV*	
1410	*1400 - 1600 Renaissance*	
1413	*Henry V* LANCASTER	
1420		
1422	*1436 - 1476 Regiomontanus*	
1430		
1440	*Henry VI*	
1450		C15 - Abacus no longer used in Spain and Italy
1460		1449 - First patent for new invention, John of Utynam, glass
1461		
1470	*Edward IV* YORK	
1471		1477 - Caxtons Printing Press
1480		1489 - Symbols '+' and '-' first used for add & subtract
1483	*Richard III*	
1485		1492 - Columbus sails to America
1490	*Henry VII* COPERNICUS	1473 - 1543, Earth moves around the Sun.
1500		
1509		
1510		1525 - G. Rudolf first uses Square root sign (√)
1520		
1530	*Henry VIII* TUDOR	
1540		
1547	*Edward VI*	
1550		1552 - Second Patent, John Smythe, Normandy Glass
1553	*Mary*	1568 - Earliest Sector - Guidobaldi del Monte, made instruments for Galileo
1558		
1560		
1564	*1616 William Shakespeare*	
1570	*Elizabeth I*	
1580		
1590		1597 - Invention of SECTOR by Galileo
1600	*1605-Gunpowder Plot*	
1603	*James I*	**1614 - Invention of LOGARITHMS by John Napier**
1610		**1617 - Logs to Base 10, NAPIER'S BONES, Patent Docket Book started**
1620	*1625 - Hand Division, Calipers, etc*	**1620 - Interpretation of Logarithms into a straight line - Edmund Gunter**
1625	*Charles I*	**1625 - Invention of Slide Rule by Rev. William Oughtred.**
1630		
1640	*1642 - Start of First Civil War*	
1649	*Cromwell's NEW PARLIAMENT.*	
1650		**1662 - 1st use of term "Sliding Rule". 1663 - Pepys mentions Slide Rule**
1660	*Newton's Laws of Gravity, Great Fire.*	
1666	*Charles II* STUARTS	1665 - Hooke's Gear Cutting Engine. 1676 - Abacal calculation used in English Exchequer, also "Tally"
1670		
1680		1677 - Moxon : *Mechanick Exercises.* Henry Coggeshall's Slide Rule
1685	*James II*	**1683 - Thomas Everard's Slide Rule**
1688		
1690	*William II & Mary*	
1700		
1702	*Anne*	
1710		1723 - Nicolas Bion: *Construction & Principal uses of Mathematical Instruments.*
1714		
1720	*George I*	

Index

Chapter 6, Appendix 3, and Table 6 (p.83) are not included in the Index,
as they are alphabetically listed.

301